LEADING-EDGE ENVIRONMENTAL BIODEGRADATION RESEARCH

LEADING-EDGE ENVIRONMENTAL BIODEGRADATION RESEARCH

LYMAN E. PAWLEY
EDITOR

Nova Science Publishers, Inc.
New York

For permission to use material from this book please contact us:
Telephone 631-231-7269; Fax 631-231-8175
Web Site: http://www.novapublishers.com

LIBRARY OF CONGRESS CATALOGING-IN-PUBLICATION DATA

Leading-edge environmental biodegradation research / Lymen E. Pawley, editor.
p. cm
Includes index.
ISBN-13: 978-1-60021-903-0 (hardcover)
ISBN-10: 1-60021-903-9 (hardcover)
1. Biodegradation--Research. 2. Bioremediation--Research. I. Pawley, Lyman E.
QH530.5.L43 2008
628.5--dc22

2007030877

Published by Nova Science Publishers, Inc. ✣ New York

CONTENTS

PREFACE

This volume presents new leading-edge research on environmental biodegradation which is the destruction of organic compounds by microorganisms. Microorganisms, particularly bacteria, are responsible for the decomposition of both natural and synthetic organic compounds in nature. Mineralization results in complete conversion of a compound to its inorganic mineral constituents (for example, carbon dioxide from carbon, sulfate or sulfide from organic sulfur, nitrate or ammonium from organic nitrogen, phosphate from organophosphates, or chloride from organochlorine). Since carbon comprises the greatest mass of organic compounds, mineralization can be considered in terms of CO2 evolution. Radioactive carbon-14 (14C) isotopes enable scientists to distinguish between mineralization arising from contaminants and soil organic matter. However, mineralization of any compound is never 100% because some of it (10–40% of the total amount degraded) is incorporated into the cell mass or products that become part of the amorphous soil organic matter, commonly referred to as humus. Thus, biodegradation comprises mineralization and conversion to innocuous products, namely biomass and humus. Primary biodegradation is more limited in scope and refers to the disappearance of the compound as a result of its biotransformation to another product.Compounds that are readily biodegradable are generally utilized as growth substrates by single microorganisms. Many of the components of petroleum products (and frequent ground-water contaminants), such as benzene, toluene, ethylbenzene, and xylene, are utilized by many genera of bacteria as sole carbon sources for growth and energy.

Short Communication - Chitosan, a biodegradable and biocompatible polysaccharide, is a potentially useful material in various fields. The authors developed a simple chitosan sheet and examined the possibility of using an adriamycin-containing chitosan sheet as a drug carrier for controlled release. To prepare a carrier consisting only of chitosan, a chitosan suspension was subjected to acid-alkaline treatment, mixed with adriamycin, frozen and freeze-dried. The adriamycin-containing chitosan sheet was inserted into the peritoneal cavity of mice in order to investigate its biodegradation. The appearance of decomposition of chitosan was observed using scanning electron microscopy, and adriamycin in urine and liver were detected for one and two weeks, respectively. Adriamycin metabolites were detected in plasma for two weeks. Furthermore, adriamycin remained in the chitosan sheet without being metabolized after two months. These results suggested that the chitosan sheet prepared in this study might improve therapeutic efficacy in topical lesions as a carriers of sustained-release drugs.

Chapter 1 - The review is devoted to up-to-date trends in application of biocatalytic systems, based on various immobilized cells of microorganisms, for environmental bioremediation. Mostly, attention is paid to the treatment of wastewaters produced by different branches of industry and agriculture. Three main aspects concerning bioremediation processes with immobilized cells are under consideration. The first one is connected with application of microorganisms in an immobilized state merely for environment protection owing to deep destruction of dangerous pollutants present in wastewaters. The use of immobilized cells for biodegradation of neurotoxic organophosphorous pesticides and fat-containing compounds as components of wastewaters is discussed. The second aspect of environmental bioremediation deals with conversion of agricultural wastes to some desired products. Mainly, the interest is concentrated on production of fuel bioethanol by immobilized yeast and filamentous fungi and obtaining of L(+)-lactic acid in the processes with immobilized bacteria and filamentous fungus cells. The efficiency of conversion of various enzymatically-pretreated wastes to the target products by immobilized cells is analyzed. Oppositely, the third direction of bioremediation is focused on the prevention of formation of stable systems with immobilized cells that can do damage to environment. The approaches, allowing completely avoiding or decreasing the appearance of corrosive biofilms composed of self-immobilized individual or mixed bacterial cultures in oil industry, are pointed in the review.

Chapter 2 - There has been an increasing concern among the regulatory bodies, scientific community and public at large regarding the potential threat of certain groups of chemicals (natural and man-made) on human, wildlife, and the environments. Although these chemicals have existed for a long time, their potential environmental threats have been realised only in the last few decades. These so called emerging contaminants (ECs) are mostly organic in nature and quit a few of them possess estrogenic activity and have been found to cause abnormalities in reproductive function of wildlife and some aquatic species. For most emerging contaminants, there is currently little information regarding their potential toxicological significance in ecosystems- particularly effects from long-term, low-level environmental exposures. These ECs include prescription and over-the-counter veterinary and human pharmaceuticals (antibiotics, steroids and reproductive hormones), as well as many chemicals found in personal care products and others (antioxidants, detergents, antimicrobials, fragrances). In addition, some pesticides, natural toxins, biomedicine, nanoparticles, flame retardants and perfluorated hydrocarbons also fall under ECs. The environmental fate of these ECs influences their exposure and, hence, is a key parameter for estimating the risk of effects on the biota. Biological degradation plays a key role in their ultimate fate in the environment. In this chapter, the authors provide an overview of past and recent studies conducted on the biodegradation of selected ECs in various environmental matrices (e.g. soils, sediments, waters, animal manure), and the factors influencing their degradation behavior in these matrices are also discussed. ECs are often present in the environment at low concentrations that might affect the ability of micro-organism to degrade these compounds. Examples of ECs considered in this chapter include steroid hormones (natural and synthetic), some commonly used veterinary antibiotics, plasticizers, and some industrial surfactants. Finally some recommendations on future research direction on environmental biodegradation of emerging organic contaminants are presented.

Chapter 3 - Since the 1980s, composting has been used as an alternative for bioremediation in open sites, with the microbial communities involved being a focal point of study. The use of this innovative bioremediation approach has been necessary due to recent changes in waste management regulations around the world, which focus towards more integrated, practical, sustainable and economic schemes than landfilling.

Currently, not much literature on the feasibility of bioremediation technologies to soils contaminated with mixtures of organic compounds exists. Even less information on the application of composting as a bioremediation technology is available. This chapter will include a case study on the successful application of in-vessel composting of a fuel-contaminated (diesel and gasoline) desert-mining soil/sawdust in the Atacama Desert (Chile). The physical and chemical factors that influence the distribution, abundance, richness diversity and evenness of microbial communities during composting of a fuel-contaminated soil during bioremediation will be discussed.

This chapter will also provide a review of the research conducted to date using molecular tools for the biomonitoring of changes in microbial communities as an indicator of remediation and restoration of oil-contaminated soils. In particular, the investigation of microbial community structure changes based on molecular tools including terminal restriction fragment length polymorphism (T-RFLP), BIOLOG EcoPlate™ and the COMPOCHIP microarray will be presented. The advantages and disadvantages of certain molecular tools for environmental management will also be discussed. Future needs for bio-monitoring of microbial community structure changes during bioremediation of oil-contaminated soils will be identified, and a discussion of the technological management, regulatory and sustainable aspects of using composting as a bioremediation technology will be included.

Chapter 4 - Polychlorinated dibenzo-*p*-dioxins (PCDDs) and polychlorinated dibenzofurans (PCDFs), commonly known as dioxins (PCDD/Fs), are toxic environmental pollutants, unintentionally formed from various sources. Due to their persistent and ubiquitous natures elimination of these pollutants from the environment is a difficult task. Removal of dioxins by biological degradation (biodegradation) is considered a reasonable method as an alternative to other expensive physico-chemical approaches. Several bacteria utilizing dioxins as sole carbon energy sources have been isolated from the environment. Biodegradation of dioxins has been extensively studied in several microorganisms, and details concerning biodiversity, biodegradation, biochemistry and molecular biology of this process have accumulated during the last three decades. There are several microbial mechanisms responsible for biodegradation of dioxins, including oxidative degradation by dioxygenase-containing aerobic bacteria, bacterial and fungal cytochrome P-450, fungal lignolytic enzymes, reductive dechlorination by anaerobic bacteria, and direct ether ring cleavage by fungi containing etherase-like enzymes. Many attempts have been made to bioremediate PCDD/Fs using this basic knowledge of microbial dioxin degradation. This review highlights the current knowledge and recent advancements in the microbial biotransformation, biodegradation, biosorption, and bioremediation of dioxins.

Chapter 5 - Microalgae-based environemntal technologies offer attractive and reliable alternatives to conventional energy-demanding and green house gases emitting wastewater treatment processes. For instance, photosynthetic oxygenation is more economical (no need for external aeration) and safer (no CO_2 emission and low risks of pollutant and microorganism spreading). In addition, nutrient removal can be coupled with the mineralization of organic comounds in a single stage process, with a subsequent reduction in plant construction and operation costs. Microalgae also support cost-efficient technologies for the removal of toxic heavy metals or in the fixation of CO_2 and the biomass produced from algae-based process is a potential source of valuable chemicals, fertilizers or biofuels (methane, hydrocarbon or biodiesel). Unfortunately, despite the merits of this approach, the limitations imposed by the light distribution to the photosynthetic cells, together with the high cost and scale-up problems of photobioreactors have constrained the use of algae-based technologies in environmental applications. However, the massive demand for algal-based products in the food, biotechnology and agricultural sectors has contributed to generate a considerable knowledge on these processes. More recently, increasing energy prices and CO_2-mitigation policies even boosted research and development on algal-based processes for bioenergy production, an application still economically unviable a decade ago. With mature technologies now becoming available and a highly favorable political and energetic conjuncture, time has come to reevaluate algal-based photobioreactors as energy-efficient and waste-to-resource pollution control processes.

Chapter 6 - Since long ago the men have been using plants, animals and also microorganisms as source of chemicals and products for his interest. With the recent development in the exploration of the microbiological and genetical world, this resource became still more promising, and stimulated a new area of research, the bioprospection. Roughly speaking it comprises the search of new economically viable compounds, produced via catabolic reactions, which can be efficiently manipulated and applied to some chemical process of interest in the industry, health or environment. As an example of new trends on Ecology Microbiology, this chapter is concerned on the bioprospection of bacteria, looking for biocatalysts to prepare pure enantiomerically compounds, to be applied on pharmacology, and hydrocarbonoclastic consortia, to be applied on bioremediation of oil contaminated environments. 106 bacteria strains, with potential for synthesis of chiral molecules, were obtained using 8 different carbon sources (4 nitriles, 2 alcohols and 2 ketones). Moreover, 5 bacteria strains, capable of degrading hydrocarbon compounds (between 16 - 45%) in laboratory conditions, were also isolated from a hydrocarbonoclastic consortium. Unexpectedly, when this consortium was applied to some *in situ* microcosms (planted on a mangrove ecosystem) the oil compounds were not reduced significantly. This result shows the necessity of making *in situ* experiments to test the real potentiality of the microorganisms for remediation of contaminated sites. It also reinforces that natural ecological systems are very complex and hard to be controlled, and that our knowledge about the biotechnological potential of microorganisms has much to be developed. In the meanwhile, the authors discuss the importance of selecting a suitable culture media (carbon source) and the modern tools for studying the biodiversity of the samples and the chemical potential of the isolates.

Chapter 7 - Composting is one of the most successful tools for recycling organic wastes, but it is also a process with a broad diversity from a microbial point of view. The nutritional nature of the materials liable to suffer this kind of biotransformation, as well as the environmental conditions that prevail throughout the process, favour the presence of microbial populations adapted to the specific circumstances of every stage. From thermophilic to xenobiotic-degrading microorganisms, there are very different microorganisms in compost, many of which exhibit interesting capabilities, not only for the composting process, but for other environmental applications. Among these communities, one of the most attention-grabbing is that of lignocellulolytic microorganisms, which play a starring role in the global carbon cycle since lignocellulose is the major biopolymer in nature. Nowadays, lignocellulolytic microorganisms are becoming more and more important on account of their implication in industrial processes. Thus, in the paper industry there are an increasing number of factories that have substituted the chemical processes by biological treatments, which are friendlier from an environmental perspective. On the other side, the exhaustion of fossil fuels has turned the eyes to alternative sources, biomass bioconversion to ethanol being one of the most promising. And it is expected in the future these activities to be engaged to a greater extent since both citizens and governments demand for a bigger environmental commitment by industries. Now it is time for scientists to answer this demand and make all these processes both environmental and economically viable.

Chapter 8 - The European Council Directive on the Landfill of Wastes 1999/31/EEC (LFD), places a requirement on member states to draw up strategies to ensure that the amount of biodegradable municipal waste deposited at landfill progressively reduces over a 15-year period to only 35% of the total amount produced in 1995. An especially problematic organic waste is the pig hair due to the high production generated in the slaughterhouses (34000 tons per year in Spain). In order to reduce this problem, according to the European Council Directive, a study about the transformation of pig hair from slaughterhouses in potential feedstuffs for animals is carried out. Because keratin is the raw material of the hair, proteolysis by commercial protease enzymes is investigated. Trials are conducted in a batch reactor at different operating conditions. Degrees of hydrolysis and amino acid profiles in the hydrolysates are determined. The essential amino acid index, the protein efficiency ratio (PER) and the biological value (BV) in the hydrolysates show the hair waste could be managed as alternative feedstuffs in swine diets, reducing the amount of biodegradable municipal waste going to landfills. From the results a general kinetic equation for the enzymatic hydrolysis of hair proteins is suggested.

Chapter 9 - Synthetic dyes are employed as coloring agents in the textile, food, and pharmaceutical industries, which are also integrated into many industrial products. The release of dyes into the wastewater is highly visible in small quantities due to their brilliance and creates major environmental problems. Tightening government legislation is forcing industries to treat their waste effluent to control water pollution. Dyestuffs in the industrial effluents are highly stable and recalcitrant to conventional treatment. Currently, removal of dyes from effluents is by physical and chemical methods. Such methods, although offering some solutions to the problems, are often very costly and, although the dyes are removed, accumulation of concentrated sludge creates a disposal problem. Biological degradation is recognized as the most effective method for the degrading the dye present in the wastewater from textile industry. Research over the past two decades has provided insight into the various aspects of biological degradation of dyes. Detailed and extensive studies have been made for treatment of dye containing wastewaters by white rot fungi and their enzyme systems. It is observed that white rot fungi have a non-specific enzyme system, which oxidizes the recalcitrant dyes. This chapter reviews the recent progress in the research on understanding the biodegradation mechanism of white rot fungi and their enzymes involved in dye decolorization. Biocatalysis and characteristics of their enzymes and emerging knowledge of the enzymology of these organisms are discussed. Study of fungal degradation products of dyes, espically azo dyes, is reviewed for its importance in the understanding of the nature of bioremediation with fungi. Some of most recent results on biodegradation mechanism of azo dyes and metabolites analysis from authors' group are reported herein. Perspectives on how to use white rot fungi as an effective, cheaper alternative for dye removal and detoxification are also discussed.

Chapter 10 - There is growing potential for the adoption of biodegradable polymers in packaging applications due to municipal solid waste concerns and packaging sustainability initiatives. Since fossil fuel based polymers are difficult to recycle when soiled with contaminants, the use of biodegradable polymers opens the opportunity to successful recycling by composting them along with other compostable materials. However, to better understand this alternative, it is necessary to investigate the compostability performance of biodegradable polymers and packages in real composting conditions as a function of time. This information would provide a time-frame for composting these packages and in general provide a better representation of biodegradation trend. This chapter discusses the current ASTM international and European standards for evaluating biodegradation of plastics in simulated composting conditions, where the biodegradation is correlated to the conversion of carbon content in the polymer into carbon dioxide. There are different methods of recording carbon dioxide such as cumulative measurement (which is recommended by ASTM/ISO), direct measurement (using a computerized control), and gravimetric measurement (weight change due to CO_2 absorption). All three methods and associated results are discussed in this article. Actual testing of biodegradable packages in real composting conditions is also described.

Chapter 11 - In nature, chlorine does not occur only as chloride or bound in substances of anthropogenic origin in the polluted environment but also in many hundreds of compounds of natural origin. Chloride deposited in the forest ecosystem from the atmosphere reacts with soil organic matter (SOM) under the mediation by enzymes and/or microorganisms (to a smaller degree also abiotically), forming chloroacetic acids (CAA), chloromethane, chloroform, other aliphatic and aromatic compounds and chlorinated humic substances. Chloride, to which Norway spruce is sensitive, is thus mostly blocked in organochlorines, which even prevail over chloride in soil. Chlorine is thus involved in SOM degradation, leading to smaller SOM decay products and finally to their mineralization. Closely related microbial processes were met in methodically similar soil bioremediation studies conducted before. The aim of the investigations was first (i) to elucidate the uptake, effects and fate of phytotoxic trichloroacetic acid (TCA, previously considered to be only a secondary atmospheric pollutant) in the Norway spruce/soil-system, (ii) to address the role of CAA, and finally (iii) to consider the global role of chlorine in coniferous forest ecosystems. Radiotracer methods were used with advantage.

Using carbon 14, the fastest uptake of TCA was found to occur in the youngest spruce shoots due to a higher transpiration stream; TCA is then rapidly biodegraded in the forest soil. This microbial degradation depends on TCA concentration, soil humidity and temperature as well as on the soil character and microorganisms. The biodegradation of TCA took place also in the phyllosphere of the needles; TCA elimination in needles may be caused by slow decarboxylation. On the other hand, TCA may be formed by biooxidation of absorbed atmospheric tetrachloroethylene in needle chloroplasts.

Using chlorine 36 it was shown for the first time that chlorination of SOM yields TCA and also dichloroacetic acid (DCA), which is biodegraded even faster than TCA. The chlorination process was shown to proceed more microbially than abiotically, basidiomycetes being suggested as the responsible microorganisms. It is not clear whether the process is mediated by extracellular enzymes or intracellularly, to what degree the abiotic chlorination proceeds, whether dehalogenation of chlorinated substances takes place, and what further influences affect the ecosystem under study.

A connection between carbon and chlorine cycle was clearly shown: products of chlorination of SOM are further mineralized and contribute to the litter decay and a part of chlorine is released from the forest ecosystem again. The previous chlorine level is thus restored. Moreover, the monitoring of chloride, TCA, AOX and other substances in forest soil, needles and soil solution in forest stands, especially near salted roads, showed an adverse effect of these substances on coniferous trees, especially spruce.

In: Leading-Edge Environmental Biodegradation
Editor: L. E. Pawley, pp. 1-10
ISBN: 978-1-60021-903-0

Short Communication

BIODEGRADATION AND FEASIBILITY OF SIMPLE CHITOSAN SHEET AS DRUG DELIVERY CARRIER

Kieko Saito*
Institute for Environmental Sciences, University of Shizuoka, 52-1 Yada Suruga Shizuoka, 422-8526, Japan

ABSTRACT

Chitosan, a biodegradable and biocompatible polysaccharide, is a potentially useful material in various fields. We developed a simple chitosan sheet and examined the possibility of using an adriamycin-containing chitosan sheet as a drug carrier for controlled release. To prepare a carrier consisting only of chitosan, a chitosan suspension was subjected to acid-alkaline treatment, mixed with adriamycin, frozen and freeze-dried. The adriamycin-containing chitosan sheet was inserted into the peritoneal cavity of mice in order to investigate its biodegradation. The appearance of decomposition of chitosan was observed using scanning electron microscopy, and adriamycin in urine and liver were detected for one and two weeks, respectively. Adriamycin metabolites were detected in plasma for two weeks. Furthermore, adriamycin remained in the chitosan sheet without being metabolized after two months. These results suggested that the chitosan sheet prepared in this study might improve therapeutic efficacy in topical lesions as a carriers of sustained-release drugs.

INTRODUCTION

Chitosan, a biodegradable polysaccharide, comprising glucosamine and *N*-acetylglucosamine, is an alkaline-deacetylated chitin derived from the exoskeletons of insects and shells of crustaceans. Chitin and chitosan have been used in a wide variety of biomedical applications [1-7] since Prudden et al. reported the acceleration of wound healing by chitin

* Tel/Fax: +81 54 264 5719; E-mail: saitok@smail.u-shizuoka-ken.ac.jp

and chitosan. In addition to its physiological activity [8-10], chitin and chitosan are biocompatible and biodegradable. Thus, the medical applications of chitin and chitosan have recently been the subjects of numerous studies.

Several carriers derived from chitosan have been developed as drug delivery systems [11-15]; however, the clinical applications remain under investigation. A wound dressing developed commercially, which is employed externally, has been shown to be clinically effective [4]. However, there was concern regarding the side effects induced by an excessive immune reaction to the ingredients contained in the biomedical product; chitosan is digested by activated macrophages. If a chitosan carrier containing drugs can be applied to the focuses of a disease, drugs can be released gradually and improve therapeutic efficiency. In order to exert notable drug activity in topical lesions without any side effects induced by biomedical materials, the quality and form of the material is very important. In the present study, we used a simple technique to develop a pure chitosan sheet with a flexible flat shape, and furthermore investigated chitosan biodegradation and subsequent drug (adriamycin; ADM) release from the sheet in order to explore the feasibility of chitosan as a drug carrier in topical lesions.

Materials and Methods

1. Chemicals

Chitosan (PSH 80, Mw. 5×10^5-1×10^6, 80% deacetylated), and ADM metabolites (adriamycinol and adriamycinon) were kind gifts from Yaizu Fishery Co. (Shizuoka, Japan) and Pharmacia (Tokyo, Japan) respectively. ADM was purchased from Kyowa Hakko Co., Ltd. (Tokyo, Japan) and other chemicals were obtained from Wako Pure Chemical Industries, Ltd. (Tokyo, Japan).

2. Preparation of Chitosan Sheets

Chitosan (1% w/v) was dissolved in hydrochloric acid (0.05 N), with agitation, overnight at room temperature, and sodium hydroxide (0.05 N) was added to neutralize the chitosan solution, which was then gelled. The gelled solution was filtered with a glass filter (25G P5), washed with distilled water several times, and homogenized using a homogenizer (KIINEMATICA, PT 10-35, Switzerland) after the addition of distilled water up to give a 1% chitosan concentration. Homogenized gelling solution (20 ml) was cooled gradually to -20°C and was then maintained for over 12 h in a plastic schale (ϕ 9 cm). Chitosan sheet thickness was dependent on the volume of the solution. The frozen solution was lyophilized and the chitosan sheet was cut into suitable shapes before the experiment. Sheet permeability to water and air were 0.8-10 $cm^3/cm^2/h$ (measured using water column at room temperature) and 2×10^3-8×10^3 $cm^3/cm^2/min$ (measured with air pressure of 10-15 kgf/cm^2 at room temperature), respectively.

3. Animal Experiments

Female ddY mice (age; 5-6 weeks) were obtained from SRL (Shizuoka, Japan) and housed 5 or 1 per cage, with a stainless steel feeder under standard conditions with a 12-h light/dark cycle. Animals were allowed free access to water and standard diet. To investigate the biodegradation of chitosan sheets *in vivo,* a piece of circular sheet (13 mg, 5-10 mm) was placed in the peritoneal cavity of mouse, and was subsequently removed for observation by scanning electron microscopy. To measure ADM metabolism, urine was collected daily and mice were sacrificed for liver and blood samples after sheet placement in the cavity. The experimental protocols were conducted in accordance with the Guidelines for Animal Experiments at the University of Shizuoka.

4. Scanning Electron Microscopy

Scanning electron microscopy (SEM) was performed using a Hitachi S-2500 (Tokyo, Japan). Sheets removed from the peritoneal cavity were dried using a critical-point drying technique, following treatment with 50, 70, 90 and 100% ethanol and immersion in ethanol-isoamyl acetate (1:1). Dried samples were Pt-Pd coated before SEM observation.

5. Preparation of Adriamycin-containing Chitosan Sheet

Adriamycin (ADM) was added at 1 mg per 100 ml of the homogenized gelling solution described in 2.2 above, and this solution was mixed for 1 h at room temperature. The solution was frozen at -20°C for over 12 h in a plastic schale (ϕ 9 cm), and thawed. The watery chitosan sheet was washed with distilled water several times to remove excess ADM, and was then lyophilized.

6. ADM Determination

ADM concentration was measured by modifying the reported high-performance liquid chromatography (HPLC) method [16]. An HPLC column with a fluorescence detector (Hitachi F1050, Tokyo, Japan) was used for analysis. A normal phase column (Mightysil Si60, Kanto Kagaku, Tokyo, Japan) was used as the stationary phase. The mobile phase consisted of chloroform-isopropanol-acetic acid-water-sodium acetate buffer (100:100:14:14:1, pH 4.5) at a flow rate of 1.0 ml/min. Fluorescence signals were monitored at Ex 470 nm and Em 585 nm. ADM was extracted by chloroform/methanol (4:1) and was then centrifuged at 15000 rpm for 15 min. The phase-separated chloroform/methanol layer was analyzed by HPLC. Pure ADM was used as a standard. The content of ADM adsorbed by the chitosan sheets was estimated to be 1 μg/mg dry chitosan. Fluorescence microscopy was used to confirm the adsorption of ADM.

RESULTS

1. Morphological Characteristics of Chitosan Sheet

The structure of chitosan sheet as observed by microscopy is shown in Figure 1. The sheet was symmetric, and on SEM, was found to consist of multiple porous layers and an uneven surface with apertures (Figure 1A and B). The shape of the pores was not uniform, with the diameter varying between 50 and 500 μm. Under a phase-contrast microscope, the sheet appeared to have a knitted stitch and fine fibrous structure (Figure 1C). The presence of uneven apertures is reflected in the difference between the light and dark regions on the micrograph. These structures enabled the permeation of water and air. The texture of the sheet was smooth and flexible, similar to a sponge or gauze, and would thus be easy to handle in medical applications. In addition, chitosan sheet with adsorbed ADM was observed by both SEM and phase-contrast microscopy, and no significant changes in morphological characteristics were noted (data not shown).

2. Biodegradation of an ADM-containing Chitosan Sheet in Mouse

An ADM-containing chitosan sheet was inserted into the peritoneal cavity of a mouse for several days in order to observe its biodegradation. Figure 2 shows SEM images of the surface structure of the chitosan sheet after biodegradation for 7 days, as initial stage of the degradation was seen at 24 h. At 7 days, the surface was considerably more irregular and the difference between light and dark regions was clearer than with the original, which appears to be a series of small cracks at x100 (Figure 2B). Furthermore, biodegradation of macromolecular substances entangled in the fibers was observed both at the surface and within the sheet at x10000 (Figure 2C). The sheet gradually decomposed and became very fragile, and for several months a lump of particulates was visible; its appearance was substantially different from the original sheet, providing a visual indication of the chitosan's degradation.

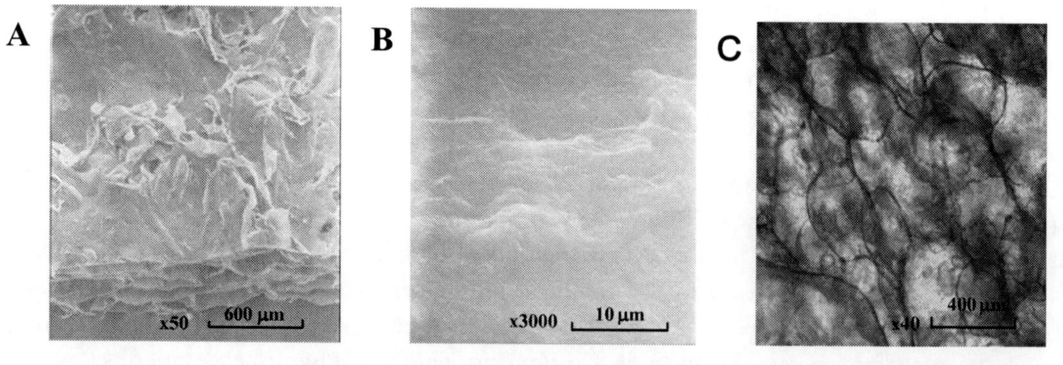

Figure 1. Photomicrographs of chitosan sheet. (A and B) Sanning electron microscopy, (C) phase-contrast microscopy.

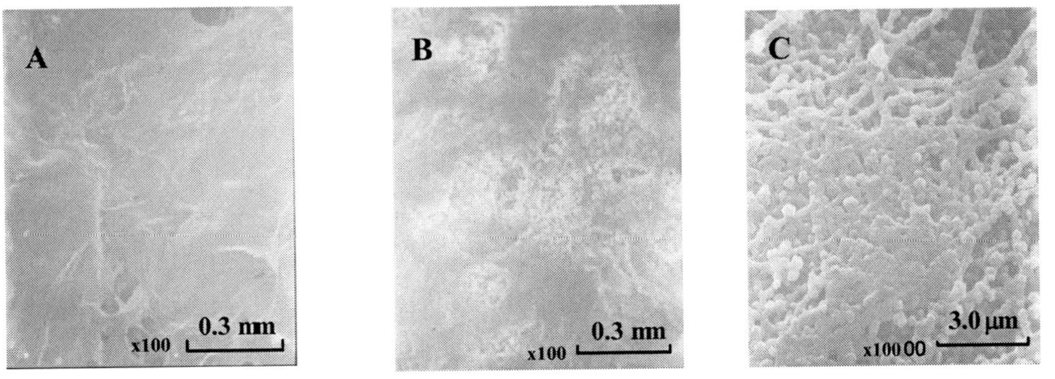

Figure 2. Photomicrographs (scanning electron microscope) of adriamycin-containing chitosan sheet after biodegradation in mouse peritoneal cavity for 7 days. (A) Before biodegradation, (B and C) after biodegradation.

3. Feasibility of ADM-containing Chitosan Sheet as Drug Carrier

ADM released from the sheet was measured in liver, blood and urine using HPLC (Figure 3). ADM was detected in urine and liver for 1 and 2 weeks, respectively, after the ADM-containing chitosan sheet was placed in the peritoneal cavity. In contrast, no ADM was detected in blood; however, ADM metabolites (adriamycinol and adriamycinone) were released into blood for 2 weeks. In particular, adriamycinone exhibited steady release from 1 to 2 weeks, before subsequently decreasing, In addition, the residual ADM on chitosan sheet was determined after application in vivo (Figure 4. A). The ADM content of the chitosan sheets decreased gradually, which indicates that it was released into tissues. ADM was released from chitosan sheet embedded in the abdominal cavity. Furthermore, besides its metabolites, over 0.5 μg/g of adriamycin remained in the chitosan sheet without being metabolized after two months in vivo (Figure 4. B), showing that the chitosan sheet might maintain and exhibit the medicinal effects of adsorbed drugs for several months.

Discussion

We developed a drug carrier that composed of 100% chitosan that was able to adsorb ADM and was subsequently biodegraded in mice. Chitosan is a slightly cationic natural polysaccharide with numerous reactive amino groups in its molecular structure that possesses many useful characteristics; viscosity, antibacterial activity, humectant, biocompatibility, biodegradation and physiological activity in animals. Various products derived from chitosan have been developed in a variety of fields, and products with useful biological properties attract the attention of medical fields. However, most such products are not produced from 100% chitosan, but rather contain chemical modifications or included impurities, which may adversely influence the living body. The chitosan sheet in this study was produced using a simple technique based on agglutination (described in Materials and Methods).

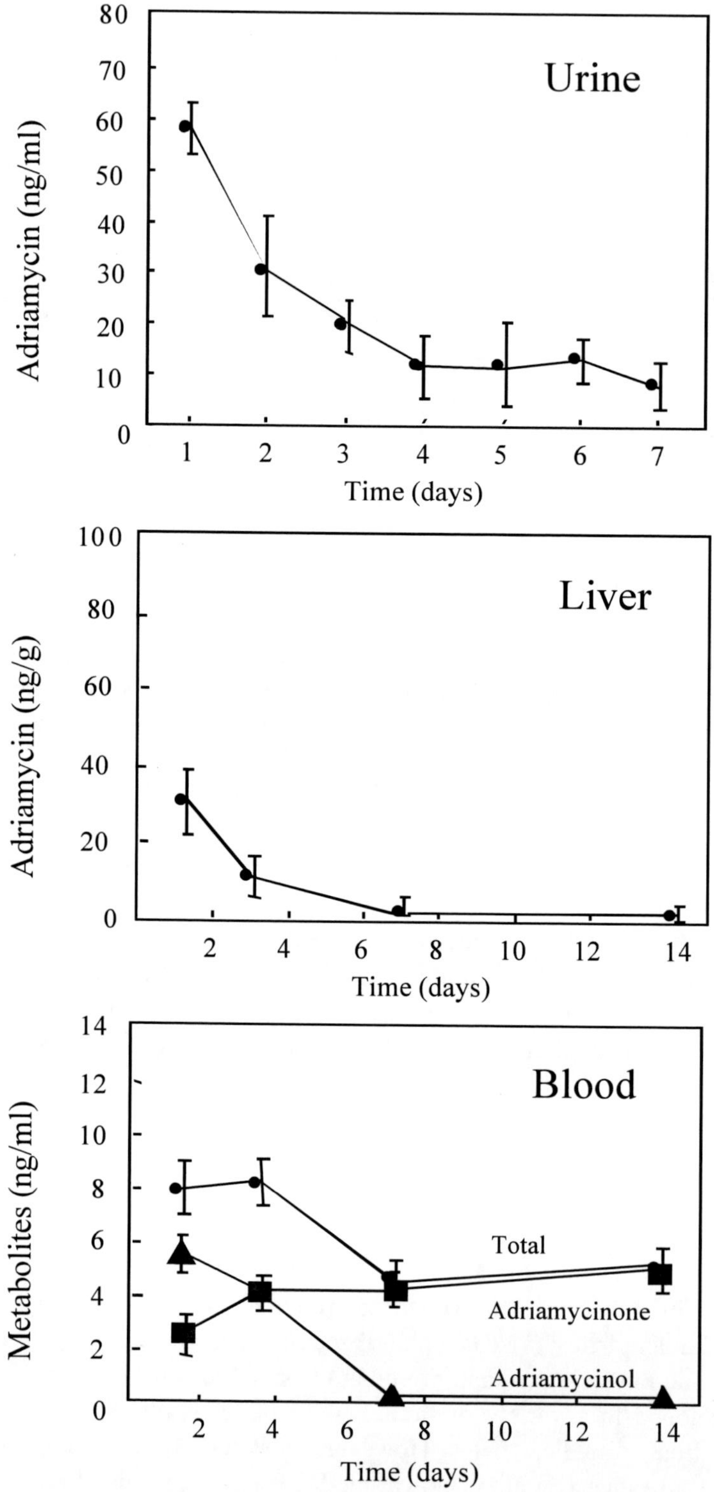

Figure 3. Release kinetics of adriamycin and its metabolites from chitosan sheet in mice. Data are presented as mean ±SD (n=4).

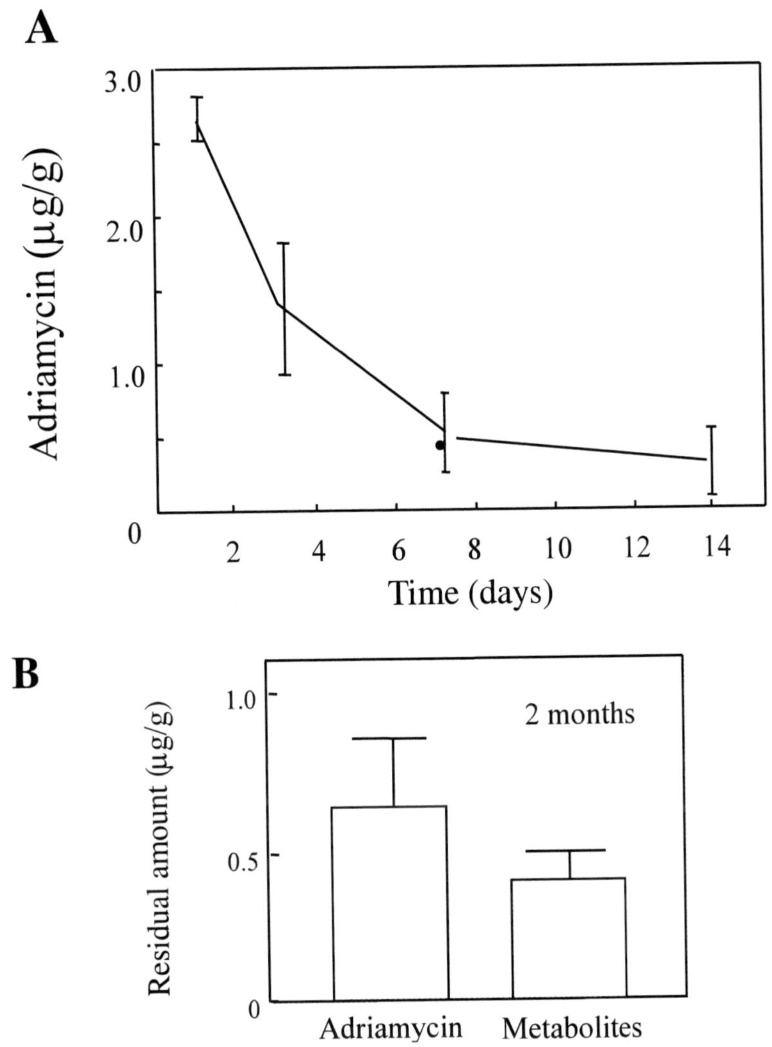

Figure 4. Decline of adriamycin in the chitosan sheet after the application to abdominal cavity of mice. (A) Residual adriamycin for 2 weeks, (B) residual adriamycin and its metabolites after 2 months. Data are presented as mean ±SD (n=4).

Chitosan was originally used as a flocculant to coagulate negatively charged suspended particles founded in turbid natural water [17] and as a water-purifying agent to remove heavy metals in wastewater treatment [18, 19]. Chitosan is also a known sorbent, effective in the sorption of transition metal ions because the amino and hydroxyl groups on the chitosan chains can serve as coordination and electrostatic interaction sites [20-22]. Chitosan is probably dominated simultaneously by adsorption, ion exchange and chelation. These characteristics may enable chitosan to adsorb ADM and coagulate uniformly under freezing conditions.

Fluorescence microscopy confirmed ADM adsorption by chitosan, even though the sheet was washed firmly with water. There were no significant differences between the ADM-containing chitosan sheet and non-ADM-containing chitosan sheet on SEM observation. In addition, no differences in biodegradability were seen between the two, which suggests that the presence of ADM did not affect chitosan sheet conformation (data not shown). However, the conformation of the chitosan molecules in the sheet is unknown, both in the presence and absence of ADM. Further experiments on the chemical behavior of chitosan molecules with ADM are necessary for subsequent drug design and clinical applications.

The biodegradation images shown in Figure 2 has the same appearance as those obtained after degradation by lysozyme (data not shown), showing that the decomposition of the sheet in mice is mediated by lysozyme, which is known to be ubiquitous in the body and to play an important role in degrading chitosan into oligomers that are further hydrolyzed into simple sugars *in vivo* [23]. Lysozyme release from specific cells or organs into blood is induced by chitosan. The enhancement of lysozyme is probably a biological defensive function, followed by activation of the macrophage system in animals [24]. We observed that cells like macrophage adhered to the surface of the chitosan sheet after application (data not shown). The chitosan sheet was stable in water and degraded with lysozyme or an acidic solution in vitro, and ADM was extracted from the sheet with an organic solvent (chloroform/methanol), however, ADM was not detected from the sheet that reacted only with lysozyme in vitro. These results suggest that biodegradation of the chitosan sheet that releases ADM in vivo is mediated not only by lysozyme but also by biofunctions related to metabolism, which might be affected by the interaction between ADM and the chitosan molecules. Although we were not certain of the rate of ADM release and the mechanism of degradation of chitosan, the biodegradable chitosan sheet appeared to decompose and release ADM. We know this because ADM was detected in liver and urine for 1 and 2 weeks, respectively. In addition, ADM metabolites, adriamycinone, and adriamycinol were present in the blood for two weeks; these were measured by HPLC as shown Figure 3. The half-life of ADM in blood is so short that it disappears within several hours. It is difficult to detect actual ADM in blood; however its metabolites exhibit medicinal effects *in vivo* [25]. For ADM, metabolites are effective in medical treatment; however, it is necessary to further investigate drugs that are not readily metabolized and their effectiveness in direct application using in chitosan sheets.

The ADM content of the chitosan sheets embedded in the abdominal cavity decreased in time dependency as shown in Figure 4, and it appeared that ADM was released into tissues as a result of the degradation of the chitosan sheet. The sheet decomposed and became very fragile, and particulates were visible after several months. Furthermore, over 1 μg/g of adriamycin and its metabolites still remained in the sheet even though the appearance of the sheet was disfigured after two months in vivo. This illustrates that the chitosan sheet was able to release drugs without being ruined.

In addition, several studies have demonstrated the high biocompatibility, low immunogenicity and biodegradability of chitosan, suggesting that chitosan can be considered as a sustained drug release material, as controlling degradation is possible by utilizing the biodegradability properties.

Multi-porous beads prepared from chitosan have been tested for physical and immunological properties as an anticancer chemotherapeutic carrier, but antitumor activity was found to be less than that of free ADM [26]. Recently, 100% chitosan particles containing a drug were developed [1,2], but were not efficacious in the affected region because mobility prevented continuous of drug activity during medical treatment. Furthermore, small particles with drugs may be digested by macrophages before exerting sufficient therapeutic effects, even if macrophages are activated. With regard to therapeutic efficiency, when producing a chitosan drug carrier for controlled release, material quality and form, in addition to function, must be carefully considered. A non-woven fabric of 100% chitosan is suitable for preventing undesirable side effects, such as inflammation and allergies.

The present chitosan sheet is permeable to water and air, and is flexible, in addition to possessing the properties typically associated with chitosan, which shows a useful material for medical use.

Conclusion

The biodegradable simple chitosan sheet, which was developed in the present study, might be an efficacious material as a carrier of sustained-release drugs and more advanced clinical applications are expected in both animals and humans.

References

[1] M. Nakajima, K. Atsumi, K. Kifune, K. Miura, H. Kanamaru, Chitin is an effective material for sutures, *Jap. J. Surg.* 16 (1986) 418-424.

[2] G.Biagini, A. Bertani, R. Muzzarelli, A. Damadei, G. DiBenedetto, A. Belligolli, G.Riccotti, C. Zucchini, C. Rizzoli, Wound management with N-carboxybutyl chitosan, *Biomaterials* 12 (1991) 281-286.

[3] S. Tanioka, Y. Okamoto, S. Minami, A. Matsuhashi, S. Tokura, H. Sashiwa, H. Saimoto, Y. Shigemasa, Development of chion and chitosan biomaterials, in: M. Yalpani (Eds.), *Carbohydtrates and Carbohydrate Polymer,* ATL Press, Mt. Prospect U.S., 1993, pp. 153-164.

[4] S. Minami, D. R. Khadal, K. Miyatake, Y. Okamoto, H. Ohta, S. Sasaki, Y. Shigemasa, Biological effect of chitosan-lactose mixure and it's clinical application. *Chitin and chitosan Reseaches* 7 (2001) 268-272.

[5] Y. Shigemasa, S. Minami, Applications of chitin and chitosan for biomaterials, Biotechnol. *Genet. Eng. Rev.* 13 (1996) 383-420.

[6] J. F. Prudden, P. Migel, P. Hanson, L. Friedrich, L. Balassa, The discovery of a potent pure chemical wound-healing accelerator, *Am J Surg.* 119 (1970) 560-564.

[7] F. L. Mi, S. S. Shyu, Y. B. Wu, S.T. Lee, J. Y. Shyong, R. N. Huang, Fabrication and characterization of a sponge-like asymmetric chitosan membrane as a wound dressing, *Biomaterials* 22 (2001) 165-173.

[8] S. Hirano, Chitin and chitosan, in: Wiley-VCH (Ed.), *Ullmann's Encyclopedia of Industrial Chemistry,* Wiley-VCH, Weinheim, Germany, 2001.

[9] T. Mori, M. Okumura, M. Matsuura, K. Ueno, S. Tokura, Y. S, Minami, T. Fujinaga, Effects of chitin and its derivatives on the proliferation and cytokine production of fibroblasts in vitro, *Biomaterials* 18 (1997) 947-951.

[10] Y.Okamoto, M. Watanabe, K. Miyatake, M. Morimoto, Y. Shigemasa, S. Minami, Effects of chitin/chitosan and their oligomers/monomers on migrations of fibroblasts and vascular endothelium, *Biomaterials* 23 (2002) 1975-1979.

[11] S. Miyazaki, H. Yamaguchi, C. Yokouchi, M. Takada,W. M. Hou, Sustained release of indomethacin from chitosan granules in beagle dogs, *J. Pharm. Pharmacol.* 40 (1988) 642-643.

[12] H. Tamura, S. Nishikawa, S. Tokura, Controlled release of model drug by using biodegradable polysaccharides, *Chitin and Chitosan Research* 7 (2001) 253-256.

[13] K. A. Janes, M. P. Fresneau, A. Marazuela, A. Fabra, M. J. Alonso, Chitosan nanoparticles as delivery systems for doxorubicin, *J. Control. Release* 73 (2001) 255-67.

[14] S. Mitra, U. Gaur, P. C. Ghosh, A.N. Maitra, Tumour targeted delivery of encapsulated dextran-doxorubicin conjugate using chitosan nanoparticles as carrier, *J. Control. Release* 74 (2001) 317-323.

[15] F. L. Mi, Y. B. Wu, S. S. Shyu, J. Y. Schoung, Y. B. Huang, Y. H. Tsai, J. Y. Hao, Control of wound infections using a bilayer chitosan wound dressing with sustainable antibiotic delivery, *J. Biomed. Mater. Res.* 59 (2002) 438-449.

[16] S. Shinozawa, T. Oda, Determination of ADM (doxorubicin) and related fluorescent compounds in rat lymph and gall by high-performance liquid chromatography, *J. Chromatogr.* 212 (1981) 323-330.

[17] Q. P. Peniston, E. L. Johnson, *Treatment an aqueous medium with chitosan and derivatives of chitin to remove an impurity,* U.S. Patent 3533940 (1970).

[18] S. E. Bailey, T. J.Olin, R. M. Bricka, D.D. Adrian, A review of potentially low-cost sorbents for heavy metals, *Water Res.* 33 (1999) 2649-2479.

[19] M.S. El-Geundi, Adsorbents for industrial pollution control, *Adsorpt. Sci. Technol,* 15 (1997) 777-787.

[20] E. Guibal, I. Saucedo, M. Jansson-Charrier, B. Delanghe, P. Le CloirecUranium and vanadium sorption by chitosan and derivatives. *Water Sci. Technol.* 30 (1994) 183-190.

[21] E. Onsoyen, O. Skaugrud, Metal recovery using chitosan, *J. Chem. Technol. Biotechnol.* 49 (1990) 395-404.

[22] O. A. Monteiro Jr., C. Airoldi, Some thermodynamic data on copper-chitoson and copper chitosan biopolymer interaction. *J. Colloid Interface Sci.* 212 (1999) 212-219.

[23] S. H. Pangburn, P.V. Trescony, Heller J. Lysozyme degradation of partially deacetylated chitin, its films and hydrogels, *Biomaterials* 3 (1982) 105-108.

[24] S. Hirano, M. Iwata, K. Yamanaka, H. Tanaka, T. Toda, H. Inui. Enhancement of serum lysozyme activity by injecting a mixture of chitosan oligosaccarides intravenously in rabbits, *Agric. Biol. Chem.* 55 (1991) 2623-2625.

[25] K. Mross, P. Maessen, W. J. van der Vijgh, H.Gall, E. Boven, H. M. Pinedo, Pharmacokinetics and metabolism of epidoxorubicin and doxorubicin in humans, *J. Clin. Oncol.* 6 (1988) 517-526.

[26] K.Nishimura, S. Nishimura, H. Seo, N. Nishi, S. Tokura, I. Azuma, Macrophage activation with multi-porous beads prepared from partially deacetylated chitin, *J. Biomed. Mater. Res.* 9 (1986) 1359-1372.

In: Leading-Edge Environmental Biodegradation
Editor: L. E. Pawley, pp. 11-51

ISBN: 978-1-60021-903-0

Chapter 1

MODERN TRENDS IN APPLICATION OF IMMOBILIZED CELLS FOR ENVIRONMENTAL BIOREMEDIATION

Elena N. Efremenko*[*]*, Ilya V. Lyagin, Olga V. Senko, Nikolay A. Stepanov, Olga V. Spiricheva and Rufat E. Azizov

The M.V. Lomonosov Moscow State University, Moscow, Russia

ABSTRACT

The review is devoted to up-to-date trends in application of biocatalytic systems, based on various immobilized cells of microorganisms, for environmental bioremediation. Mostly, attention is paid to the treatment of wastewaters produced by different branches of industry and agriculture. Three main aspects concerning bioremediation processes with immobilized cells are under consideration. The first one is connected with application of microorganisms in an immobilized state merely for environment protection owing to deep destruction of dangerous pollutants present in wastewaters. The use of immobilized cells for biodegradation of neurotoxic organophosphorous pesticides and fat-containing compounds as components of wastewaters is discussed. The second aspect of environmental bioremediation deals with conversion of agricultural wastes to some desired products. Mainly, the interest is concentrated on production of fuel bioethanol by immobilized yeast and filamentous fungi and obtaining of L(+)-lactic acid in the processes with immobilized bacteria and filamentous fungus cells. The efficiency of conversion of various enzymatically-pretreated wastes to the target products by immobilized cells is analyzed. Oppositely, the third direction of bioremediation is focused on the prevention of formation of stable systems with immobilized cells that can do damage to environment. The approaches, allowing completely avoiding or decreasing the appearance of corrosive biofilms composed of self-immobilized individual or mixed bacterial cultures in oil industry, are pointed in the review.

[*] efremenko@enzyme.chem.msu.ru

Keywords: immobilized cells, organophosphorous pesticides, chemical warfare agents, wastewaters, lactic acid, bioethanol, corrosive biofilms

INTRODUCTION

The rational use of environmental sources as well as effective treatment of wastes, accumulated as a result of functioning of modern technologies intensively applied in industry and agriculture, is a problem of primary importance. In general, the humanity survival depends on effective solution of the problem, since the amounts of wastes increases annually, whereas the evident decrease in nonrenewable natural resources is observed. So, modern biotechnology should offer a number of efficient methods and approaches to the problem solution, allowing renewing and remaining a balance in environment.

1. APPLICATION OF IMMOBILIZED MICROORGANISMS FOR ENVIRONMENT PROTECTION

1.1. Immobilized Cells in the Biodegradation of Neurotoxic Organophosphorous Compounds

1.1.1. Modern Situation Confirming Necessity of Detoxification of Organophosphorous Pesticides and Warfare Agents

According to the expert estimations made by analytical specialists from Frost & Sullivan, the world volume of applied organophosphorous pesticides, possessing neurotoxic and carcinogenic action, increases per 11.7% annually. In general, 1.4 billion hectares is treated by pesticides in the world, and organophosphorous compounds (OPC) compose the major part of used insecticides. OPC are widely used in the cultivation of various crops: grains, fruits, vegetables, tobacco and cotton [1]. The acute toxicity of these broad-range neurotoxins is of great environmental concern, and stringent regulations have been implemented worldwide, limiting the agricultural targets as well as the maximum levels of permissible residues in food for human and animal consumption. Nevertheless, the application of OPC along with their low water solubility results in the accumulation of the compounds in soil, ground, river and waste agricultural waters. All OPC are characterized by acute delayed neurotoxicity, but the mutagenic effect of cumulative low OPC quantities appears to be even more dangerous than neurotoxic one. The need for effective detoxification systems, which can be used in an environmentally responsible manner, is generally recognized.

There is a lot of information concerning the chromosome aberrations revealed among the population of territories treated with the OPC as well as participating in the OPC production. Along with this, OPC could penetrate the human body over skin, inhalation and with water and foods.

The scale application of useful organophosphorous pesticides as well as necessity of destruction of toxic pesticides, obsolete or already prohibited from the agricultural use, activates development of appropriate preparations with high enough detoxifying effect and creation of technologies based on employment of detoxifiers for the destruction of resistant organophosphorous pollutants. It should be noted, that 500 thousand tons of unfit organophosphorous pesticides are stored today in the stockpiles all over the world and should be destroyed.

Another one problem, concerning destruction of OPC and aimed at the reduction of environmental risk, is formed by the realization of Chemical Weapons Conversion, according to which thousand tons of stored organophosphorous warfare agents (Vx, sarin, soman) should be destroyed during next five years [2]. The technologies accepted for the destruction of chemical warfare agents (CWA) have various disadvantages (non-destructed residual amounts of toxic CWA, corrosive or flammable wastes, etc.) and all means enabling their overcoming are of high scientific and practical importance. The analysis of technologies offered for the destruction of CWA and based on various chemical and physical methods showed that there would be lots of solid, liquid and gaseous wastes [3]. The residual concentrations of CWA (up to 0.001%) in reaction mixtures, obtained after alkaline neutralization of the OPC, and hydrolytic products containing very stable C-P bond, namely phosphonic acid and its esters, constitute the main threat for human beings and nature.

The biotechnological methods based on the application of microorganisms look very attractive for their use in the degradation of organophosphorous pollutants since biological approaches are characterized by a high degree of OPC destruction, effective assimilation of products and thereby guarantee the ecological safety of processes.

Immobilization of cells provides an increase in efficiency of microbial treatment of wastes with neurotoxic OPC.

1.1.2. Main Factors Effecting on Characteristics of Immobilized Microbial Biocatalysts for OPC Degradation

To date a number of various approaches to the immobilization of cells catalyzing destruction of OPC are known (Table 1). The immobilized biocatalysts are differed in methods, microbial cells and supports used for their production. The variation in methods of cell immobilization was directed at the finding of conditions guarantying a less decrease in cell metabolic activity and long operation stability. Nevertheless, efficiency of action of immobilized biocatalyst mostly depends on the hydrolytic characteristics of used cells possessing by them in relation to OPC.

The known data testified to the fact that degradation of OPC by immobilized biocatalysts can be successfully realized under batch and continuous conditions. It is necessary to note that one of the first works related to this field was carried out in continuous reactor [13]. The next immobilized biocatalyst tried under similar conditions was developed only ten years later. Certainly, it was predetermined by difficulties appearing throughout the long-term exploitation of immobilized cells. Most difficulties were caused by the disadvantages of supports used for cell entrapment. To confirm the supposition, it should be noted that immobilized biocatalyst obtained by entrapment of bacterial consortium into Manville R-635 diatomaceous earth pellets [13] lost 50% of its initial metabolic activity for 60 days (Table 1). The posterior elaborations on the base of natural and artificial bacterial consortia immobilized into cryogel of poly(vinyl alcohol) (PVA cryogel) [11-12] were characterized by a high stability of functioning of developed systems. The 5%-loss of initial metabolic activity was established for discussed biocatalysts during twice longer period of their investigation compared to biocatalyst prepared on the base of Manville R-635.

Table 1. Degradation of OPC by immobilized biocatalysts based on various microorganisms

<table>
<tr><th>Process</th><th>Cells</th><th>Carrier</th><th>Immobilization</th><th>Residual activity (%) after usage period (d)</th><th>Substrate</th><th>Process productivity, mg/L/h</th><th>Ref.</th></tr>
<tr><td rowspan="12">Batch</td><td>Pseudomonas A3</td><td rowspan="3">Ca-alginate gel</td><td rowspan="3">Entrapment</td><td>73% after 12 d</td><td>Methyl-parathion</td><td>12.8</td><td>[4]</td></tr>
<tr><td>Bacillus sp.</td><td>n/a*</td><td>Methyl-parathion</td><td>1.62</td><td>[5]</td></tr>
<tr><td rowspan="8">E. coli</td><td>n/a</td><td>Chlorpyrifos</td><td>0.00129</td><td>[6]</td></tr>
<tr><td rowspan="4">Polypropylene</td><td rowspan="4">Adsorption</td><td rowspan="4">97% after 19 d</td><td>Paraoxon</td><td>116</td><td rowspan="4">[7]</td></tr>
<tr><td>Coumaphos</td><td>2.08</td></tr>
<tr><td>Methyl-parathion</td><td>18.0</td></tr>
<tr><td>Diazinon</td><td>197</td></tr>
<tr><td>Cellulose fiber</td><td>Specific attachment via cellulose-binding domen</td><td>90% after 45 d</td><td>Paraoxon</td><td>60.6</td><td>[8]</td></tr>
<tr><td rowspan="4">PVA cryogel</td><td rowspan="5">Entrapment</td><td>80% after 90 d</td><td>Paraoxon</td><td>26.0</td><td>[9]</td></tr>
<tr><td>95% after 120d</td><td>Coumaphos</td><td>614</td><td>[10]</td></tr>
<tr><td rowspan="3">Bacterial consortium</td><td>n/a</td><td>IMPA</td><td>1.13</td><td>[11]</td></tr>
<tr><td>n/a</td><td>Parathion</td><td>21.0</td><td>[12]</td></tr>
<tr><td rowspan="4">Continuous</td><td>Manville R-635</td><td>50% after 60 d</td><td>Glyphosate</td><td>45.0</td><td>[13]</td></tr>
<tr><td rowspan="3">E. coli</td><td rowspan="2">Porous glass</td><td rowspan="2">Chemical cross-linking</td><td rowspan="2">n/a</td><td>Coumaphos</td><td>36.3</td><td rowspan="2">[14]</td></tr>
<tr><td>Paraoxon</td><td>275</td></tr>
<tr><td>Cellulose fiber</td><td>Specific attachment via cellulose-binding domen</td><td>90% after 48 d</td><td>Paraoxon</td><td>95.6</td><td>[15]</td></tr>
</table>

* n/a – not available.

Various microorganisms are used to obtain immobilized biocatalysts for OPC degradation, but *E. coli* cells are applied more often than others. Mainly, that is due to a wide use of the cells as gene-engineered producers of recombinant enzymes catalyzing the hydrolysis of OPC [7-10, 14-16]. The application of such cells as active parts of immobilized biocatalysts guarantee the realization of individual intracellular enzymatic activities sighted on the hydrolysis of certain substrates, whereas the following transformation and utilization of products of the enzymatic reactions is not happened. Since then, the application of bacterial consortia including microorganisms with various enzymatic activities, catalyzing consequent reactions of OPC transformation, is more attractive and efficient from point of view of environmental safety and a high degree of biodegradation [11-13]. In some cases, the successful involvement of recombinant *E. coli* cells into multicomponent microbial associations of OPC destructors appeared to be very positive for general process [12]. The cells with single but key enzymatic activity, guaranteeing the detoxification of mostly toxic substrate, provide access to other members of complicated microbial system to products easily metabolized by them. Thereby, the introduction of some cells with certain catalytic feature into complex biodegradation system usually widens its potential substrate specificity and improves complete breakdown of pollutants. Immobilization plays an important role in this case since it allows to keep integrity of used multicomponent biosystem.

Definitely, the methods of immobilization of living cells are critical to the operation of biocatalysts. There are several known methods of cell immobilization tried out in the development of biodegradation systems for OPC destruction: cell entrapment to the support structure [4-6, 9-13]; adsorption of cells on support surface [7]; crosslinking of cells with glutaric aldehyde [14] and affine binding of genetically modified cells to support via fusion protein localized on the cell surface [8, 15]. All these methods are characterized by convincing merits and demerits.

The inclusion of cells into the volume of porous non-soluble supports as well as adsorption of cells on a carrier surface are physical methods of immobilization, that have no negative effect on viability of microorganisms, can be easily reproduced, scaled up and technically realized in industry. At the same time, both absorption and adsorption of cells are reversible processes, and therefore slow withdrawal of cells from biocatalyst can be observed. From technological point of view, the phenomenon usually results in decrease in total efficiency of action of immobilized biocatalyst and reduction of its functionality. From ecological point of view, the transition of recombinant cells, being a part of immobilized consortium, from the biocatalyst under environmental conditions can have unpredictable consequences, since genetically engineered cells usually contain mobile elements such as plasmids with genes, encoding resistance to antibiotics [9-11]. In this connection, the wide application of recombinant cells for realization of biotechnological processes in environment is still under discussion. Sure, the process of cell release from granules of immobilized biocatalysts should be controlled. First of all, that can be done by use carriers possessing a high enough mechanical and chemical stability and by organization of certain regimes of biocatalyst exploitation, allowing the reduction of cell desorption.

Several materials were known to use as carriers for physical immobilization of microorganisms participating in OPC degradation: Ca-alginate gel [4-6], PVA cryogel [9-12], polypropylene [7] and Manville R-635 diatomaceous earth pellets [13]. The lowest level of cell desorption was revealed in the case of PVA cryogel, since it is known that the polymer partially penetrate to cell wall during cryostructuring [17] and plays a role of additional anchor for immobilized cells.

The crosslinking of cells with glutaric aldehyde is another well-known method, based upon chemical interactions of bifunctional agent with functional groups localized on a surface of immobilized cells [14]. The method allows to completely avoid the release of cells from immobilized biocatalyst, but it brings a lot of other problems. As usual, this approach to cell immobilization, based on application of toxic compound such as glutaric aldehyde, is accompanied by dramatic decrease in viability of microorganisms and thereby the biodegradation features of biocatalysts. Both toxicity of immobilizing agent and its high cost considerably limit the distribution of the method for immobilization of OPC destructors and their use in environment. So, it can be correctly recommended only for processes of laboratory scale.

The modern approach to obtain the biocatalysts, based on immobilization of cells owing to multiple interactions between carrier and bacteria containing genetically introduced surface affine proteins, has an intermediate position among physical and chemical immobilization methods. On the one hand, this approach guarantees the absence of cell release from biocatalyst, and on the other hand, it does not require the use of toxic binding agents. The *E. coli* cells with surface-expressed cellulose-binding protein were used to elaborate immobilized biocatalysts on the base of a carrier such as cellulose fiber [8, 15].

The method is very attractive and simple in implementation, but it can be realized only with genetically modified cells. It is necessary to note that use of recombinant cells for obtaining and exploitation of immobilized biocatalysts is a justified act only if the accumulation of cell biomass with genetically changed properties could be done without application of expensive inductors, providing a high level of biosynthesis of target affine proteins. Since then, the serious genetic investigation, connected with creation of a new cell destructor of OPC, should precede the development of immobilized biocatalyst by this method. Moreover, the carrier used under environmental conditions should possess a resistance to biodegradation, and cellulose fiber has no that characteristic. So, all that facts testify to existence of definite limitations for the application of discussed method of immobilization of cells degrading the OPC.

Obviously, this method can become more attractive for its biotechnological application if mentioned limitations will be overcame via development of highly efficient expression systems on the base of *E. coli* cells and another carriers with a high resistance to biodegradation will be offered.

Thus, the support, employed for cell immobilization, microbial content of immobilized biocatalyst, enabling realization of various useful transformation activities, and the method of cell immobilization are the main factors influencing on the effectiveness of developed biocatalyst in the processes of OPC biodegradation. Analyzing all foresaid facts, it is possible to conclude, that use of microbial consortia with cells, producing key enzymes for OPC biodegradation, and the use of physical methods for cell immobilization, allowing control of cell release, compose the main modern tendencies of development of immobilized biocatalysts for OPC destruction.

1.1.3. Effectiveness of Action of Immobilized Biocatalysts in Biodegradation of Organophosphorous Pesticides

The comparison of different immobilized biocatalysts, recently developed for OPC destruction, appeared to be very difficult, because catalytic efficiency of their action was estimated using one or maximum two substrates. An exception to the rule was composes by a single work [7], in which four organophosphorous substrates were tested for biodegradation by immobilized cells.

Nevertheless, we tried to analyze the biocatalytic systems by process productivity determined in relation to degradation of most toxic substrate (Table 1). It should be emphasize, that comparison of productivity of processes carried out by immobilized cells under batch and continuous conditions is not correct, since the settings of the processes are absolutely different.

The general analysis of characteristics of biodegradation processes undertaken with immobilized biocatalysts (Table 1) allowed to conclude that loading of reactor with biomass of immobilized cells notably effects the total productivity of process. Furthermore, the concentration of cells inside of granules of biocatalysts also plays an important role in the efficiency of biosystem functioning. Usually, the concentration of cells in the content of wet biocatalyst is close to 10% (w/w) [9-12], but in some cases it can be less (down to 2-6% [4-6, 8]) or higher (up to 20% [7]) than that value.

Comparison of published results [4-5, 7] reflecting the application of various immobilized bacterial cells in the biodegradation of methyl-parathion evidently confirmed that use of genetically improved cells allowed to have significant increase in the productivity of process as compared to native bacterial strains (Table 1). Since in all three discussed biosystems the cells were immobilized by similar physical methods, so the established improvement of process characteristics obviously depended on the type of used cells.

As it was previously mentioned, the method of immobilization of cells significantly influences on the characteristics of biocatalyst and thereby on the productivity of process. So, the data obtained by different research groups [7, 9-10] confirmed this fact in experiments with same substrates (Paraoxon and Coumaphos). It is necessary to note, that in fact the combination of immobilization method with features of used carrier predetermines the final results in many cases.

The analysis of data, obtained in realization of batch and continuous approaches to the organization of processes with immobilized cells, showed that the latter approach guarantees more efficient functioning of the biocatalysts (Table 1). The productivity of the process operated under continuous regime was 1.6-times higher compared to the productivity of batch process with the same immobilized biocatalysts [8, 15].

As a final point, it should be mentioned that the regime of functioning of definite immobilized biocatalytic system always is chosen reasoning from the main tasks of the biodegradation technology and depends on the technological realization of biotechnological process.

The interest to the biodegradation of organophosphorous herbicide such as Glyphosate, being derivative of phosphonic acid and containing C-P bond, existed since beginning of its application in practice, namely for 20 years [18]. This compound inhibits the biosynthesis of aromatic amino acids in plant cells, completely terminating their growth [19]. The information concerning toxicity of Glyphosate for human beings is still absent. Very high resistance of C-P bond to the chemical cleavage provokes the intensive search for microbiological sources capable of its degradation.

Today, only one immobilized biocatalyst developed for the biodegradation of Glyphosate is known [13]. The application of bacterial consortium in immobilized form enabled the complete degradation of the herbicide. It was established, that cells utilized the substance not only as a source of phosphorus, but also as sources of carbon and nitrogen.

1.1.4. Immobilized Biocatalysts for the Degradation of Products of Chemical Neutralization of Organophosphorous Warfare Agents

The biotechnological approach to the destruction of products of chemical neutralization of orhanophosphorous warfare agents (Vx, soman, sarin) looks very attractive, since the use of microorganisms, catalyzing the cleavage of C-P bond, appeared to be the most effective for realization of the process.

The biocatalyst based on immobilized bacterial mixed cultures and PVA cryogel was developed for the biodetoxification of product of alkaline neutralization of sarin such as O-isopropyl-methylphosphonic acid (IMPA) [11]. The mixed cultures initially isolated from four different natural sources were used in the work. There were cells from: 1) water samples collected from ERDEC, Aberdeen Proving Ground (APG), Maryland; 2) soil samples collected from a cotton field where organophosphate pesticides has been used; 3) activated sludge sampled from a municipal wastewater treatment plant; 4) activated sludge from an APG wastewater treatment facility.

Before immobilization, all samples of mixed cultures were adapted to the nutrient medium containing IMPA for a long period of time (totally 35 days). The final concentration of IMPA at the end of cell adaptation was 0.35 g/L. The granules of PVA cryogel (2 mm in diameter) with immobilized bacterial cells were obtained as a result of entrapment of mixed cultures into polymer matrix. The formation of spherical granules of hydrogel was carried out in the hydrophobic medium formed by hexane. To remove the residual amounts of hexane from the surface of biocatalysts, the granules were washed by 15 mM HEPES buffer.

The biodestruction of IMPA by immobilized cells (5 g of wet biocatalyst/L) was investigated, when four biocatalysts with various mixed cultures were used in parallel. The best results on biodegradation were obtained with cells from the first isolate.

It was established, that it is possible to reach 91% destruction of 85 mg/L IMPA during 42.5 h, but methylphosphonic acid (MPA), being the product of IMPA hydrolysis, accumulated in the investigated system up to 0.29 mM and 30-35% of the amount was detected inside of immobilized cells. Further prolongation of biocatalyst application did not result in increase in degree of MPA degradation.

Under same conditions, the 85% hydrolysis of 350 mg/L IMPA was reached for 400 h (~17 days) with the immobilized biocatalyst obtained on the base of the first isolate. The process also was followed by the accumulation of high enough concentrations of MPA and by absence of their further destruction even with elongation of cultivation time of immobilized cells.

Thus, the cells selected for the destruction of IMPA in this work obviously had esterase activity and catalyzed the hydrolysis of IMPA till the MPA and isopropyl alcohol, but the further destruction of MPA with the cleavage of C-P bond was seriously complicated. Probably, the latter phenomenon was a main factor negatively influenced on the functioning of whole immobilized system in general.

The specialists from U.S. Army Edgewood Chemical Biological Center and AlliedSignal Inc. [20-21] used their joint forces to develop some technological realization of a process with immobilized cells, aimed at the destruction of alkaline hydrolysates of Vx. The microbial consortium GB2 [22-23], containing various types of bacterial cells, was immobilized inside of polyurethane foam and placed into 1 L flow reactor

The hydrolysates of Vx were introduced to the cultivation medium in 0.05-0.1% concentrations. Thus, there was a 1000-fold dilution of initially obtained alkaline hydrolysates. The O-ethyl-methylphosphonic acid (EMPA) present in Vx-hydrolysates as a main hydrolytic product was considered as a source of phosphorus in the medium for immobilized cells in the cultivation medium.

Isopropyl alcohol, glycerol, molasses and the mixture of two latter substrates were tested as main carbon sources in the work with immobilized cells. The ratio "C:P" was varied from 100:1 to 64:1 in the medium. The main idea of the variations was connected with the maximal increase in yield of cell biomass in the volume of bioreactor, since the intensive growth and metabolic activity of immobilized cells should contributed to enhancement of degree of EMPA biodegradation. The nutrition medium (600-800 mL) was pumped into reactor with the rate 50 mL/day.

It was revealed, that the maximal degree of EMPA biodegradation, that can be reached in such reactor with used immobilized biocatalyst, was 30-40%, and it did not depend on the initial concentration of carbon-containing substrate in the medium and the duration of reactor functioning. The total operational time of the system was 190 days.

The authors of the results supposed, that the use of two subsequently combined bioreactors with similar amounts of immobilized biocatalyst could improve the situation with EMPA biodegradation. However, in practice they preferred to use another approach to the process intensification based on the improvement of characteristics of the immobilized biocatalyst with GB2 consortium [23]. Finally, the 1.25 mm cubic granules of polyurethane foam, containing both bacterial cells and parts of activated charcoal, were prepared. Then, the aerobic biodegradation of diluted hydrolysates of Vx was carried out with immobilized cells (680 mL/L) at 25°C and pH 6.0-9.0. The total volume of medium with granules of immobilized biocatalyst was 740 ml/L in the 1 L reactor.

The 96% decrease in chemical oxygen demand (COD) and 75% removal of OPC such as EMPA and MPA from the medium were demonstrated with this biocatalytic system after its functioning for 15 days.

In spite of the fact, that deep enough degradation of OPC in the hydrolysates of Vx was reached in the work with immobilized cells, it was not total. Since then, the investigations processes of OPC degradation should be continued.

1.2. Biotreatment of Food Wastewaters by Immobilized Cells of Microorganisms

The biotreatment of various wastewaters of different fields of food industry is a problem of current importance, since most of them are characterized by the presence of a variety of organic compounds in high enough concentrations that should be eliminated. There are several well-known approaches to the problem solution: physical separation of wastes, application of chemical reagents for dirt flocculation or destruction and biotechnological treatment. The latter method supposes the use of wastewaters with a mixture of components as nutrient media for cultivation of microorganisms capable of utilizing their main ingredients (Table 2). The review of recent publications, analyzing content of wastewaters from food industry and domestic wastes, showed that lipids, proteins, carbohydrates and salts are among the most abundant components.

Recently, a number of biotechnological processes aimed at the effective treatment of wastewaters were developed on the base of diverse microorganisms [41-45], but realization most of them was entailed by necessity of utilization of large amounts of biomass actively accumulated as a result of degradation of organic wastes [46-47]. Usually, these consequences take place, since the processes are based on the application of actively growing cells.

Immobilization allows to overcome the problem, because it enables creation of high cell concentrations in the reactors from the beginning of process and to use the same biomass for wastes treatment. In addition, the application of cells in immobilized form looks more attractive for biological treatment of wastewaters as compared to suspended form of microorganisms, since it guarantees simple separation of cultural medium from cells and opportunity of carrying out of the process under continuous regime [48]. The long-term exploitation of immobilized cells in discussed systems is possible owing to their higher resistance to various unfavorable conditions as compared to free cells.

There are several investigations, where the authors estimated various carriers for immobilization of cells, used in the biotreatment of wastewaters [49-52]. The researches were guided by following main criteria, when they choose an appropriate carrier for the development of immobilized microbial biocatalysts: high mechanical stability and porosity of structure, low biodegradability, and simplicity of cell immobilization and reliability of coupling with cells.

Table 2. The content and main characteristics of food industry wastewater

Wastewater	Organic compound concentration, g/L			pH	COD, g/L	Ref.
	lipids	carbohydrates	proteins			
Starch processing	0.52-5.0	1.52-44.87	1.2-1.5	3.2-4.4	8.80-18.90	[24-26]
Meat processing	0.05-1.02	0.06-3.5	1.0	4.1-4.5	0.78-14.10	[27-29]
Wastewater from abattoirs	0.15-1.00	0.5-13.0	0.25-1.7	4.9-7.0	0.89-8.63	[30-31]
Whey processing plant	n/d*	0.36-0.68	2.2-14.8	5.0-9.0	1.16-2.12	[32]
Dairy wastewater	0.035-0.04	n/d	n/d	8.2-8.8	4.0-4.42	[33]
Olive mill wastewater	n/d	21.0-33.0	n/d	5.3-5.6	1.8-174	[34-36]
Oil-containing wastewater (kitchen, restaurant)	0.2-0.9	0.2-21.0	0.3-9.0	5.0-8.0	1.3-11.0	[37]
Food processing wastewater	n/d	0.005-9.28	0.27	4.5-11.7	0.15-21.0	[38-40]

* n/d – not detected.

Polyurethane is widely tried as a polymeric carrier for immobilized cells actively participating in the wastewater biotreatment [49], but poly(vinyl alcohol) hydrogels appeared to be the most suitable polymeric carriers for immobilization of various cells used for wastewater biotreatment since they usually have high-porous structure [49-51]. Natural polymeric gels such as Ca-alginate and k-carrageenan cannot compete with synthetic polymers as cell carriers in the wastewater treatment due to their biodegradability [53].

The biotechnological treatment of wastes with high concentrations of organic compounds, carrying out under aerobic conditions, usually appeared to be more effective compared to anaerobic operational systems. Since then, further discussion of advanced technological approaches developed in the field of wastewater biotreatment will be focused upon aerobic functioning of biocatalytic systems with immobilized cells.

An approach, based on the use of self-immobilized cells (biofilms) for the biotreatment of fat-containing waters, is well known and widely applied. To form active biofilms, capable of degrading various components present in wastewaters, free cells are usually inoculated into reactor, loaded with particles of various adsorbents (gravel, haydite, plastic, asbestos cement) [54] (Table 3). Later, the cells grown in the reactor form a biofilm covered the surface of porous carriers. Under natural conditions, the content of formed biofilms can differ from the initially introduced monoculture and comprises association of microorganisms.

Unfortunately, all such systems are characterized by two main limitations: 1) the biodegradation of fat-substances occurs only on the surface of biofilms; 2) the velocity of biodegradation of fat-containing components is low enough. For example, the degradation of 6-8 g/L fatty ingredients was established after 4-5 days of wastewater treatment by biofilms with complex of immobilized cells: *Pseudomonas, Proteobacteria, Flavobacteria, Acinetobacter, Microlunatus, Haliscomenobacter, Actinomycetes, Nitrosospira, Planctomycetes* [57-58]. The limitations in application of biofilms exist due to a low porosity of adsorbents used as carriers and low biodegradable activity of cells accumulating in the biocatalysts.

Another modern approach to the obtaining of active biocatalysts with immobilized cells proposes recycling pumping of activated sludge through a layer of polyurethane foam loaded into reactor [55]. The procedure allows to get an immobilized activated sludge in the porous structure of carrier. It was shown, that the biological system could be utilized in the treatment of 0.8 g/L fat-containing wastewaters of food industry for 100 days, but it was necessary to constantly add 10 g/L anionic detergents, to emulsify the fats. The fat-particles should not be more that 100 μm, since only the size guaranteed the bioavailability of fats for cells immobilized into the carrier. According to authors' data, the increase in size of fat-particles resulted in significant reduction of the process efficiency.

Table 3. The effectiveness of usage of activated sludge in the treatment of wastewaters of food industry under aerobic conditions

Carrier	Wastewater components and their concentrations, g/L	Process efficiency, %	Time of exploitation, d	Ref.
Gravel, haydite, etc.	Grease, 6.0-8.0 g/L	98-100	4-5	[54]
Polyurethane	Mixture of various oils, 30 g/L	98	100	[55]
Poly(vinyl-chloride)	Volatile fatty acids, 0.65 g/L	95	100	[56]

Another one polymer, poly(vinyl chloride), was used as carrier for the immobilization of activated sludge in the process providing biotreatment of wastewaters with a very wide spectrum of organic compounds [56]. The process was going on under continuous conditions in the reactor, loaded with polymer blocks containing immobilized cells. The wastewater treatment by this biosystem was conducted for 100 days. It should be noted, that there was no decrease in COD for the first 10 days of reactor functioning at all. Finally, the immobilized cells guaranteed 95%-decrease in COD. Probably, an adaptation of immobilized cells to the complicated combination of substrates took place.

Special consortium of microorganisms immobilized on ceramic carrier was offered for the treatment of wastewaters of food processing factory. Bacterial cells of genus *Acinetobacter* constituted the base of the consortium [38] (Table 4), which initially was isolated from the same wastewaters. So, there was need in special cell adaptation. The COD level (1.1 g/L) of wastewaters (pH 8.2) treated by the immobilized biocatalyst was low enough as compared to general level of COD known for wastes taken from different fields of food industry (Table 2). The maximal decrease in COD was 87%. The total operational period of the biocatalyst was 162 days.

Table 4. The effectiveness of usage of biocatalysts based on various immobilized microorganisms in the treatment of wastewaters of food industry

Microorganism	Carrier	Concentration of components, g/L	Process efficiency, %	Time of exploitation, d	Ref.
Consortium with bacteria of genus *Acinetobacter*	Cheramics	Carbohydrates (up to 9.28)	87	162	[38]
Mucor miehei	PVA	Grease (1.0-1.5)	100	30	[59]
Rhodococcus sp	Cheramics	Volatile fatty acids	100	2.5	[60]
Aspergillus niger	Polyurethane	Sugars (20-29)	85	6	[34]
Coriolus versicolor	Ca-alginate gel	Sugars (4-6)	60-100	15	[35]
Funalia trogii					
Phanerochaete chrysosporium	Loofa	Glucose (10.5)	39-88	8	[36]
Rhodobacter shaeroide	Ca-alginate and agar gels	Mixture of canola and soybean oils (5)	55.5	57	[61]
Rhizopus oryzae	PVA cryogel	Lipids (up to 21)	95-100	82	[37]
		Lipids (up to 21), carbohydrates (up to 50), proteins (up to 20)	95-100	82	[62]

The use of microorganisms secreting various hydrolytic extracellular enzymes seems to be the most effective way to obtain highly active immobilized biocatalyst for the water biotreatment. In this case, it is possible to apply single culture instead of a number of microorganisms with different substrate specificities. The fewer components are in the content of immobilized microbial biocatalyst, the easier control of the system is needed.

So, the immobilized biocatalyst was prepared on the base of filamentous fungi *Mucor miehei* being active producer of extracellular thermophilic lipases [59] (Table 4). The film of poly(vinyl alcohol), covered the internal part of flow reactor, was used as a carrier for the adsorption of spores and their further growth up to developed mycelium. The formation of immobilized mycelium was conducted at 40-42°C for 4 days by pumping of a nutrient medium into reactor.

The biotreatment of wastewater, containing 1.0-1.5 g/L cotton oil, was carried out at 40°C, when special additive of certain content (malt extract – 7.0 g/L, $(NH_4)_2SO_4$ – 0.3 g/L, $CaCO_3$ – 0.1 g/L) was introduced into the flow. The biocatalyst was exploited for 30 days. Obviously, the constant introduction of necessary additive into the treated water flow was a serious disadvantage of the system functioning.

A new biocatalyst was developed using *Rhodococcus sp.* cells immobilized on ceramic granules for biodegradation of various fatty acids in the content of municipal wastewaters (pH 8.0) [60]. The biotreatment was carried out at 37°C in the flow reactor. The complete removal of acetic and propionic acids from water was reached for 48 h of biocatalyst functioning. The 100% degradation of butyric and valeric acids was established after 60 h of biotreatment of water by immobilized cells. It should be noted, that using the same cells in suspended form guaranteed only 20%-destruction of all these acids after cultivation of bacteria for 72 h.

Newly developed biocatalyst was used to treat wastewaters of sugar industry. It was obtained on the base of immobilized bacteria and applied in flow reactor supplied with boiling bed [63]. The maximal conversion of sugar was 52 kg/m^3/day.

The olive mill wastewater (pH 3.9-6.3) is considered to have complicated content: 20-29 g/L sugars, 6 g/L phenols, 0.19-0.2 g/L phosphates and various microelements. The biocatalyst with fungus cells *Aspergillus niger* immobilized in polyurethane foam was efficiently used to treat the wastes. It was established, that it is possible to have 85%-decrease in COD after biotreatment of wastes for 6 days, but it was realized only after separation of olive oil from the wastewater and introduction of additional amounts of phosphates (up to 3 g/L) to the treated medium. Same biotreatment, undertaken with application of white rot fungi *Coriolus versicolor* and *Funalia trogii*, immobilized in Ca-alginate gels, was effective only after 5-fold dilution of olive mill wastewater [34, 35] (Table 4).

Recently, loofa-immobilized cells *Phanerochaete chrysosporium* were used to develop an empirical model useful for result prognosis in biotreatment of olive oil mill wastewater (OMW) [36]. The OMW had the following composition: 174 g/L COD, 5.2 g/L total phenols and pH 5.38. Various dilutions of OMW were used as basis for preparation of media for the biotreatment. Glucose and nitrogen sources were added into the treated media at different concentrations. The pH of the media at the end of the treatment was almost constant at 4.5-5.0. Loofa sponge used in the study was obtained from the matured dried fruit of *Luffa cylindrica*. For the experimental use, the sponge was cut into appropriate segments. The height and diameter of each segment were about 1 cm and 7 cm, respectively. The immobilization of *P. chrysosporium* cells within the loofa sponge disc was carried out as follows: the spore suspension was inoculated into Erlenmeyer flask containing diluted OMW solution and was incubated at 30°C for 8 days without agitation.

The removal of total phenols decreased as the OMW concentration increased from 20% to 40%. The COD decreased as the concentration of the added nitrogen source increased from 0.5 to 0.75 g/L. The COD removal increased as glucose concentration increased to the high level (7 g/L), while nitrogen source concentration was at 0.25 g/L. The TP removal was at its lowest level when the concentrations of the OMW and nitrogen source were 40% and 0.25 g/L, respectively. The optimum conditions for this treatment was achieved by setting the experiment with OMW concentration at 40%, while the nitrogen source concentration and glucose level were at 0.5 and 5 g/L, respectively. The mathematical equations presented in this study are in agreement with the results of other relevant OMW studies, regarding the correlation between the major pollutant removal profiles, i.e., COD and total phenols. Nevertheless, more information is needed for presenting fractionation procedure rapid and reliable enough, to give clear quantitative and informative relation in this area.

It was shown, that biocatalysts developed on the base of photosynthetic bacteria *Rhodobacter shaeroide* S and NR-3, entrapped into Ca-alginate or agar gels, could be applied for the treatment of oil-containing sewage wastewater [61]. The biotreatment of wastewater guaranteed the 3.5- and 3.0-fold decrease in initial concentration of oil, in fact, mixture of canola and soybean oils, for 6 days for the biocatalyst prepared with Ca-alginate and agar gels, respectively. It was possible to exploit both biocatalysts for 57 days. The maximal COD lowering was 55.5%, the level was reached for 54 days.

A new biocatalyst was developed for the effective biodegradation of various fats present as ingredients in wastewaters of different fields of food industry. Immobilized biocatalyst was prepared, using cells of filamentous fungi *Rhizopus oryzae* and PVA cryogel [37, 62]. It was established that period of half-life of the biocatalyst is 41 day in the media, containing up to 21 g/L lipids. The initial level of COD typical for wastewaters treated by the immobilized biocatalyst was in the range 2.7-11.0. The biocatalyst provided 95-100%-decrease in COD for 24 h (during single working cycle of batch process).

2. Conversion of Agricultural Wastes to Desired Products by Immobilized Microbial Biocatalysts

Microorganisms are widely distributed in nature, because they are capable of destructing various chemical compounds. Since then, the use of biocatalytic systems based on application of microbial cells allows to efficiently solve the problems concerning organization of ecologically-safe production of different substances and utilization of industrial and agricultural wastes with complicated chemical contents. To intensify the existing technologies, the development of advanced biocatalytic systems, involving perspective producers and alternative technological approaches, is definitely needed. The elaboration of biocatalysts comprising microbial cells, immobilized by various methods, and appropriate carriers looks like a very attractive and up-to-date method of innovative activity that can bring a highly efficient result.

2.1. Production of L(+)-lactic Acid by Immobilized Biocatalysts

2.1.1. General Situation with Biotechnological Production of Lactic Acid

Lactic acid (LA) is widely used in various fields of industrial production, i.e., in food and cosmetic industry, pharmaceutical and chemical industry, agriculture and production of biodegradable polymers [64-65]. A high level of LA production in the world could stimulate its wide use as a monomer for the chemical synthesis of polymers with high mechanical stability and plasticity that can be applied for the production of various goods of different shape and size (Table 5).

Table 5. Biodegradable polymer materials based on polylactates [64, 66]

Material	Producer	Material type	Application
Bellfree, Lactron	Kanebo	Granules, films	Labels, tapes, wares, packaging, garbage bags, shrinkable filmes, hawker's stands, rigid containers
EcoPla, NatureWorks	Cargill Dow Polymers		
Lacea	Mitsui Chemical		
Biorhan	Trespaphan	Films	
Ecoloju	Mitsubishi Plastics		
Terramac	Unitica		
Palgreen	Tohcello		
Nature-Works PLA	Nature-Works LLC		
Ingeo		Fiber	Textile and clothes
Bellfree, Lactron	Kanebo		Health and beauty aids, protective clothes and filters
Terramac	Unitica		
Biotech Color	Dainipon Inc & Chemical	Varnish-and-paint materials on the base of polylactates and starch	Paints

The current worldwide demand for LA is estimated roughly 150 thousant tons per year. However, the global consumption of LA for the production of polymers, containing poly(lactic acid) may increase the annual need in the monomer up to 500 thousand tons by 2010 [67].

Today, the production of LA is realized by both chemical and microbiological ways. The first one is preferable for the production of LA of low grade, whereas the microbiological process is used for the production of L(+)-form of LA appropriate for food industry, medicine and polymer synthesis. The microbiological approach to the production of LA also allows to utilize various substrates that in fact are wastes of various branches of industry. Of cause, they can be used in LA production after a special enzymatic pretreatment (Figure 1).

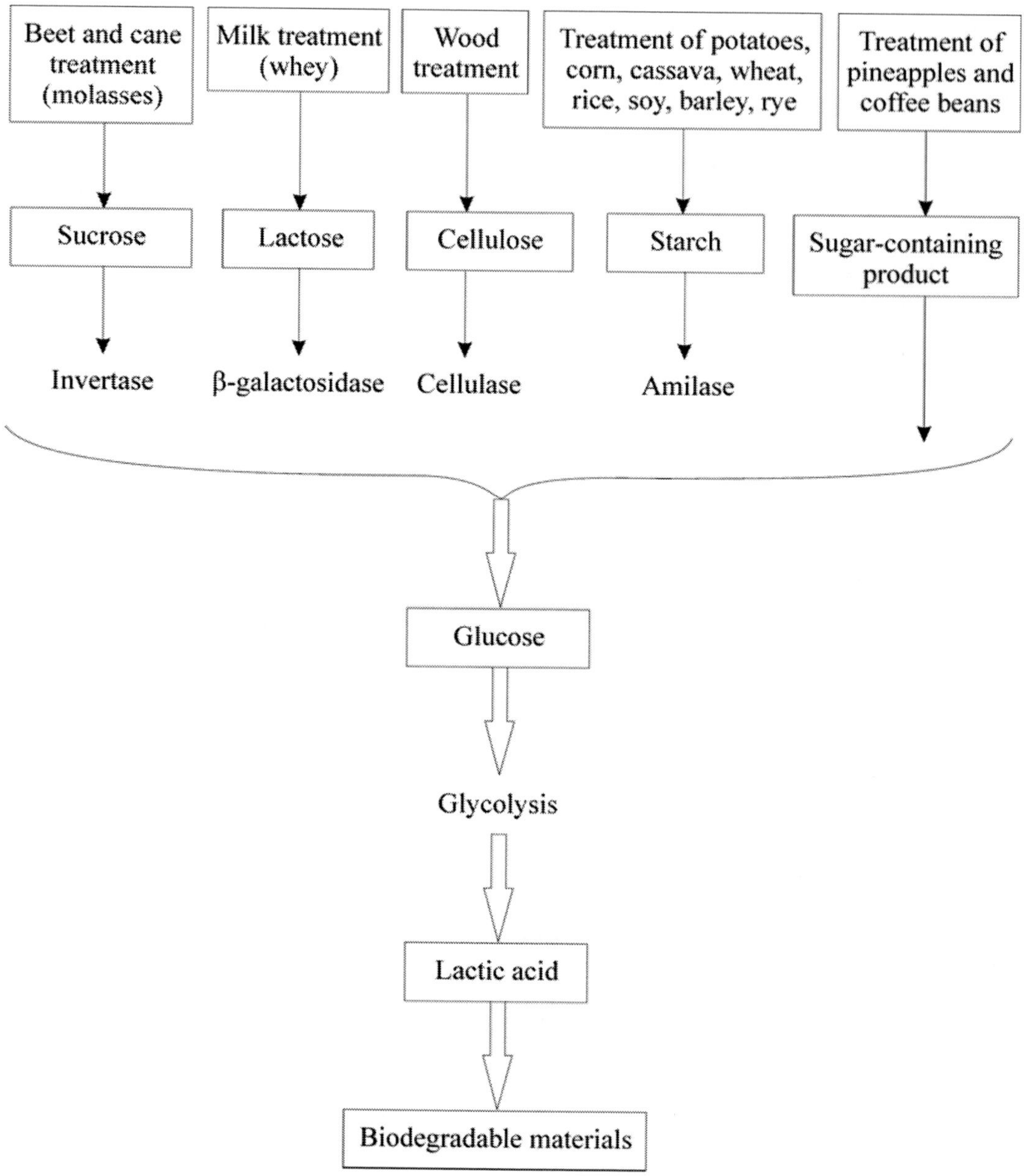

Figure 1. Various industrial wastes usable as raw materials for biotechnological production of lactic acid by microorganisms.

The annual volume of different agricultural wastes is about 3.5 billion tons [68], and most of the wastes could be converted into LA. For example, the wastes obtained after treatment of sugar beet [69-71], cassava [72-76], wheat [68, 77-79], rye [80], barley [81], rice [82], oats [83], corn [84-86], potatoes [71, 87-89], sugar-cane [90], pineapples [91], cellulose-containing materials [92-93], and even coffee [94] were already tried as substrates for LA production.

The homoenzymatic lactic acid bacteria such as *Lactococcus lactis* [95], *Lactobacillus delbrueckii* [75, 96], *L. helveticus* [97] and *L. casei* [74, 98] are considered to be very attractive for their use in practice for L(+)-LA production. The strains *L. amylophilus*, *L. manihotivorans* [79, 99] and *L. amylovorus* ATCC 33622 [100] are capable of secreting of enzymatic complexes catalyzing hydrolysis of complicated starch-containing substrates. However, it was established, that fermentation of starch and pentoses, i.e., xylose, is low effective, when the bacterial cells are used. The main reason is the formation of a racemic mixture of L(+)- and D(-)-forms of LA.

The representatives of filamentous fungi of genus such as *Mucor*, *Monilia* and *Rhizopus* also produce L(+)-LA. The fungus cells *Rhizopus oryzae* seems to be the most perspective LA-producer among others [82].

The homoenzymatic fermentation realized with bacterial cells requires the introduction of complicated nitrogen sources, containing various amino acids, and vitamins into the medium. Thereby, the latter ingredients increase the cost of the last product and create complications in LA purification [72, 101].

The filamentous fungi require utilization of media with a narrow list of components. The fungus cells appeared to be more resistant to the low pH values, possess the tolerance to inhibition effect of high enough concentrations of LA [102]. In addition, the filamentous fungus cells are capable of secreting complexes of hydrolytic enzymes. The latter feature of fungi enables the application of a wide spectrum of substrates for LA production and combination of substrate conversion to glucose with its transformation to the target product in the same process [66, 82, 85, 87-88].

However, it should be noted that fungus cells are sensitive to oxygen concentrations in the medium, so the cultivation of fungi should be organized under aerobic conditions in the reactors with satisfied mass-transfer characteristics. The application of mechanical stirrers usually causes the mechanical cell stress and results in the significant decrease in LA production by fungus cells.

Thus, immobilization of both bacterial and fungus cells could play an important role in the intensification of existing biotechnological processes of LA production.

2.1.2. Immobilized Cells in Biotechnological Production of L(+)-lactic Acid

The use of microorganisms in immobilized state in the production of L(+)-LA should improve their resistance to the action of negative factors (i.e., pH, LA high concentration, inhibitors present in various wastes used as substrates) and allows to increase the operational time of viable cells and their concentrations in the reactors.

At present, the use of immobilized cells in the production of L(+)-LA is known only at a level of lab-scale reactors. The results obtained with immobilized cells in the processes of conversion of various wastes and substrates are well known (Table 6), but they did not still realized in practice.

For example, it was shown, that molasses solution, being a waste of sugar-cane industry, could be easily converted into L(+)-LA by *Lactobacillus casei,* immobilized onto granules of activated charcoal by adsorption. The continuous process was carried out at 42°C with the immobilized cells, and the specific productivity of the process was 0.31 g/L/h [103].

Table 6. Production of lactic acid from food industry wastes with immobilized biocatalysts

<table>
<tr><th>Carrier</th><th>Microorganism</th><th>Substrate</th><th>[LA], g/L</th><th>Yield, %</th><th>Productivity, g/L/h</th><th>Ref.</th></tr>
<tr><td>Delignified cellulose</td><td>Lactobacillus brevis</td><td>Dairy whey</td><td>36</td><td>72</td><td>2 ***</td><td>[66]</td></tr>
<tr><td>Ca-alginate gel</td><td>Lactobacillus delbrueckii</td><td>Beet molasses*</td><td>70</td><td>82</td><td>13.92 **</td><td>[70]</td></tr>
<tr><td>Glass granules “Porabact A”</td><td>Lactobacillus casei</td><td>Dairy whey*</td><td>27</td><td>30</td><td>4.5 ***</td><td>[77]</td></tr>
<tr><td>Activated Alumina</td><td>Lactobacillus helvetikus</td><td>Lactose</td><td>75</td><td>75</td><td>2 **</td><td>[97]</td></tr>
<tr><td>Activated charcoal</td><td>Lactobacillus casei</td><td>Cane molasses</td><td>65</td><td>100</td><td>0.31 **</td><td>[103]</td></tr>
<tr><td>Polyethylene/polypro-pylene filmes</td><td rowspan="2">Lactobacillus casei</td><td>Lactose</td><td>30</td><td>100</td><td>1.2 **</td><td>[104]</td></tr>
<tr><td>Glass granules “Poraver”</td><td rowspan="6">Dairy whey*</td><td>93</td><td>93</td><td>3 ***</td><td>[105]</td></tr>
<tr><td rowspan="5">k-carrageenan gel</td><td>Lactococcus lactis</td><td>4</td><td>10</td><td>0.7 ***</td><td rowspan="3">[106]</td></tr>
<tr><td>Lactobacillus delbrueckii</td><td>12</td><td>30</td><td>2 ***</td></tr>
<tr><td>Streptococcus salivarius</td><td>20</td><td>50</td><td>2.5 ***</td></tr>
<tr><td rowspan="2">Lactobacillus helvetikus</td><td>43</td><td>70</td><td>21.5 **</td><td>[107]</td></tr>
<tr><td>41</td><td>70</td><td>14 **</td><td>[108]</td></tr>
<tr><td>Agarose gel</td><td>Lactobacillus casei</td><td>Dairy whey</td><td>33</td><td>65</td><td>0.7 ***</td><td>[109]</td></tr>
<tr><td>Ca-alginate gel</td><td>Lactobacillus helvetikus</td><td>Pineapple cannery wastes</td><td>30</td><td>94</td><td>0.4 ***</td><td>[110]</td></tr>
<tr><td>Glass granules “Siran beads”</td><td>Lactobacillus rhamnosus</td><td>Beer industry wastes*</td><td>15</td><td>98</td><td>0.2 ***</td><td>[111]</td></tr>
<tr><td>Loofa / Ca-alginate gel</td><td>Aspergillus awamori, Lactococcus lactis</td><td>Cassava</td><td>120</td><td>60</td><td>1.6 ***</td><td>[112]</td></tr>
<tr><td rowspan="2">Ca-alginate gel</td><td>Aspergillus awamori, Streptococcus lactis</td><td>Potatoes starch</td><td>45</td><td>70</td><td>2 ***</td><td>[113]</td></tr>
<tr><td>Lactobacillus casei, Lactococcus lactis</td><td>Whey*</td><td>40</td><td>50</td><td>1.1 ***</td><td>[114]</td></tr>
<tr><td>Cotton fiber</td><td rowspan="2">Rhizopus oryzae</td><td>Corn starch</td><td>127</td><td>100</td><td>1.7 ***</td><td>[115]</td></tr>
<tr><td>PVA cryogel</td><td>Potatoes starch</td><td>15</td><td>30</td><td>0.6 ***</td><td>[116]</td></tr>
</table>

* Process with substrate pretreatment.

** Continuous process.

*** Batch process.

It was established, that molasses being a waste of beet-sugar factories could be used by *Lactobacillus delbrueckii* IFO 3202 cells entrapped into Ca-alginate gel as a substrate for L(+)-LA production under batch and continuous conditions. Herein, the acidic hydrolysis of wastes was conducted before their use in LA production. The maximal yield of the product was 82% when the initial substrate (sucrose) concentration was 78 g/L/h [70]. The immobilized biocatalyst possessed the highest productivity in the continuous process such as 13.92 g/L/h.

Both lactose and lactose-containing sources was established to be good substrates for the L(+)-LA obtaining. The cells *Lactobacillus helvetikus*, adsorbed on the Activated Alumina granules used lactose to produce LA. The conversion of process reached the value about 3.9 g/L/h after 30-36 h of cultivation of immobilized cells in reactor [97].

The cells *Lactobacillus casei* immobilized using polyethylene carrier Raschell also converted lactose (30 g/L) into L(+)-LA [104]. The productivity of process was 1.2 g/L/h, whereas the product yield was close to 100%.

Same bacterial producers adsorbed on the surface of glass particles of Pora-bact A were applied for the production of LA under batch conditions from whey of milk pretreated with proteases. The final concentration of LA in the medium was 27 g/L [77]. The use of another glass beads Poraver as carrier for same cells allowed to considerably increase the concentration of LA up to 93 g/L [105].

It was shown that whey of milk could be used by a number of bacterial cells, i.e., *Lactobacillus helvetikus*, *Lactococcus lactis*, *Lactobacillus delbrueckii* and *Streptococcus salivarius*, immobilized into k-carrageenan gel, as substrate for its conversion into L(+)-LA [106-108].

There are some investigations where the conversion of non-pretreated milk whey was tried for the production of LA, but all of them were characterized by low enough process productivity and product yield. For example, the *Lactobacillus casei* cells immobilized in agarose carrier were used under mentioned conditions. The process productivity was 0.7 g/L/h, and the product yield was 65% [109].

Much better results were obtained in the batch process carried out with *Lactobacillus brevis* cells when delignified cellulose carrier was used. The productivity of the process was 2 g/L/h, and the yield of LA was 72% [68].

The wastes obtained after fruits and vegetables processing are considered to be very perspective substrates for their conversion into L(+)-LA. For example, the wastewaters from a pineapple cannary factory contained high enough concentration of glucose were used in LA production. The process was carried out with *Lactobacillus delbrueckii* cells immobilized in Ca-alginate gel. The solution with 30 g/L LA was obtained in the fermentation of wastes for 72h. The degree of conversion of glucose into LA was 94% [110].

The brewing industry needs to utilize up to 1 million tons of used grain annually. It was revealed, that the wastes liquefied and saccharified by amylolytic enzymes could be used in the LA production. The *Lactobacillus rhamnosus* cells adsorbed on the Siran glass beads provided the 98%-yield of L(+)-LA from the pretreated wastes of brewing industry [111].

Thus, it was clearly demonstrated, that application of bacterial cells in immobilized state is possible for the conversion of various wastes to LA under continuous and batch conditions. However, obligatory waste pretreatment is needed in the most mentioned processes. Since then, two alternative approaches to the production of L(+)-LA from various raw materials were offered:

i. to use combination of several cultures in the same process, allowing to use one culture as producer of extracellular hydrolytic enzymes useful for biodegradation of complicated polysaccharide substrates, and another culture converting accumulated glucose into L(+)-LA;
ii. to use cultures possessing both hydrolytic activities in relation to various substrates and producing L(+)-LA.

There are several examples of processes where co-immobilized microorganisms were used for the LA production according to the first approach.

The joint immobilization and utilization of *Lactobacillus casei* and *Lactococcus lactis* cells immobilized into Ca-alginate gel guaranteed the productivity of process equal to 1.5 g/L/h under batch conditions [114].

The biocatalyst containing aerobic filamentous fungus cells *Aspergillus awamori* immobilized in the near-surface parts of Ca-alginate granules, and anaerobic lactic acid bacteria *Streptococcus lactis* immobilized in the inner parts of same Ca-alginate granules was developed [113]. The use of the biocatalyst in the process of LA production provided 80% conversion of potatoes starch into the target product. However, production of the granules with so complicated content required the use of special equipment and conditions for cell immobilization.

The co-immobilization of fungi *Aspergillus awamori* and bacteria *Lactococcus lactis* in the complex carrier composed of loofa and Ca-alginate gel enables the 76%-conversion of cassava starch into L(+)-LA [112]. The productivity of process with this biocatalyst was 1.6 g/L/h and the maximal concentration of LA in cultural medium was 120 g/L.

Unfortunately, the realization of the offered approach based on co-immobilization of cells of different cultures was characterized by a number of technical difficulties and needed in further optimization.

The processing of potatoes and corn usually results in the accumulation of starch-containing wastewaters. The use of fungi *Rhizopus oryzae* immobilized on the cotton fiber in direct transformation of native corn starch into LA gave 100% -yield of required product [115]. So, the productivity of the process where non-pretreated starch was converted into LA was 1.65 g/L/h and the final concentration of LA was 127 g/L/h.

It was established that immobilized biocatalyst prepared on the base of the same cells, possessing extracellular amylolytic activity, and PVA cryogel, could convert the native potatoes starch into L(+)-LA during 480 h with only 15%-decrease in process characteristics (both productivity and yield of product) [116]. The maximal productivities of both process and biocatalyst were obtained with 50 g/L starch in the medium. The maximal concentration of L(+)-LA was 15 g/L. The last product contained up to 98% of L(+)-isomer.

The use of immobilized fungus cells with extracellular enzymatic activity for the production of LA looks like more effective approach than the use of co-immobilized cultures. The main reason is the absence of difficulties in immobilization procedure and usage of biocatalysts.

To date, the processes aimed at the production of L(+)-LA with application of immobilized fungus cells still have no industrial application despite of their attractive properties as compared to free and immobilized bacterial and fungus cells (Tables 6, 7). One of the reasons is the appearance of technological difficulties in the processes with immobilized fungi, provided by uncontrollable growth of free fungus mycelium on various internal surfaces of reactor (stirrers, electrodes, etc.) [117]. A decrease in process productivity, necessary suspension of LA-production and additional inputs required for reactor cleaning constitute the negative aspects of application of immobilized cells.

Table 7. Free cells of different microorganisms in the production of lactic acid from various wastes of food industry

Microorganism	Substrate	[LA], g/L	Yield, %	Productivity, g/L/h	Ref.
Lactobacillus delbrueckii	Sugar-cane molasses*	90	58	3.80	[69]
Lactobacillus amylovorus ATCC 33620	Cassava	4.8	40	0.20	[73]
Rhizopus oryzae	Sugar-beet molasses	94	78	1.38	[71]
		137	76	1.43	
	Barley	25	41	0.30	[82]
	Corn starch	40	55	0.50	
	Rice	41	56	0.60	
	Cassava	32	74	0.40	
	Oats	11	23	0.10	
		51.7	44	0.52	[83]
	Potatoes starch*	98	82	1.4	[85]
	Corn starch	36	44	0.40	[86]
	Potatoes starch	21	84	0.40	[88]

* Process with substrate pretreatment.

Obviously, to overcome the problem it is necessary to control the initial concentration of spores introduced with biocatalyst into reactor and to use carrier allowing good retention of fungus cells on its surface or inside of its porous structure [102, 118-119].

Finally, it is necessary to note that comparison of characteristics of processes of LA production organized under batch and continuous regimes obviously testified to the benefits of the latter regime, especially in relation to process productivity (Table 6). The organization of process with constant parameters and significantly simplified management are additional advantages of continuously functioning systems with immobilized cells as compared to batch processes. However, it should be noted that diluted solutions of LA also is one of the specific and not pleasant characteristics of the continuous processes, since it results in additional energetic inputs during LA isolation and concentration. So, the choice of process organization should be done upon the process target, namely, the production of low or high concentrated LA.

2.2. Production of Fuel Ethanol by Immobilized Cells of Various Microorganisms

2.2.1. Bioethanol as a Source of Renewable Energy

A very fast growth of interest to the production of ethanol is observed during the last 5-6 years [120]. Actually, there are several main reasons of the situation:

1. a considerable growth of world oil price;
2. a rapid economic development of Asian countries, i.e., China and India, provoking intensive growth of need in energy sources;
3. the potential of ethanol as a source of renewable energy and its possible wide application as 10%-part of petrol [121], enabling 40-50%-decrease in toxicity of exhaust.

The annual world need in ethanol is about 2 billion tons [122]. According to expert data, the distribution of ethanol among different fields of its application generally is following: 15% of ethanol is used in food industry, 25% of ethanol is utilized by chemical industry, and the rest 60 % of produced ethanol applied as a fuel. It should be noted here, that 7% of ethanol is produced by chemical synthesis and the rest 93% of it is a product of biotechnological fermentation.

A wide diversity of agricultural crops is used as a biorenewable sources for ethanol production. Sugar beet, sugar cane, wheat, corn, potatoes, sweet sorghum and cassava are among sources mostly used for ethanol fermentation (Table 8).

Table 8. Productivity and cost of ethanol obtained from various agricultural materials

Raw material	Productivity, tones/ha	Cost, $\$/m^3$
Sugar-beet	2.5-3.0	300-400
Sugar-cane	3.5-5.0	160
Corn	2.5	250-400
Wheat	0.5-2.0	380-400
Potatoes	1.2-2.7	800-900
Sorghum	3.0-5.0	200-300
Barley	7.0-7.5	250-350
Oat	5.5-6.0	350-400
Rice	10.0-11.5	200-300
Cassava	1.5-6.0	700

However, the question concerning the expediency of production of fuel ethanol from various types of grain that could be used in food industry is still under serious discussion. Both starch- and sugar-containing wastes, i.e., molasses, beet bagasse, sugar cane bagasse, corn-cob, accruing in processing of foresaid agricultural crops, could be effectively applied as alternative sources for ethanol production [123-125].

Various cellulose-containing materials being wastes of industrial fields and agriculture also could be used as a source for ethanol production [126-128]. However, obligatory strong physical-chemical and enzymatic treatment of such substrates should be carried out before their use in the production of ethanol. Therefore, the cost of enzymatic preparations used for the treatment is of a high importance.

The choice of ethanol producer is critical part of general process. Bacterial and yeast cells cannot ferment pentoses such as xylose, arabinose and ribose, that usually present in the hydrolysates of cellulose-containing materials in high enough concentrations (up to 25-30%). The filamentous fungi can produce ethanol from pentoses, but they are very sensitive to the concentration of ethanol accumulated in the medium and cannot produce ethanol in concentrations higher than 50 g/L [129].

Bacterial cells *Zymomonas mobilis* were previously mentioned as producers with a high perspective in industrial ethanol production due to their unique opportunity to realize the fermentation process via Enter-Doudoroff pathway [130-132]. The metabolic way guarantees low enough accumulation of cell biomass and twice less accumulation of intracellular ATP concentration as compared to glycolytic way used by others microorganisms. Concurrently, same pathway provided higher yield of CO_2 as compared to yeast cells. Moreover, bacteria possess tolerance to ethanol in concentrations less than 120 g/L and could produce ethanol using only three main sugars: glucose, fructose, and sucrose. Since then, only yeast cells *Saccharomyces cerevisiae* are seriously discussed as industrial producers of fuel ethanol. Only the cells could produce ethanol in concentrations up to 18% (v/v). Nevertheless, the yeast cells are also sensitive to various fermentation inhibitors present in the agricultural raw materials after their hydrolytic treatment, i.e., phenols, furfural-containing compounds, acetic acid, tannins, formic acid, polyaromatic compounds.

The immobilization of cells producing ethanol is an important step towards improvement of the situation with the features of the microorganisms and enrichment of results that can be obtained in the field.

2.2.2. Immobilized Cells in the Bioethanol Production

The investigation of potential use of immobilized cells in the ethanol production started 15-20 years ago [131-134]. Even the first semicommercial production of ethanol from the wastes of sugar-cane processing was offered in Japan in 1984, and the process was based on the application of immobilized yeast cells [135].

It was shown, that immobilization of ethanol producers allows:

- to use continuous processes,
- to reduce working volumes of utilized fermentation equipment,
- to carry out complete and rapid separation of fermentation medium from the cells,
- to improve the tolerance of cells to main product and medium acidity and thereby to increase the yield of ethanol.

A wide variety of methods and carriers were tested in the experiments with immobilized cells being producers of ethanol [136]. Finally, it was established that adsorption of cells is not an appropriate method to be used in industrial fermentation processes. It appeared that sorption capacity of most carriers considerably reduced under exploitation conditions being affected by different factors (temperature, presence of various compounds in the fermentation medium, pH, etc.). So, characteristics of the processes realized with cells immobilized by adsorption were not stable, and that was a main argument against use of such a type of cell immobilization. Nevertheless, the method of cell immobilization by adsorption is very simple and can be easily realized in practice, so, it continues to be widely tried in modern investigations [137-140] (Table 9).

The entrapment of cells producing ethanol into gels of natural polymers such as Ca-alginate, k-carrageenan and agar was tried very widely many years ago [141-144].

It was shown that all of these carriers are characterized by significant defects in mechanical stability becoming more pronounced during biocatalyst testing in the reactors. Moreover, the biodegradable polysaccharides underlie the chemical structure of the carriers, since then microbial destruction of the carriers can take place in the reactors, when various agricultural wastes contaminated with natural microbiota are utilized for ethanol fermentation. The heterogeneous porosity of natural gels was established to create the diffusion limitations for substrates and products inside of granules of immobilized biocatalyst, and further, the mass-transfer problems resulted in the reduction of process productivity.

Table 9. Production of ethanol by immobilized cells from various sources

Microorganism	Carrier	Substrate	Reducing sugars, g/L	[Ethanol], g/L	Ref.
S. pastorianuse, A. awamori	Cellulose bead	Starch	100	44.5	[137]
Fleischmann yeast	Sugar-cane stalks	Cane molasses	112	47	[140]
S. cerevisiae	Ca-alginate gel	Mahula flowers mash	281-363 g/kg	152-205 g/kg	[147]
	PVA cryogel	whey	60	24	[152]
	Agar gel	Cane molasses	195	79.5	[143]
K. marxianus	PVA Ca-alginate gel		140	52-53 57	[150]
Z. mobilis	k-carrageenan gel	Corn starch	162-172	70	[144]
S. cerevisiae		Pineapple cannery waste	82.3	37	[155]
A. niger, Z. mobilis	Ca-alginate gel	Starch	50	18.9	[156]
S. cerevisiae		Beet molasses	109	46.2	[146]
Z. mobilis, S. diastaticus		Cassava starch	150	46.9	[157]
Kefir yeast	Delignified cellulose	whey	21.2-52.5	20.3-41.8	[158]

Discussing the characteristics of carriers obtained on the base of Ca-alginate gel, it is necessary to note that Ca-alginate gel has one additional but very important disadvantage as compare to agar and k-carrageenan, namely, the gel is very chemically unstable in the media containing phosphate-ions and organic acids [141, 145].

Surprisingly, it appeared that all well-known negative characteristics of the polymers did not restrain investigators from their use in up-to-date developments of new immobilized biocatalysts for ethanol production [146-147]. It should be recognized that Ca-alginate gels continues to be very popular among specialists working with immobilized cells producing ethanol.

Obviously, the use of synthetic carriers for immobilization of ethanol producers looks very attractive and rational. Resistance to different mechanical actions (stirring, abrasive damage, pressing, etc.) as well as hydrolysis by various enzymes characterizes the carriers. Now, when the second tide of interest to use of immobilized systems in fermentation industry is observed, the application of synthetic polymers for the development of new immobilized biocatalysts is evidently preferable [148-152].

Previously, polyacrylamide gels were tried as carriers for ethanol producing cells [142, 153]. It was revealed that the toxicity of main monomer (acrylic acid) used for the polymer preparation significantly decreased the viability and metabolic activity of immobilized cells, so, the number of living cells possessing high levels of ethanol productivity considerably reduced in the content of immobilized biocatalysts. The heterogeneous porosity of formed gel matrices negatively influenced on supply of immobilized cells with substrates as well as on release of metabolites. As a result, the death of cells localized in the inner parts of granules was determined.

Alternatively to polyacrylamide gels, the use of PVA cryogels as carriers for cell immobilization was offered [152]. PVA cryogels was shown to provide satisfied results with bacterial and immobilized cells used in ethanol production [17, 154]. The mechanical and chemical resistance of the polymer macroporous structure, already mentioned above, played an important role in the productive functioning of immobilized biocatalysts in the processes of ethanol fermentation.

The Table 9 contains the data concerning application of immobilized cells in modern investigations aimed at the development of advanced fermentation processes. All of them propose the use of different sugar- or starch-containing wastes for ethanol production by immobilized cells; however, the use of cellulose-containing wastes for same processes obviously is on agenda.

Analysis of the data (Table 9) testified to the fact, that today preferences are given to yeast cells as producers of ethanol and to cell entrapment as immobilization method.

3. Prevention of Formation of Dangerous Systems with Immobilized Cells in Environment

For a long time it was considered, that corrosion processes presumably have the chemical nature, since there are a lot of different aggressive substances present in oil wastewaters. The analysis of recent results of multiple investigations of different corrosion objects [159-161] confirms that various microorganisms provoke 30% of all corrosive situations.

The term biocorrosion or microbially influenced corrosion refers to the accelerated deterioration of metals due to the presence of biofilms on their surfaces. Biofilms are formed by self-immobilized cells of different taxonomic groups [162-163].

Recent investigations of biocorrosion were focused on the study of electrochemical reactions, planktonic cultures, biomineralization processes, extracellular enzymes present in the biofilm matrix, phylogenetic characteristics of corrosive consortia and concentrations of dissolved oxygen in corrosive media [164-166].

According to recent literature data [162], biocorrosion occurs in aquatic and terrestrial habitats differed in nutrient content, temperature, pressure and pH. Biofilms promote interfacial physicochemical reactions, not normally favoured under abiotic conditions. Mechanisms of biocorrosion processes reflect the range of physiological activities carried out by diverse types of microorganisms found within biofilms. These mechanisms evidently vary depending on microbial species colonising metal surfaces.

Biocorrosion provoked by cells of sulphate-reducing bacteria (SRB) was a subject of numerous detailed investigations [162, 167-168]. The main results of the investigations were followings:

- SRB stimulate development of corrosion as a consequence of their activities within microbial consortia existing in the form of biofilms adherent to the mild steel;
- the biofilms with SRB have a oxic/anoxic interface;
- the primary role of SRB comprises the production of sulphide;
- the iron dissolution takes place at the electrochemical anode, it combines with biogenic sulphide to give precipitated iron sulphides as corrosion products;
- the corrosive biofilms have heterogeneous microbial content (consortia of mixed microbial species) but always contain SRB.

Despite considerable research efforts aimed at the investigation of biocorrosion phenomenon the detailed mechanisms of biocorrosion developed under the action of biofilms are still poorly understood.

The information concerning factors provoking formation of corrosive biofilms is of a high importance, since it helps to found effective approaches to the prevention of corrosion development.

It was recently established that iron sulphide produced by SRB in the frame of biofilms plays a critical role in the formation of corrosive self-immobilized consortia [169].

According to recent data (Fg.2 and 3), the formation of corrosive biofilm is going on quicker on surfaces covered with iron sulfide as compared to clear metal surfaces. The presence of sulfide on a surface mainly effects log phase of biofilm growth, so specific rate of biofilm growth increases concurrently with increase in sulfide concentration on a metal surface. The influence of sulfide precipitated on a metal surface on the growth of corrosive biofilms appeared to be more pronounced at low concentrations, approximately up to 1 $\mu g/cm^2$.

It was shown that porous sulfide layer formed anoxic zone even under aerobic conditions, which accelerates formation of heterogeneous biofilm [166]. So, metal surface covered by sulfide is more accessible for adhesion of bacterial cells from various wastewaters to form anaerobic part of biofilms.

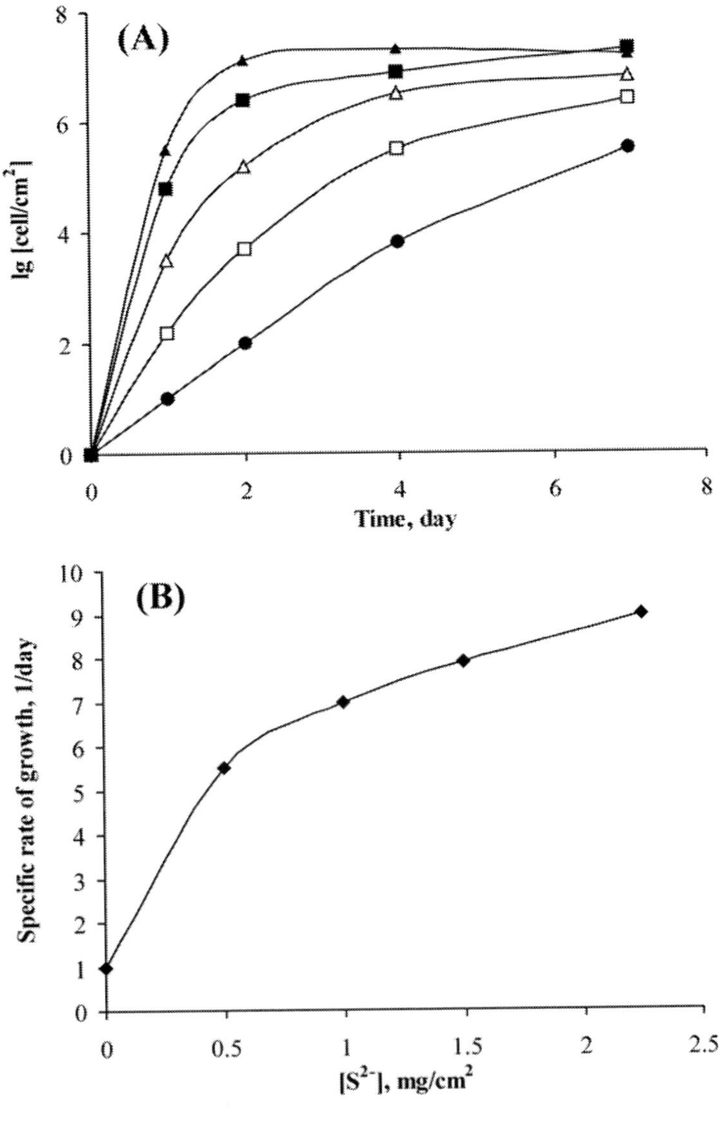

Figure 2. Growth kinetics (A) and specific rate of growth (B)of biofilms on a metal surfaces covered by different sulphide concentrations: (●) 0 μg/cm^2, (□) 0.45 μg/cm^2, (Δ) 0.92 μg/cm^2, (■) 1.41 μg/cm^2, (▲) 2.20 μg/cm^2.

Since then, it is necessary to prevent the precipitation of sulphide, mostly producing by planktonic cells in the wastewater systems, on the surface of metal pipelines and equipment, to avoid the development of corrosive biofilms.

Although microstructure and chemical composition of metal used as parts of modern pipelines, reservoirs and equipments is considered to be crucial factors in initiation of biocorrosion development [170]. It should be noted that all mentioned metallic constructions are made from metals differed in chemical composition.

Mild, carbon and stainless steel coupons were used to analyze biofilm formation depending on chemical composition of metal. Kinetic curves of growth were established for biofilms on each kind of metal. The growth rates of biofilms formed on the tested metal surfaces evidently were different. The investigation of metal corrosion carried out concurrently with the study of biofilm formation showed that chemical corrosion accelerated growth of self-immobilized cells. The differences in the colonization behavior of investigated steel surfaces testified to the influence of chemistry of used metals on the attachment and proliferation of bacterial cells.

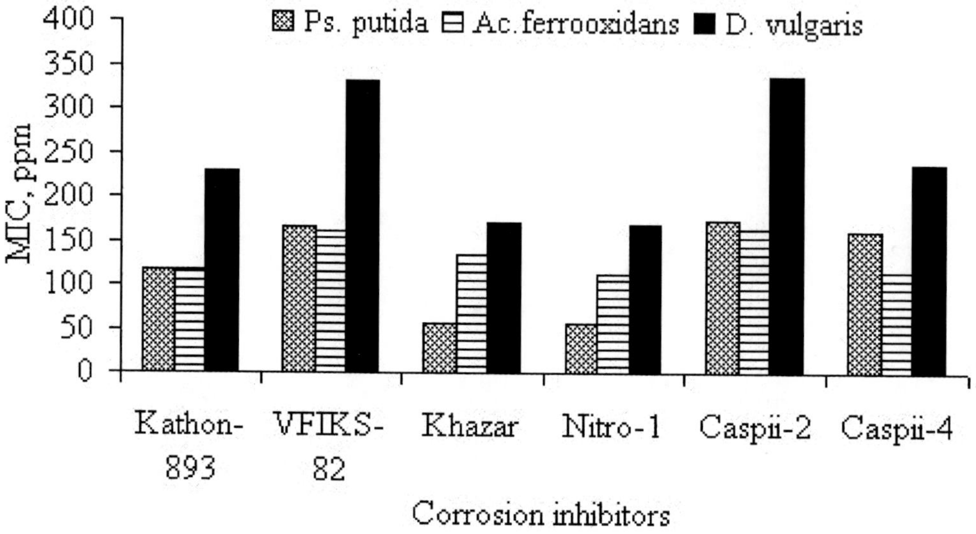

Figure 3. Minimal inhibition concentration of various CI widely used in practice.

Since notable corrosion was observed in the case of mild steel [171-172], it is possible to conclude that the surface non-covered with hard oxide film appeared to be opened for biofilm formation. It is well-known that release of iron ions from stainless steel should be at a minimum, as the surface covered with Cr-rich oxide film. Cr-rich oxide film possesses high metal passivity in different media and stability to pitting corrosion. Since then, the formation of corrosive biofilms on the surface of stainless steel did not observe for at least month of continued experiments.

Obviously, both biological and chemical types of corrosion operate conjointly, therefore it is necessary to undertake activities directed to the simultaneous counteraction against the chemical corrosion as well as destruction of microorganisms promoting biocorrosion processes. Since then, the corrosion inhibitors (CI), chemically synthesized in many laboratories, having biocide activity are of special interest [173-174].

To provide the ecological safety of CI application as biocides, the Minimal Inhibitory Concentrations (MIC) causing the death of all cells of certain type of microorganisms are defined. The physiological solution (PS) (0.9% NaCl) is used as main medium for cell exposition in the presence of different CI concentrations and MIC determination. In experiments, PS plays a role of model medium which content simulates the oil waste waters. Usually, the cells at concentration equal to 10^5-10^7 cell/mL are left in contact with CI for 6-7 days at 28°C. The residual quantities of living cells in probes with different CI concentrations are defined by the microbiological method, whereas the last one is enough time and labour consuming.

The data of Figure 3 illustrate the situation, when various CI have biocide influence on the different microorganisms, participating in the formation of biofilms, in non-equal concentrations. Mechanisms of cell death provoked by compounds with biocide activity can be different, but in all cases MIC is the most important parameter.

The results testify to a high tolerance of SRB to action of CI as biocides. That fact should be taken into account in practice when CI are used to prevent the formation of corrosive biofilms.

Conclusion

The problem concerning OPC biodegradation has tremendous environmental, social and even political meaning, especially when the modern history of humanity knows the cases when organophosphorous agents were used in terrorist acts or used as a mean of threat. All facts mentioned herein make the elaboration and application of new men-friendly microbial immobilized biocatalysts, decontaminating various wastewaters containing OPC, very actual.

Talking about the treatment of wastewaters of enterprises of food industry containing high enough concentrations of organic components, it is necessary to note, that huge volumes of the wastes should be treated daily. The methods of biotreatment based on the application of immobilized cells should guarantee high velocities of biodegradation of various organic components at a high enough flow rate in reactors. Search for new carriers enabling execution of effective biotechnological processes with immobilized microorganisms seems to be the main direction of advanced developments in this field.

New biotechnological processes based on the application of cells of various microorganisms could constitute an adequate alternative for the technologies with free cells. The use of immobilized cells in the processes of production of some required products concurrently with waste utilization makes the usage of biocatalysts as well as the processes as they are economically pronounced and ecologically safe.

Summarizing the data foresaid in connection with ethanol production by immobilized cells and discussing the future of immobilized biocatalysts in the field, it is necessary to say, that further search for new carriers, possessing high chemical and mechanical stability in complicated media such as hydrolysates of industrial and agricultural wastes, should be continued. New synthetic polymers being products of innovative processes realized in chemical industry should enable the development of effective biocatalysts, containing immobilized cells of ethanol production. The use of improved ethanol producers in immobilized form is expected to allow the increase in yield of target product. The immobilization of new gene engineered yeast strains possessing new ability in pentose fermentation should guarantee the stability of recombinant biosystems and their safe use under industrial conditions.

To prevent the development of self-immobilized cells in the systems of wastewater treatment with further acceleration of corrosion situation, it is necessary to apply methods enabling elimination of main factors provoking formation of corrosive biofilms. The information concerning evident influence of chemical characteristics of used type of steel on adhesion of bacterial cells and formation of corrosive biofilms should be taken into account. All reviewed results are significant in the context of steel choice related to the exploitation media and emphasize the importance of necessary removal of sulfide from wastewaters. It is necessary to remember that processes of sulfide accumulation, formation of biofilm constituted by self-immobilized bacteria and chemical corrosion of metals depend on each other.

ACKNOWLEDGEMENTS

The work was financially supported by Russian Foundation of Basic Research (Project ofi_a 05-04-08018), the Program of Basic researches of Presidium RAS "Basic problems of energetic" and Federal Agency of Science and Innovations of Russian Federation (State contract No. 02.515.11.5002).

REFERENCES

[1] Costa, L.G. (2006) Current issues in organophosphate toxicology. *Clin. Chim. Acta, V.366*, p.1-13.

[2] Organisation for the Prohibition of Chemical Weapons. (2005) Convention on the prohibition of the development, production, stockpiling and use of chemical weapons and on their destruction. Available from: http://www.opcw.org/

[3] Committee on Review and Evaluation of International Technologies for the Destruction of Non-Stockpile Chemical Materiel, National Research Council. (2006) *Review of International Technologies for Destruction of Recovered Chemical Warfare Materiel.* Washington, DC: National Academies Press.

[4] Ramanathan, M.P. & Lalithakumari, D. (1996) Methylparathion degradation by *Pseudomonas* sp. A3 immobilized in sodium alginate beads. *World J. Microb. Biot., V.12*, p.107-108.

[5] Screenivasuulu, C. & Aparna, Y. (2001) Bioremediation of methylparathion by free and immobilized cells of *Bacillus* sp. isolated from soil. *Bull. Environ. Contam. Toxicol., V.67*, p.98-105.

[6] Qiao, Ch.-L., Yan, Y.-Ch., Shang, H.Y., Zhou, X.T., Zhang, Y. (2003) Biodegradation of pesticides by immobilized recombinant *Escherichia coli. Bull. Environ. Contam. Toxicol., V.71*, p.370-374.

[7] Mulchandani, A., Kaneva, I., Chen, W. (1999) Detoxification of organophosphate nerve agents by immobilized *Escherichia coli* with surface-expressed organophosphorus hydrolase. *Biotechnol. Bioeng., V.63(2)*, p.216-223.

[8] Wang, A.A., Mulchandani, A., Chen, W. (2002) Specific adhesion to cellulose and hydrolysis of organophosphate nerve agents by a genetically engineered *Escherichia coli* strain with a surface-expressed cellulose-binding domain and organophosphorus hydrolase. *Appl. Environ. Microb., V.68(4)*, p.1684-1689.

[9] Hong, M.S., Rainina, E., Grimsley, J.K., Dale, B.E., Wild, J.R. (1998) Neurotoxic organophosphate degradation with polyvinyl alcohol gel-immobilized microbial cells. *Bioremed. J., V.2(2)*, p.145-157.

[10] Kim, J.-W., Rainina, E.I., Mulbry, W.W., Engler, C.R., Wild, J.R. (2002) Enhanced-rate biodegradation of organophosphate neurotoxins by immobilized nongrowing bacteria. *Biotechnol. Progr.*, *V.18*, p.429-436.

[11] Zhang, Y., Autenrieth, R.L., Bonner, J.S., Harvey, S.P., Wild, J.R. (1999) Biodegradation of neutralized sarin. *Biotechnol. Bioeng.*, *V.64(2)*, p.221-233.

[12] Efremenko, E.N. (2006) Biocatalytic systems based upon the immobilized cells degrading parathion and p-nitrophenol. In: G.E. Zaikov et al. (Eds.), *Biocatalysis and Biocatalytic Technologies* (p.23-37). Hauppauge, NY: Nova Science Publishers Inc.

[13] Hallas, L.E., Adams, W.J., Heitkamp, M.A.. (1992) Glyphosate degradation by immobilized bacteria: field studies with industrial wastewater effluent. *Appl. Environ. Microb.*, *V.58(4)*, p.1215-1219.

[14] Mansee, A.H., Chen, W., Mulchandani, A. (2000) Biodetoxification of coumaphos insecticide using immobilized *Escherichia coli* expressing organophosphorus hydrolase enzyme on cell surface. *Biotechnol. Bioproc. Eng.*, *V.5*, p.436-440.

[15] Wang, A.A., Chen, W., Mulchandani, A. (2005) Detoxification of organophosphate nerve agents by immobilized dual functional biocatalysts in a cellulose hollow fiber bioreactor. *Biotechnol. Bioeng.*, *V.91(3)*, p.379-386.

[16] Efremenko, E.N., Votchitseva, Yu.V., Aliev, T.K., Varfolomeyev, S.D. (2005) Recombinant plasmid DNA pTES-His-OPH encoding synthesis of polypeptide with properties of organophosphate hydrolase, and strain *E.coli* – producer of polypeptide with properties of organophosphate hydrolase. *Patent RU 2255975.*

[17] Lozinsky, V.I. & Plieva, F.M. (1998) Poly(vinyl alcohol) cryogels employed as matrices for cell immobilization. 3. Overview of recent research and developments. *Enzyme Microb. Tech.*, *V.23*, p.227-242.

[18] Kononova, S.V. & Nesmeyanova, M. A. (2002) Phosphonates and their degradation by microorganisms. *Biochemistry (Moscow)*, *V.67(2)*, p.184-195.

[19] Cox, C. (2004) Glyphosate. *J. Pest. Ref.*, *V.24(4)*, p.10-15.

[20] DeFrank, J.J., Guelta, M., Harvey, S., Fry, I.J., Earley, J.P., Lupton, F.S. (2000) Biodegradation of hydrolyzed chemical warfare agents by bacterial consortia. In: B.Zwanenburg et al. (Eds.), *Enzymes in Action: Green Solution for Chemical Problems* (p.193-209). Netherlands: Kluwer Academic Publishers.

[21] Guelta, M.A. & DeFrank, J.J. (1998) Performance of immobilized cell bioreactors for treatment of HD and VX hydrolysates. *U.S. Army Edgewood Research, Development and Engeneering Center Technical Report*, ERDEC-TR-497.

[22] Fry, I.J., Earley, J.P., DeFrank, J.J. (1998) Process biodegradation of hydrolyzed sarin by the GB2 microbial consortium. *Proceedings of the 1997 ERDEC Scientific Conference on Chemical and Biological Defense Research*, p.615-636.

[23] Fry, I.J., DeFrank, J.J., Earley, J.P. (2006) Microbial biodegradation of phosphonates. *Patent US 7001758B1.*

[24] Tung, T.Q., Miyata, N., Iwahori, K. (2004) Growth of *Aspergillus oryzae* during treatment of cassava starch processing wastewater with high content of suspended solids. *J. Biosci. Boieng.*, *V.97(5)*, p.329-335.

[25] Jin, B., van Leeuwen, H.J., Yu, Q., Patel, B. (1999) Screening and selection of microfungi for microbial biomass protein production and water reclamation from starch processing wastewater. *J. Chem. Tecnol. Biot.*, *V.74(2)*, p.106-110.

[26] Jin, B., van Leeuwen, H.J., Patel, B., Doelle, H.W., Yu, Q. (1999) Production of fungal protein and glucoamylase by *Rhizopus oligosporus* from starch processing wastewater. *Process Biochem., V.34,* p.59-65.

[27] Johns, M.R. (1995) Developments in wastewater treatment in the meat processing industry: a review. *Bioresourse Technol., V.54*, p.203-216.

[28] Grabas, M. (2000) Organic matter removal from meat processing wastewater using moving bed biofilm reactors. *Environ. Prot. Eng., V.26(1),* p.55-62.

[29] Thayalakumaran, N., Bhamidimarri, R., Bickers, P.O. (2003) Biological nutrient removal from meat processing wastewater using a sequencing batch reactor. *Water Sci. Technol., V.47(10)*, p.101-108.

[30] Mittal, G.S. (2004) Characterization of the effluent wastewater from abattoirs for land application. *Food Rev. Int., V.20(3),* p.229-256.

[31] Mittal, G.S. (2006) Treatment of wastewater from abattoirs before land application – a review. *Bioresource Technol., V.97*, p.1119-1135.

[32] Fang, H.H.P. (1990) Treatment of wastewater from a whey processing plant using activated sludge and anaerobic process. *J. Dairy Sci., V.74*, p.2015-2019.

[33] Demirel, B. & Yenigun, O. (2004) Anaerobic acidogenesis of dairy wastewater: the effects of variations in hydraulic retention time with no pH control. *J. Chem. Technol. Biotechnol., V.79*, p.755-760.

[34] Vassilev, N., Fenice, M., Federici, F., Azcon, R. (1997) Olive mill waste water treatment by immobilized cells of *Aspergillus niger* and its enrichment with soluble phosphate. *Process Biochem., V.32(7)*, p. 617-620.

[35] Yesilada, O., Sik, S., Sam, M. (1998) Biodegradation of olive oil mill wastewater by *Coriolus versicolor* and *Funalia trogii*: effects of agitation, initial COD concentration, inoculum size and immobilization. *World J. Microb. Biot., V.14,* p.37-42.

[36] Ahmadi, M., Vahabzadeh, F., Bonakdarpour, B., Mehranian, M. (2006) Empirical modeling of olive oil mill wastewater treatment using loofa-immobilized *Phanerochaete chrysosporium*. *Process Biochem., V.41*, p.1148–1154.

[37] Sen'ko, O.V., Spiricheva, O.V., Lozinsky, V.I., Efremenko, E.N. (2007) Immobilized biocatalyst for treatment of fat-containing wastewaters at enterprises of food industry. *Catalysis in Industry, V.1*, p.55-61.

[38] Kariminiaae-Hamedaani, H-R., Kanda, K., Kato, F. (2003) Wastewater treatment with bacteria immobilized onto a ceramic carrier in an aerated system. *J. Biosci. Bioeng., V.95(2)*, p.128-132.

[39] Van Ginkel, S.W., Oh, S.-E., Logan, B.E. (2005) Biohydrogen gas production from food processing and domestic wastewaters. *Int. J. Hydrogen Energ., V.30(15),* p.1535-1542.

[40] Oh, S.-E. & Logan, B.E. (2005) Hydrogen and electricity production from a food processing wastewater using fermentation and microbial fuel cell technologies. *Water Res., V.39,* p.4673-4682.

[41] Wakilin, N.G. & Forster, C.F. (1997) An investigation into microbial removal of fats, oils and greases. *Bioresource Technol., V.59(1),* p.37-43.

[42] Becker, P., Köster, D., Popov, M.N., Markossian, S., Antranikian, G., Märkl, H. (1999) The biodegradation of olive oil and the treatment of lipid-rich wool scouring wastewater under aerobic thermophilic conditions. *Water Res., V.33(3),* p.653-660.

[43] Mongkolthanaruk, W., Dharmsthiti, S. (2002) Biodegradation of lipid-rich wastewater by a mixed bacterial consortium. *Int. Biodeter. Biodegr., V.50(2)*, p.101-105.

[44] Chipasa, K.B. & Mędrzycka, K. (2006) Behavior of lipids in biological wastewater treatment process. *J. Ind. Microbiol. Biot., V.33(8)*, p.635-645.

[45] Deng, S.B., Bai, R.B., Hu, X.M., Luo, Q. (2003) Characteristics of a bioflocculant production by *Bacillus mucilaginosus* and its use in starch wastewater treatment. *Appl. Microbiol. Biotechnol., V.60*, p.588-593.

[46] Crognale, S., D'Annibale, A., Federici, F., Fenice, M., Quaratino, D., Petruccioli, M. (2006) Olive oil mill wastewater valorization by fungi. *J. Chem. Technol. Biot., V.81,* p.1547-1555.

[47] Lanciotti, R., Gianotti, A., Baldi, D., Angrisani, R., Suzzi, G., Mastrocola, D., Guerzoni, M.E. (2005) Use of *Yarrowia lipolytica* strains for the treatment of olive mill wastewater. *Bioresource Technol., V.96,* p.317-322.

[48] Coulibaly, L., Gourene, G., Agathos, N.S. (2003) Utilization of fungi for biotreatment of raw wastewaters. *Afr. J. Biotech., V.2(12),* p.620-630.

[49] Leenen, E.J.T.M., Dos Santos, V.A.P., Grolle, K.C.F., Trampler, J., Wijffels, P.H. (1996) Characteristics of and selection criteria for support materials for cell immobilization in wastewater treatment. *Water. Res., V.30(12),* p.2985-2996.

[50] Okazaki, M., Hamada, T., Fujii, H., Kusudo, O., Mizobe, A., Matsuzawa, S. (1995) Development of poly(vinyl alcohol) hydrogel for waste water cleaning. II. Treatment of *N,N*-dimethylformamide in waste water with poly(vinyl alcohol) gel with immobilized microorganisms. *J. Appl. Polym. Sci., V.58(12),* p.2243-2249.

[51] Chang, I.-S., Kim, C.-I., Nam, B.-U. (2005) The influence of poly-vinyl-alcohol (PVA) characteristics on the physical stability of encapsulated immobilization media for advanced wastewater treatment. *Process Biochem., V.40(9),* p.3050-3054.

[52] Rostron, W.M., Stuckey, D.C., Young, A.A. (2001) Nitrification of high strength ammonia wastewaters: comparative study of immobilization media. *Water Res., V.35(5),* p.1169-1178.

[53] Van der Sluis, C., Mulder, A.N.T., Grolle, K.C.F., Engbers, G.H.M., ter Schure, E.G., Tramper, J., Wijffels, R.H. (2000) Immobilized soy-saurce yeast development and characterization of a new polyethylene-oxide support. *J. Biotechnol., V.80(2),* p.179-188.

[54] Koshkina, L.Yu., Sirotkin, A.S., Kulikov, Yu.M., Nabatkin, A.N., Vazharova, T.A., Shaginurova, G. I. (2001) Complex technology of industrial wastewaters treatment containing grease and SSAC. *Ind. Chem., V.7,* p.5-8 (*Rus.)*.

[55] Matsui, T., Miura, A., Iiyama, T., Shinzato, N., Matsuda, H., Fuuhashi, K. (2005) Effect of fatty oil dispersion on oil-containing wastewater treatment. *J. Hazard. Mater. B, V.118*, p.255-258.

[56] Di Berardino, S., Bersi, R., Converti, A., Rovatti, M. (1997) Starting-up an anaerobic hybrid filter for the fermentation of wastewater from food industry. *Bioprocess Biosyst. Eng., V.16*, p.65-70.

[57] Schmid, M., Thill, A., Purkhold, U., Walcher, M., Bottero, J.Y., Ginestet, P., Nielsen, P.H., Wuertz, S., Wagner, M. (2003) Characterization of activated sludge flocks by confocal laser scanning microscopy and image analysis. *Water Res., V.37,* p.2043-2052.

[58] Eschenhagen, M., Schuppler, M., Röske, I. (2003) Molecular characterization of the microbial community structure in two activated sludge system for the advanced treatment of domestic effluents. *Water Res., V.37,* p.3224-3232.

[59] Mirzaev, T.S., Sattarov, A.S., Davranov, K.D. (2000) Preparation and application of immobilized cells from the fungus *Mucor miehei. Chem. Nat. Compd., V.36(4),* p.402-403.

[60] Yun, S.I. & Ohta, Y. (2005) Removal of volatile fatty acids with immobilized *Rhodococcus sp.* B261. *Bioresource Technol., V.96(10),* p.41-46.

[61] Takeno, K., Yamaoka, Y., Sasaki, K. (2005) Treatment of oil-containing sewage wastewater using immobilized photosynthetic bacteria. *World J. Microb. Biot., V.21,* p.1385-1391.

[62] Efremenko, E.N., Senko, O.V., Spiricheva, O.V., Lozinsky, V.I., Varfolomeyev, S.D. (2006) Immobilized biocatalyst for biological purification of fat-containing wastewaters and the method of biocatalyst production. *Patent Proposal RU 2006107637.*

[63] Jördening, H.J., Hausmann, K, Demuth, B, Zastrutzki, M. (2006) Use of immobilized bacteria for the wastewater treatment examples from the sugar industry. *Water Sci. Tecnol., V.53(3),* p.9-15.

[64] Datta, R., Moon, S.-H., Tsai, S.-P., Bonsignore, P., Frank, J.R. (1995) Technological and economic potential of poly(lactic acid) and lactic acid derivatives. *FEMS Microbiol. Rev., V.16(2-3),* p.221-231.

[65] Hofvendahl, K. & Hahn-Hägerdal, B. (2000) Factors affecting the fermentative lactic acid production from renewable resources. *Enzyme Microb. Tech., V.26(2-4),* p.87-100.

[66] John, R.P., Nampoothiri, K.M., Pandey, A. (2007) Fermentative production of lactic acid from biomass: an overview on process developments and future perspectives. *Appl. Microbiol. Biot., V.74,* p.524-534.

[67] Wee, Y.-J., Kim, J.-N., Ryu, H.-W. (2006) Biotechnological production of lactic acid and its recent application. *Food Technol. Biotechnol., V.44(2),* p.163-172.

[68] Elezi, O., Kourkoutas, Y., Koutinas, A.A., Kanellaki, M., Bezirtzoglou, E., Barnett, Y. A., Nigam, P. (2003) Food additive Lactic acid production by immobilized cells of *Lactobacillus brevis* on delignified cellulosic material. *J. Agr. Food Chem., V.51,* p.5285-5289.

[69] Kotzanmanidis, C., Roukas, T., Skaracis, G. (2002) Optimization of lactic acid production from beet molasses *Lactobacillus delbrueckii* NCIMB 8130. *World J. Microb. Biot., V.18(5),* p.441-448.

[70] Goksungur, Y. & Guvenc, U. (1999) Production of lactic acid from beet molasses by calcium alginate immobilized *Lactobacillus delbrueckii* IFO 3202. *J. Chem. Technol. Biot., V.74(2),* p.131-136.

[71] Soccol, C.R., Marin, B., Raimbault, M., Lebeault, J.M. (1994) Potential of solid state fermentation for production of L(+)-lactic acid by *Rhizopus oryzae. Appl. Microbiol. Biot., V.41(4),* p.286-290.

[72] Yumoto, I. & Ikeda, K. (1995) Direct fermentation of starch to L(+)-lactic acid using *Lactobacillus amylophilus. Biotechnol. Lett., V.17,* p.543-546.

[73] Xiaodong, W., Xuan, G., Rakshit, S.K. (1997) Direct fermentation of lactic acid from cassava or other starch substrates. *Biotechnol. Lett., V.19,* p.841-843.

[74] Rojan, P.J., Nampoothiri, K.M., Nair, A.S., Pandey, A. (2005) L(+)-Lactic acid production using *Lactobacillus casei* in solid-state fermentation. *Biotechnol. Lett., V.27*, p.1685-1688.

[75] John, R.P., Nampoothiri, K.M., Pandey, A. (2006) Solid-state fermentation for L-lactic acid production from agro wastes using *Lactobacillus delbrueckii. Process Biochem., V.41*, p.759-763.

[76] John, R.P., Nampoothiri, K.M., Pandey, A. (2006) Simultaneous saccharification and L-(+)-lactic acid fermentation of protease treated wheat bran using mixed culture of lactobacilli. *Biotechnol. Lett., V.28*, p.1823-1826.

[77] Senthuran, A., Senthuran, V., Mattiasson, B., Kaul, R. (1997) Lactic acid fermentation in a recycle batch reactor using immobilized *Lactobacillus casei. Biotechnol. Bioeng., V.55*, p.841-853.

[78] Garde, A., Jonsson, G., Schmidt, A.S., Ahring, B.K. (2002) Lactic acid production from wheat straw hemicellulose hydrolysate by *Lactobacillus pentosus* and *Lactobacillus brevis*. *Bioresource Technol., V.81*, p.217-223.

[79] Naveena, B.J., Altaf, M., Bhadriah, K., Reddy, G. (2005) Selection of medium components by Placket Burman design for the production of L(+)-lactic acid by *Lactobacillus amylophilus* GV6 in SSF using wheat bran. *Bioresource Technol., V.96*, p.485-490.

[80] Raccach, M. & Mamiro, T. (1997) The effect of temperature on the lactic acid fermentation of rye flour. *Food Microbiol., V.4*, p.213-220.

[81] Linko, Y.Y. & Javaneinen, P. (1996) Simultaneous liquefaction, saccharification and lactic acid fermentation on barley starch. *Enzyme Microb. Tech., V.19*, p.118-123.

[82] Yu, R. & Hang, Y.D. (1989) Kinetics of direct fermentation of agriculutural commodities to L(+)-lactic acid by *Rhizopus oryzae*. *Biotechnol. Lett., V.11(8)*, p.597-600.

[83] Koutinas, A.A., Malbranque, F., Wang, R., Campbell, G.M., Webb, C. (2007) Development of an oat-based biorefinery for the production of L(+)-lactic acid by *Rhizopus oryzae* and various value-added coproducts. *J. Agr. Food Chem., V.55*, p.1755-1761.

[84] Cheng, P., Muller, R.E., Jaeger, S., Bajpai, R., Jannotti, E.L. (1991) Lactic acid production from enzyme-thinned corn starch using *Lactobacillus amylovorus. J. Ind. Microbiol. Biot., V.7*, p.27-34.

[85] Yin, P., Nishina, N., Kosakai, Y., Yahiro, K., Park, Y., Okabe, M. (1997) Enhanced production of L(+)-lactic acid from corn starch in a culture of *Rhizopus oryzae* using an air-lift bioreactor. *J. Ferment. Bioeng., V.84(3)*, p.249-253.

[86] Hang, Y.D. (1989) Direct fermentation of corn to L(+)-lactic acid by *Rhizopus oryzae. Biotechnol. Lett., V.11(4)*, p.299-300.

[87] Saito, K., Kawamura, Y., Oda, Y. (2003) Role of the pectinolytic enzyme in lactic acid fermentation of potato pulp by *Rhizopus oryzae*. *J. Ind. Microbiol. Biot., V.30*, p.440-444.

[88] Huang, L.P., Jin, B., Lant, P., Zhou, J. (2003) Biotechnological production of lactic acid integrated with potato wastewater treatment by *Rhizopus arrhizus*. *J. Chem. Technol. Biot., V.78*, p.899-906.

[89] Anuradha R., Suresh, A.K., Venkatesh, K.V. (1999) Simultaneous saccharification and fermentation of starch to lactic acid. *Process Biochem., V.35*, p.367-375.

[90] Xavier, S. & Lonsane, B.K. (1994) Sugarcane pressmud as a novel and inexpensive substrate for production of lactic acid in a solid state fermentation system. *Appl. Microbiol. Biot.*, *V.41*, p.291-295.

[91] Jin, B., Yin, P., Ma, Y., Zhao, L. (2005) Production of lactic acid and fungal biomass by *Rhizopus* fungi from food processing waste streams. *J. Ind. Microbiol. Biot.*, *V.32(11-12)*, p.678-686.

[92] Karel, M., Jaroslav, V., Vera, H., Mojmir, R. (1997) Lactic acid production in a cell retention continuous culture using lignocellulosic hydrolysate as a substrate. *J. Biotechnol.*, *V.56*, p.25-31.

[93] Venkatesh, K.V. (1997) Simultaneous saccharification and fermentation of cellulose to lactic acid. *Bioresource Technol.*, *V.62*, p.91-98.

[94] Pandey, A. & Soccol, C.R. (2000) Economic utilization of crop residues for value addition: a future approach. *J. Sci. Ind. Res. India*, *V.59*, p.12-22.

[95] Nolasco-Hipolito, C., Matsunaka, T., Kobayashi, G., Sonomoto, K., Ishizaki, A. (2002) Synchronised fresh cell bioreactor system for continuous L(+)-lactic acid production using *Lactococcus lactis* IO-1 in hydrolysed sago starch. *J. Biosci. Bioeng.*, *V.93*, p.281-287.

[96] Kadam, S.R., Patil, S.S., Bastawde, K.B., Khire, J.M., Gokhale, D.V. (2006) Strain improvement of *Lactobacillus delbrueckii* NCIM 2365 for lactic acid production. *Process Biochem.*, *V.41*, p.120-126.

[97] Tango, M.S. & Ghaly, A.E. (2002) A continuous lactic acid production system using an immobilized packed bed of *Lactobacillus helveticus*. *Appl. Microbiol. Biot.*, *V.58*, p.712-720.

[98] Hujanen, M., Linko, S., Linko, Y.Y., Leisola, M. (2001) Optimisation of media and cultivation conditions for L(+)(S)-lactic acid production by *Lactobacillus casei* NRRL B-441. *Appl. Microbiol. Biot.*, *V.56*, p.126-130.

[99] Ohkouchi, Y. & Inoue, Y. (2006) Direct production of L(+)-lactic acid from starch and food wastes using *Lactobacillus manihotivorans* LMG18011. *Bioresource Technol.*, *V.97*, p.1554-1562.

[100] Zhang, D.X. & Cheryan, M. (1991) Direct fermentation of starch to lactic acid by *Lactobacillus amylovorus*. *Biotechnol. Lett.*, *V.10*, p.733-738.

[101] Yoon, K.Y., Woodams, E.E., Hang, Y.D. (2005) Fermentation of beet juice by beneficial lactic acid bacteria. *Lebenson. Wiss. Technol.*, *V.38(1)*, p.73-75.

[102] Efremenko, E.N., Spiricheva, O.V., Lozinsky, V.I., Varfolomeev, S.D. (2006) *Rhizopus oryzae* fungus cells producing L(+)-lactic acid: kinetic and metabolic parameters of free and PVA-cryogel-entrapped mycelium. *Appl. Microbiol. Biot.*, *V.72*, p.480-485.

[103] Day, D. & Thongwai, N. (1998) An immobilized cell system for the continuous production of lactic acid from molasses. *Ann. NY Acad. Sci.*, *V.864*, p.259-262.

[104] Daraktchiev, R., Beschkov, V., Kolev, N., Aleksandrova, T. (1997) Bioreactor with a semi-fixed packing: anaerobic lactose to lactic acid fermentation. *Bioproc. Biosyst. Eng.*, *V.16*, p.115-117.

[105] Senthuran, A., Senthuran, V., Kaul, R., Mattiasson, B. (1999) Lactic acid production by immobilized *Lactobacillus casei* in recycle batch reactor: a step towards optimization. *J. Biotechnol.*, *V.73*, p.61-70.

[106] Audet, P., Paquin, C., Lacroix, C. (1989) Sugar utilization and acid production by free and entrapped cells of *Streptococcus salivarius* subsp. *thermophilus*, *Lactobacillus*

delbrueckii subsp. *bulgaricus*, and *Lactococcus lactis* subsp. *lactis* in a whey permeat medium. *Appl. Environ. Microb.*, *V.55*, p.185-189.

[107] Schepers, A.W., Thibault, J., Lacroix, C. (2006) Continuous lactic acid production in whey permeate/yeast extract medium with immobilized *Lactobacillus helveticus* in a two-stage process: model and experiments. *Enzyme Microb. Tech.*, *V.38*, p.324-337.

[108] Norton, S., Lacroix, C., Vuillemard, J.-C. (1994) Reduction of yeast extract supplementation in lactic acid fermentation of whey permeate by immobilized cell technology. *J. Dairy Sci.*, *V.77*, p.2494-2508.

[109] Zayed, G. & Winter, J. (1995) Batch and continuous production of lactic acid from salt whey using free and immobilized cultures of lactobacilli. *Appl. Microbiol. Biot.*, *V.44*, p.362-366.

[110] Irdis, A. & Suzana, W. (2006) Effect of sodium alginate concentration, bead diameter, initial pH and temperature on lactic acid production from pineapple waste using immobilized *Lactobacillus delbrueckii*. *Process Biochem.*, *V.41*, p.1117-1123.

[111] Shindo, S. & Tachibana, T. (2004) Production of L-lactic acid from spent grain, a by-product of beer production. *J. Inst. Brew.*, *V.110*, p.347-351.

[112] Roble, N.D., Ogbonna, J.C., Tanaka, H. (2003) l-Lactic acid production from raw cassava starch in a circulating loop bioreactor with cells immobilized in loofa (*Luffa cylindrica*). *Biotechnol. Lett.*, *V.25*, p.1093-1098.

[113] Kurusava, H., Ishikawa, H., Tanaka, H. (1988) L-lactic acid production from starch by co-immobilized mixed culture system of *Aspergillus awamori* and *Streptococcus lactis*. *Biotechnol. Bioeng.*, *V.31*, p.183-187.

[114] Roukas, T. & Kotzekidou, P. (1998) Lactic acid production from deproteinized whey by mixed cultures of free and coimmobilized *Lactobacillus casei* and *Lactococcus lactis* cells using fedbatch culture. *Enzyme Microb. Tech.*, *V.22*, p.199-204.

[115] Tay, A. & Yang, S.-T. (2002) Production of L(+)-lactic acid from glucose and starch by immobilized cells of *Rhizopus oryzae* in a rotating fibrous bed bioreactor. *Biotechnol. Bioeng.*, *V.80(1)*, p.1-12.

[116] Efremenko, E.N., Spiricheva, O.V., Veremeenko, D.V., Baibak, A.V., Lozinsky, V.I. (2006) L(+)-lactic acid production using PVA-cryogel entrapped *Rhizopus oryzae* fungus cells. *J. Chem. Technol. Biot.*, *V.81*, p.519-522.

[117] Yin, P., Yahiro, K., Ishigaki, T., Park, Y., Okabe, M. (1998) L(+)-lactic acid production by repeated batch culture of *Rhizopus oryzae* in air-lift bioreactor. *J. Ferment. Bioeng.*, *V.85(1)*, p.96-100.

[118] Kosakai, Y., Park, Y.S., Okabe, M. (1997) Enhancement of L(+)-lactic acid production using mycelial flocs of *Rhizopus oryzae*. *Biotechnol. Bioeng.*, *V.55(3)*, p.462-466.

[119] Oh, H., Wee, Y-J., Yun, J-S., Han, S. H., Jung, S., Ryu, H-W. (2005) Lactic acid production from agricultural resources as cheap raw materials. *Bioresource Technol.*, *V.96(13)*, p.1492-1498.

[120] Gray, K.A., Zhao, L., Emptage, M. (2006) Bioethanol. *Curr. Opin. Chem. Biol.*, *V.10*, p.141-146.

[121] Fulton, L. (2004) Driving ahead – biofuels for transport around the world. *Renew. Energ. World*, *V.7*, p.180-189.

[122] Zervos, A., Lins, Ch., Schrafer, O. (2004) Tomorrow's world: 50% renewables scenario for 2040. *Renew. Energ. World*, *V.7*, p.238-245.

[123] Mojovic, L., Niciolic, S., Rakin, M., Vukasinovic, M. (2005) Production of bioethanol from corn meal hydrolyzates. *Fuel., V.85*, p.1750-1755.

[124] Kim, S. & Dale, B.E. (2004) Global potencial bioethanol production from wasted crops and crop residues. *Biomass Bioenerg., V.26,* p.361-375.

[125] Bebic, Z., Jakovljevic, J., Baras, J. (2000) The corn starch hydrolyzate as a fermentation substrates for ethanol production. *Hem. Ind., V.54,* p.5-9.

[126] Pimentel, D. & Patzek, T.W. (2005) Ethanol production using corn, switchgrass, and wood; biodiesel production using soybean and sunflower. *Nat. Resour. Res., V.14,* p.65-76.

[127] Sreenath, H.K. & Jeffries, T.W. (2000) Production of ethanol from wood hydrolyzate by yeasts. *Bioresource Technol., V.72*, p.253-260.

[128] Sun, Y. & Cheng, J. (2002) Hydrolysis of lignocellulosic materials for ethanol production: a review. *Bioresource Technol., V.83,* p.1-11.

[129] Karimi, K., Emtiazi, G., Taherzadeh, M.J. (2006) Ethanol production from dilute-acid pretreated rice straw by simultaneous saccharification and fermentation with *Mucor inducus*, *Rhizopus oryzae*, and *Saccharomyces cerevisiae*. *Enzyme Microb. Tech., V.40,* p.138-144.

[130] Olsson, L. & Hahn-Hägerdal, B. (1996) Fermentation of lignocellulosic hydrolysates for ethanol production. *Enzyme Microb. Tech., V.18,* p.312-331.

[131] Nellaiah, H. & Gunasekaran, P. (1992) Ethanol production from. cassava starch hydrolysate by immobilized *Zymomonas mobilis*. *Ind. J. Microbiol., V.32,* p.435-442.

[132] Gunasekaran, P. & Kamini, N.R. (1991) High ethanol productivity from lactose by immobilized cells of *Kluyveromyces fragilis* and *Zymomonas mobilis*. *World J. Microbiol. Biot., V.7*, p.551-556.

[133] Godia, F., Casas, C., Sola, C. (1987) A survey of continuous ethanol fermentation systems using immobilized cells. *Process Biochem., V.22,* p.43-48.

[134] Youssef, K.A., Ghareib, M., Khalil, A.A. (1989) Production of ethanol by alginate-entrapped *Saccharomyces cerevisiae* strain "14-12". *Indian J. Exp. Biol., V.27,* p.121-123.

[135] Samejima, H., Nagashima, M., Azuma, M. (1984) Semicommercial production of ethanol using immobilized microbial cells. *Enzyme Eng. V.434*, p.394-405.

[136] Ramakrishna, S.V. & Prakasham, R.S. (1999) Microbial fermentations with immobilized cells. *Curr. Sci., V.17,* p.87-100.

[137] Fujii, N., Oki, T., Sakurai, A., Suye, S., Sakakibara, M. (2001) Ethanol production from starch by immobilized *Aspergillus awamori* and *Saccharomuces pastorianus* using cellulose carriers. *J. Ind. Microbiol. Biot., V.27,* p.52-57.

[138] Plessas, S., Bekatorou, A., Koutinas, A.A., Soupioni, M., Banat, I.M., Marchant, R. (2007) Use of *Saccharomyces cerevisiae* cells immobilized on orange peel as biocatalyst for alcoholic fermentation. *Bioresource Technol., V.98,* p.860-865.

[139] Sakurai, A., Nishida, Y., Saito, H., Sakakibara, M. (2000) Ethanol production by repeated batch culture using yeast cells immobilized within porous cellulose carriers. *J. Biosci. Bioeng., V.90,* p.526-529.

[140] de Vasconselos, J.N., Lopes, C.E., Franca, F.P. (2004) Continuous ethanol production using yeast immobilized on shugar-cane stalks. *Brazilian J. Chem. Eng., V.21,* p.357-365.

[141] Martinsen, A., Skjak-Brak, G., Smidsrod, O. (1989) Alginate as immobilization material: I. Correlation between chemical and physical properties of alginate gel beads. *Biotechnol. Bioeng., V.33,* p.79-89.

[142] Chibata, I. (1979) Immobilized microbial cells with polyacrylamide gel and carrageenan and their industrial applications. In: K. Venkatasubramanian (Eds.), *Immobilized microbial cells* (p.187-202). Washington, DC: American Chemical Society.

[143] Nigam, J.N., Gogoi, B. K., Bezbaruah, R.L. (1998) Short communication: alcoholic fermentation by agar-immobilized yeast cells. *World J. Microbiol. Biot. V.14,* p.457-459.

[144] Krishnan, M.S., Nghiem, N.P., Davison, B.H. (1999) Ethanol production from corn starch in a fluidized-bed bioreactor. *Appl. Biochem. Biotech., V.78,* p.359-372.

[145] Amsden, B. & Turner, N. (1999) Diffusion characteristics of calcium alginate gels. *Biotechnol. Bioeng., V.65,* p.605-610.

[146] Goksungur, Y. & Zorlu, N. (2001) Production of ethanol from beet molasses by Ca-alginate immobilized yeast cells in a packed-bed bioreactor. *Turk. J. Biol., V.25,* p.265-275.

[147] Swain, M.R., Kar, S., Sahoo, A.K., Ray, R.C. (2007) Ethanol fermentation of mahula (*Madhuca latifolia* L.) flowers using free and immobilized yeast *Saccharomyces cerevisiae*. *Microbiol. Res., V.162,* p.93-98.

[148] Petre, M., Zarnea, G., Adrian, P., Gheorghiu, E. (1999) Biodegradation and bioconversion of cellulose wastes using bacterial and fungal cells immobilized in radiopolymerized hydrogels. *Resour. Conserv. Recy., V.27,* p.309-332.

[149] El-Batal, A., Farahat, L., Abd El-Rehim, H. (2000) Ethanol production by *Kluyveromyces lactis* immobilized cells in copolymer produced by radiation polymerization. *Acta Microbiol. Pol., V.49,* p.157-166.

[150] Gough, S., Barron, N., Zubov, A. L., Lozinsky, V. I., McHale, A. P. (1998) Production of ethanol from molasses at 45°C using *Kluyveromyces marxianus* IMB3 immobilized in calcium alginate gels and poly(vinyl alcohol) cryogel. *Bioproc. Biosyst. Eng., V.19(2),* p.87-90.

[151] Rebros, M., Rosenberg, M., Stloukal, R., Kristofikova, L. (2005) High efficiency ethanol fermentation by entrapment of *Zymomonas mobilis* into lentikats. *Lett. Appl. Microbiol., V.41,* p.412-416.

[152] Sviridenko, Yu.Ya., Abdullaeva, L.V., Rainina, E.I., Efremenko, E.N. (1994) Immobilized yeast for whey fermentation. *Food Ind., V.8,* p.6-7 *(Rus.).*

[153] Pundle, A., Prabhune, A., Sivaraman, H. (1988) Immobilization of *Saccharomyces uvarum* cells in porous beads of polyacrylamide gel for ethanolic fermentation. *Appl. Microbiol. Biot., V.29,* p.426-429.

[154] Lozinsky, V.I., Zubov, A.L., Titova, E.F. (1997) Poly(vinyl alcohol) cryogels employed as matrices for cell immobilization. 2. Entrapped cells resemble porous fillers in their effects on the properties of PVA-criogels carrier. *Enzyme Microb. Technol., V.20,* p.182-90.

[155] Nigam, J.N. (2000) Continuous ethanol production from pineapple cannery waste using immobilized yeast cells. *J. Biotechnol., V.80,* p.189-93.

[156] Novak, J., Czarnecki, Z., Kaminski, E. (1997) Co-immobilization of *Aspergillus niger* and *Zymomonas mobilis* for ethanol production from starch. *Pol. J. Food Nutr. Sci., V.6,* p.65-70.

[157] Amutha, R. & Gunasekaran, P. (2001) Production of ethanol from liquefied cassava starch using co-immobilized cells of *Zymomonas mobilis* and *Saccharomyces diastaticus*. *J. Biosci. Bioeng., V.92,* p.560-564.

[158] Kourkaoutas, Y., Psarianos, C., Koutinas, A., Kanellaki, M., Banat, I., Marchant, R. (2002) Corcoutas Continuous whey fermentation by kefir yeast immobilized on delignified cellulosic material. *J. Agric. Food Chem., V.50,* p.2543-2547.

[159] Flemming, H.C. & Geesey, G.G. (1991) *Proceedings of the Int. Workshop on biofouling and biocorrosion*. Stuttgart: Springer-Verlag Berlin Heidelberg, p.187-193.

[160] Da Silva, S., Basseguy, R., Bergel, A. (2002) The role of hydrogenases in the anaerobic microbiologically influenced corrosion of steels. *Bioelectrochemistry, V.56,* p.77-79.

[161] Dubey, R.S. & Upadhyay, S.N. (2001) Microbial corrosion monitoring by an amperometric microbial biosensor developed using whole cell of *Pseudomonas sp. Biosens. Bioelectron., V.16*, p.995-1000.

[162] Beech, I.W. & Sunner, J. (2004) Biocorrosion: towards understanding interactions between biofilms and metals. *Curr. Opin. Biotech., V.15*, p. 181-186.

[163] Lee, A.K., Buehler, M.G., Newman, D.K. (2006) Influence of dual-species biofilm on the corrosion of mild steel. *Corros. Sci., V.48*, p.165-178.

[164] Tributsch, H. & Rojas-Chapana, J.A. (2000) Metal sulfide semiconductor electrochemical mechanisms induced by bacterial activity. *Electrochim. Acta, V.45*, p.4705-4716.

[165] Jones, D.A. & Amy, P.S. (2002) A thermodynamic interpretation of microbiologically influenced corrosion. *Corrosion, V.58*, p.638-645.

[166] Hamilton, W.A. (2003) Microbially influenced corrosion as a model system for the study of metal microbe interactions: a unifying electron transfer hypothesis. *Biofouling, V.19*, p.65-76.

[167] Beech, I.B., Sunner, J.A., Hiraoka, K. (2005) Microbe-surface interactions in biofouling and biocorrosion processes. *Int. Microbiol., V.8*, p.157-168.

[168] Santegoeds, C.M., Ferdelman, T.G., Muyzer, G., De Beer, D. (1998) Structural and functional dynamics of sulfate-reducing populations in bacterial biofilms. *Appl. Environ. Microb., V.64(10)*, p.3731-3739.

[169] Efremenko, E.N., Azizov, R.E., Abbasov, V.M., Varfolomeyev, S.D. (2006) Analysis of bacteria self-immobilized into corrosive biofilms. In: G.E. Zaikov et al. (Eds.), *Biocatalysis and Biocatalytic Technologies* (p.39-58). Hauppauge, NY: Nova Science Publishers Inc.

[170] Hamilton, W.A. (2000) Microbially influenced corrosion in the context of metal microbe interactions. In: C.A.C. Sequira (Eds.), *Microbial corrosion* (p.3-17). London: European Federation of Corrosion.

[171] Lee, W., Lewandowski, Z., Okabe, S., Characklis, W.G., Avci, R. (1993) Corrosion of mild steel underneath aerobic biofilms containing sulfate-reducing bacteria. Part I: at low dissolved oxygen concentration. *Biofouling, V.7*, p.197-216.

[172] Lee, W., Lewandowski, Z., Okabe, S., Characklis, W.G., Avci, R., Nielsen, P.H. (1993) Corrosion of mild steel underneath aerobic biofilms containing sulfate-reducing bacteria. Part II: at high bulk oxygen concentration. *Biofouling, V.7*, p.217-239.

[173] Ramesh, S., Rajeswari, S., Maruthamuthu, S. (2003) Effects of inhibitors and biocide on corrosion control of mild steel in natural aqueous environment. *Mater. Lett., V.57*, p.4547-4554.

[174] Ramesh, S. & Rajeswari, S. (2005) Evaluation of inhibitors and biocide on the corrosion control of copper in neutral aqueous environment. *Corros. Sci., V.47*, p.151-169.

In: Leading-Edge Environmental Biodegradation
Editor: L. E. Pawley, pp. 53-93

ISBN: 978-1-60021-903-0

Chapter 2

BIODEGRADATION OF SELECTED EMERGING ORGANIC CONTAMINANTS IN THE ENVIRONMENT–AN OVERVIEW

Ajit K. Sarmah*

Soil Chemical & Biological Interactions, Soil & Landscape Team, Landcare Research New Zealand Ltd., Private Bag 3127, Hamilton, New Zealand

Bent Halling-Sørensen†

Dept of Pharmaceutics and Analytical Chemistry
The Pharmaceutical Faculty, Copenhagen University, 2 Universitetsparken, DK 2100 Copenhagen, Denmark

ABSTRACT

There has been an increasing concern among the regulatory bodies, scientific community and public at large regarding the potential threat of certain groups of chemicals (natural and man-made) on human, wildlife, and the environments. Although these chemicals have existed for a long time, their potential environmental threats have been realised only in the last few decades. These so called emerging contaminants (ECs) are mostly organic in nature and quit a few of them possess estrogenic activity and have been found to cause abnormalities in reproductive function of wildlife and some aquatic species. For most emerging contaminants, there is currently little information regarding their potential toxicological significance in ecosystems- particularly effects from long-term, low-level environmental exposures. These ECs include prescription and over-the-counter veterinary and human pharmaceuticals (antibiotics, steroids and reproductive hormones), as well as many chemicals found in personal care products and others (antioxidants, detergents, antimicrobials, fragrances). In addition, some pesticides, natural toxins, biomedicine, nanoparticles, flame retardants and perfluorated hydrocarbons also

* SarmahA@LandcareResearch.co.nz
† bhs@farma.ku.dk

fall under ECs. The environmental fate of these ECs influences their exposure and, hence, is a key parameter for estimating the risk of effects on the biota. Biological degradation plays a key role in their ultimate fate in the environment. In this chapter, we provide an overview of past and recent studies conducted on the biodegradation of selected ECs in various environmental matrices (e.g. soils, sediments, waters, animal manure), and the factors influencing their degradation behavior in these matrices are also discussed. ECs are often present in the environment at low concentrations that might affect the ability of micro-organism to degrade these compounds. Examples of ECs considered in this chapter include steroid hormones (natural and synthetic), some commonly used veterinary antibiotics, plasticizers, and some industrial surfactants. Finally some recommendations on future research direction on environmental biodegradation of emerging organic contaminants are presented.

INTRODUCTION

The last few decades have witnessed increasing concern about the potential adverse effects of the production, use, and disposal of a range of chemicals on humans, wildlife, and receiving ecosystem environments. These chemicals include not only those derived from industrial and agricultural production, but also those that are naturally released by animals and human, as well as those from household activities, medical, and health applications (Daughton & Ternes 1999). These chemicals which are mostly organic have been termed emerging contaminants (ECs) in today's environment. The United States Geological Survey and European institutions have identified prescription and over-the-counter veterinary and human pharmaceuticals (antibiotics, steroid and reproductive hormones), as well as many chemicals found in personal care products and others (antioxidants, detergents, antimicrobials, fragrances) (Brack et al. 2007, Kolpin et al. 2002). Additional compounds include pesticides, natural toxins, biomedicine, nanoparticles, flame retardants and perfluorated hydrocarbons, contributions from technological breakthroughs in chemistry, materials science and biotechnology (Kolpin et al. 2002).

Recent studies have shown that some ECs are weak estrogenic disruptors, possess other specific biological activity, persist for a long time, or are continuously released with treated sewage water at levels sufficient to potentially cause damage to aquatic species and terrestrial organisms. Degradation of organic contaminants in the environment influences the exposure and, hence, it is a key parameter for estimating the risk of effects on the biota. The biodegradation and transformations of many anthropogenic ECs in soil have gained increased interest and applications for soil bioremediation during the last decades (Alexander, 1999). The concentration levels of ECs are normally very low in both surface waters (10 ng/L to 10 μg/L) and soils (few μg/kg soil). Biodegradation of a compound in both water and soil at very low level is often difficult and not obviously predictable by simulation tests performed at much higher concentrations.

Standardized test systems can be used to produce reproducible results with general information, but improvements of the present standardized test system are necessary as they are only used for categorization of compounds as "readily" and "not readily" biodegradable (Gotvajn and Zagorc-Koncan, 1999). Estimation of biodegradation rates plays an important role in many mathematical chemical fate calculation e.g. in risk assessment of chemicals.

Given the paucity of directly applicable experimental rate data, often modelers and assessors take the easy option of educated guesses. There is a need for kinetic data to be obtained at low compound concentration, and this requires development of standard test guidelines for biodegradability studies in simulation tests with low test concentrations in both surface water and soil at relevant test conditions.

In this chapter, recent studies conducted on environmental biodegradation of some emerging contaminants (Figures 1 and 2) are reviewed. Since the list of emerging organic contaminants is long and continues to grow, and space in this chapter is limited, we select only five distinct groups of contaminants. These include steroid hormones (natural and synthetic); selected groups of veterinary antibiotics (VAs) used in the animal production; industrial surfactants such as alkylphenol polyethoxylates (APEs) and its metabolite nonylphenol (NPs); plasticiser bisphenol-A (BPA), and various phthalate esters. These compounds have been frequently detected in various environmental matrices and are a risk to the environment due to their estrogenic nature and toxicity to aquatic and some terrestrial species. Examples are provided on the degradation of these selected ECs in various environmental matrices (river water, sediment, soils, sludge and animal manure). Factors influencing their degradation behavior in these matrices are briefly discussed with examples from the available literature. Recommendations on the direction of future research on environmental biodegradation of emerging organic contaminants are presented in the concluding section.

Steroid Hormones

Livestock wastes are potential sources of steroid hormones (17α-estradiol; 17β-estradiol and its derivatives estrone and estriol; testosterone and its derivative androstenedione). These compounds are excreted by humans and animals of all species, sexes, and classes in their faeces and/or urine, and reach the environment through the release of municipal and animal wastes to receiving waters, animal waste application onto land, or by direct excretal input from grazing livestock. The estrogens (17β-estradiol, estriol and estrone) are predominantly female steroid hormones, while androgens are dominant in male species (Fotsis et al. 1980). During pregnancy, women excrete significantly higher levels of estrogens: 259 μg/person/day of 17β-estradiol, 600 μg/person/day of estrone, and 6000 μg/person/day of estriol (Fotsis et al. 1980), depending upon the term of their pregnancy (Berg & Kuss 1992). In addition to natural estrogens and androgens, a number of synthetic steroid hormones provide sources of excreted steroid hormones. For example, ethinylestradiol and mestranol are produced as active ingredients in the contraceptive, and conjugates of natural estrogens are active ingredients in female hormone replacement therapy (HRT) products.

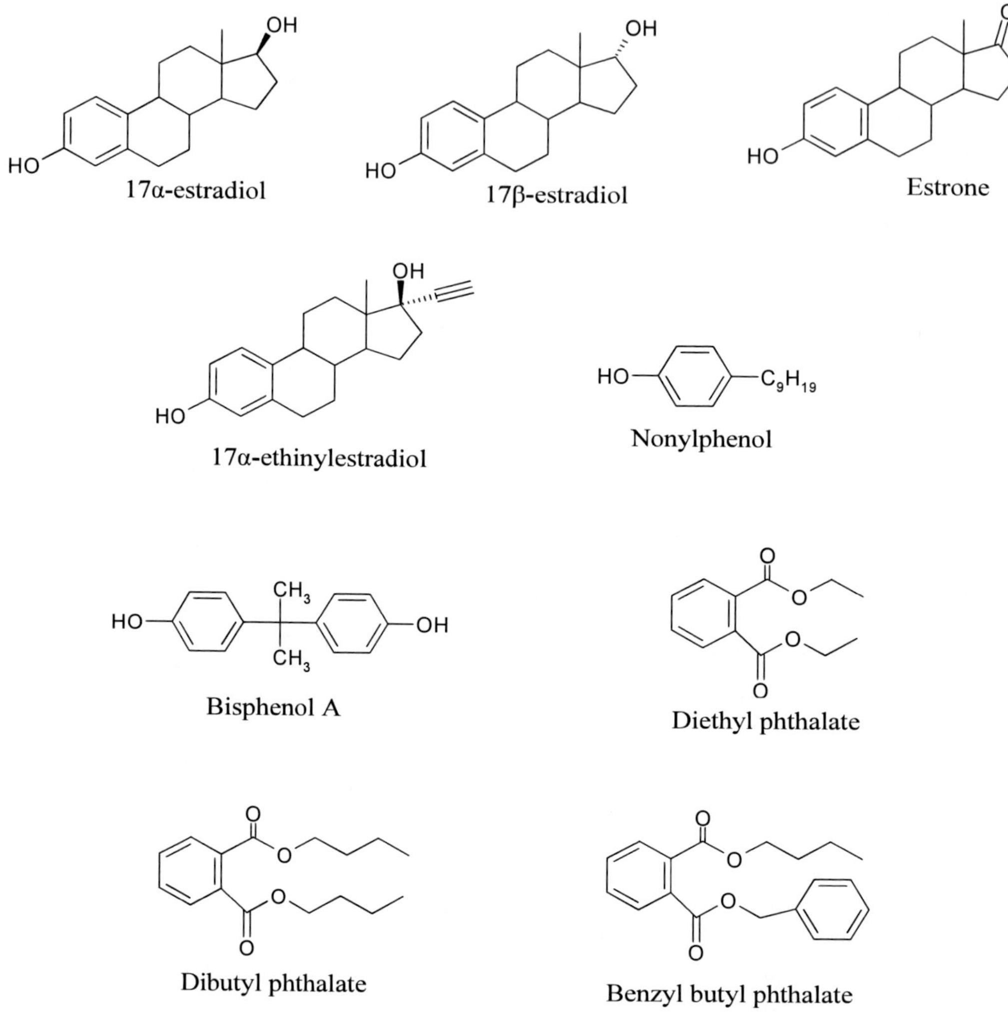

(Source: Sarmah et al. 2005)

Figure 1. Molecular structures of some emerging contaminants of concern.

Steroid hormones, originating either from animal waste or through sewage discharge, may interfere with the normal functioning of endocrine systems, and affect the reproduction and development of wildlife in receiving environments (Jobling et al. 1998; Hansen et al. 1998). The potential for such effects to be elicited is demonstrated by the induction of vitellogenin in male trout to concentrations of estradiol as low as 1 ng/l (Hansen et al. 1998), and by the formation of ova in the testis of Japanese medaka at a concentration of 4 ng/l estradiol (Metcalfe et al. 2001). A growing body of literature is now available on the occurrence and distribution of steroid hormones in waste water treatment plant (WWTP) effluents, surface water, and groundwater (Ternes et al. 1999a; Baronti et al. 2000; Peterson et al. 2001; Kolpin et al. 2002). Measured concentrations of up to 21, 52, 80 and 6 ng/L have been obtained for 17β-estradiol, estrone, estriol and 17α-ethinylestradiol respectively in raw sewage and WWTP influents in Brazil and Italy (Ternes et al; 1999a; Baronti et al. 2000).

17 ring of the molecule with estradiol (E2) oxidizing from the cyclopentane ring D at C-17 into estrone (E1) during enzymatic degradation and then further degrading into metabolite X1 and finally to carbon dioxide through a tricarboxylic acid (TCA) cycle (Figure 2). Recently, Weber et al. (2005) studied the degradation of estradiol and ethinylestradiol by activated sludge and using a defined mixed culture consisting of two strains, *Achromobacter xylosoxidans* and *Ralstonia* species, under conventional and membrane sewage treatment plants. Their study revealed that while estradiol was converted to estrone, with complete transformation within 3 days, ethinylestradiol was persistent in both type of activated sludge. The turn over rates, between 0.08 and 0.08–0.11 μg h^{-1} (g dry residue)$^{-1}$, fit with other degradation studies with activated sludge in the literature (e.g., Fujii et al. 2002; Kurisu et al. 2002; Layton et al. 2000).

Andersen et al. (2003) investigated the fate of estrone (El), 17beta-estradiol (E2), and 17alpha-ethinylestradiol (EE2) in mechanical and biological sewage treatment as well as in sewage-sludge treatment at a municipal German sewage treatment plant (STP). The main outcome of the study was that a common municipal STP with an activated sludge system for nitrification and denitrification including sludge recirculation can appreciably eliminate natural and synthetic estrogens. As a consequence, the endocrine effects of biota in the receiving waters should be significantly reduced. All estrogen concentrations decreased gradually along the treatment train. In the STP effluent, the steroid estrogen concentrations were always below the quantification limit of 1 ng/L. The elimination efficiency of the natural estrogens (E1 and E2) exceeded 98%, and EE2 was reduced by more than 90%. The natural estrogens were largely degraded biologically in the denitrifying and aerated nitrifying tanks of the activated sludge system, whereas EE2 was only degraded in the nitrifying tank. Only about 5% of the estrogens are sorbed onto digested sewage sludge. It is very likely that conjugates (glucuronides and sulfates) of the estrogens were cleaved into the parent compounds mainly in the first denitrification tank.

Because of its potency and negative effects on the hormonal system, ethinylestradiol has been studied by many researchers in an effort to isolate microbes capable of degrading the compound using laboratory studies (Haiyan et al. 2007). Shi et al. (2002) isolated *Fusarium proliferatum* sp. from a manure sample that could degrade ethinylestradiol with an initial concentration of 25 mg/L. Similarly, Yoshimoto et al. (2004) isolated two microbial species (*Rhodococcus zopfii* and *Rhodococcus equi*) using an initial concentration of 100 mg/L from sample collected from WWTP in Japan. However, neither of these two studies identified the degradation products of ethinylestradiol. The use of NAS and ammonia-oxidizing bacterium *Nitrosomonas europaea* to degrade ethynylestrdaiol, estradiol, estrone and estriol was also demonstrated by Shi et al. (2004); however, it was speculated that transformation of estrone from estradiol in NAS may have been due to the presence of other heterotrophic bacteria and not to ammonia-oxidizing bacteria which was added to the system. Recently, Haiyan et al. (2007) isolated *Sphingobacterium* sp. JCR5 from a WWTP sludge sample of a contraceptive factory in China. The authors demonstrated that at an initial concentration of 30 mg/L, the strain JCR5 could degrade 87% of ethinylestradiol under aerobic condition. It was revealed that ethinylestradiol was first oxygenized to estrone, and 2-hydroxy-2, 4-dienevaleric acid and 2-hydroxy-2,4-diene-1,6-dioic acid were the main catabolic intermediates (Figure 3).

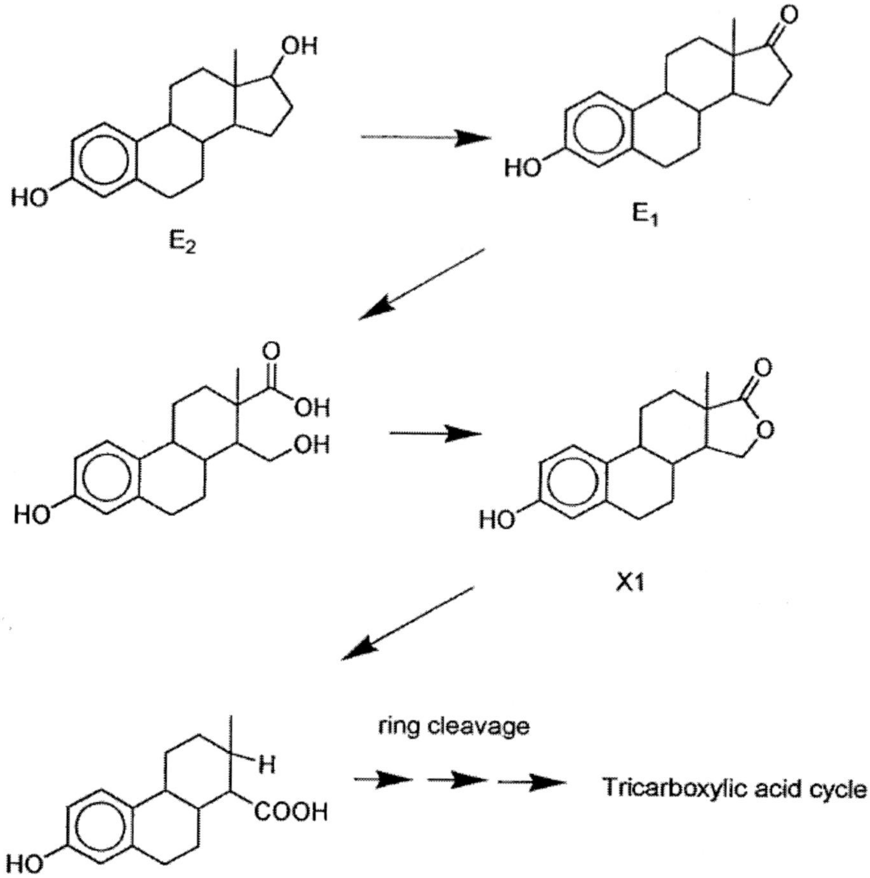

(Adapted from Lee and Liu, 2002)

Figure 2. Proposed metabolic pathway of 17*β*-estradiol by sewage bacteria.

Degradation in River Waters, River Sediments and Agricultural Soils

Biodegradation of steroid hormones in river waters, sediments or in terrestrial soils can be one of myriad of factors, such as nutrient compositions (N, P, K), bio-chemical oxygen demands (BOD), pH and the physico-chemicals properties of the media. Each of these factors could potentially contribute to overall degradation pattern for these hormones in these media.

In recent years, the biodegradability of steroid hormones has been investigated through laboratory study involving river water, agricultural soil and river sediment samples collected from urban and agricultural areas (e.g., Williams et al. 1999; Colucci et al. 2001; Colucci and Topp 2001; Jürgens et al. 2002; Lee et al. 2003; Ying and Kookana, 2003; Lorenzen et al. 2005; Jacobsen et al. 2005; Czajka and Londry, 2006; Fan et al. 2007; Sarmah and Northcott. 2007). From a laboratory study, Williams et al (1999) estimated the half-lives of estrogenic steroids to be 2–6 days in water and sediment samples collected in the UK. Jürgens et al. (2002) reported that 17β-estradiol rapidly transformed to estrone under aerobic conditions in river water with a half-life of 0.1–9 days at 20°C, with estrone further degraded at similar times. Studies conducted by the same authors also showed that 17α-ethinylestradiol is more resistant to biodegradation than 17β-estradiol in aerobic river water, with a half-life nearly 10 times higher than that of 17β-estradiol. Czajka and Londry (2006) investigated the potential for anaerobic biodegradation of 17β-estradiol and 17α-ethinylestradiol in water and sediment samples collected from Lake 260 of the experimental Lakes Area near Kenora, Ontario, Canada. These results showed that even after long incubation (3 years), anaerobic degradation of ethinylestradiol was not observed in the samples. However, estradiol was transformed to estrone under methanogenic, SO_4^{2-}, and Fe^{3+} reducing anaerobic condition, but the extent of conversion was different for each electron acceptor.

In many countries, recycled water and animal effluents containing these hormone steroids are often land-applied as an alternative for disposal, and it is therefore important to know their degradation behavior in these media. Laboratory studies of the persistence of 17β-estradiol, estrone and 17α-ethinylestradiol under a range of temperature and moisture conditions have demonstrated that 17β-estradiol is oxidised to estrone in both sterile and non-sterile agricultural soils (Colucci et al. 2001; Colucci & Topp 2001). At 30°C, following 3 days of incubation, > 56% of applied estradiol was found to be non-extractable in three Canadian agricultural soils with reported half-lives of <0.5 days in all three soils (Colucci et al. 2001). The authors did not find any contribution of any soil properties (texture, pH, organic matter) to the overall degradation rate of estradiol; however the rate of dissipation was slower when soils were air-dried or adjusted to field moisture capacity. The microbial degradation of ethinylestradiol in terrestrial soil has not been well investigated, presumably because the only source for ethinylestradiol is likely to be through the use of sewage biosolids as fertilizer. Colucci and Topp (2001), however, studied persistence of ethinylestradiol in three Canadian agricultural soils using laboratory microcosms at varied soil temperatures (4, 10, 20 and 30°C) and moisture regimes (5% – field capacity). Their study showed that ethinylestradiol was rapidly dissipated in all the soils and rate of dissipation increased with increases in incubation temperature, while the corresponding 50% dissipation times for ethinylestradiol were 8 and 3 days, at temperatures of 40°C and 30°C, respectively. Jacobsen et al (2005) investigated the fate of 17β-estradiol and testosterone in swine manure and biosolids amended soils, and observed both parent and transformation products dissipated within a few days of incubation under aerobic condition. The authors concluded that both biosolids and swine manure slurry delayed the conversion of testosterone and 17β-estradiol to less hormonally active transient products in agricultural soil. Elsewhere, from a laboratory incubation study in four soils collected in southern Australia, Ying and Kookana (2005) reported both estradiol and ethinylestradiol to degrade rapidly in four South Australian soils under aerobic condition; however, both persisted longer under anaerobic condition. Similarly, Fan et al. (2006) reported in a degradation study involving 17β-estradiol and testosterone in three US agricultural soils, that rate of mineralization for testosterone was more than 10-fold greater than for 17β-estradiol in the same soils under aerobic condition. Under anaerobic treatment, however, only 2% of testosterone degraded under methanogenic condition in the same soil. A study conducted in the UK with grassland soils amended with animal wastes in different solvents (water, artificial urine and natural sheep urine) showed that both 17β-estradiol and estrone degraded at similar rates, although this was dependent on both the soil type and solvent used (Lucas and Jones, 2006).

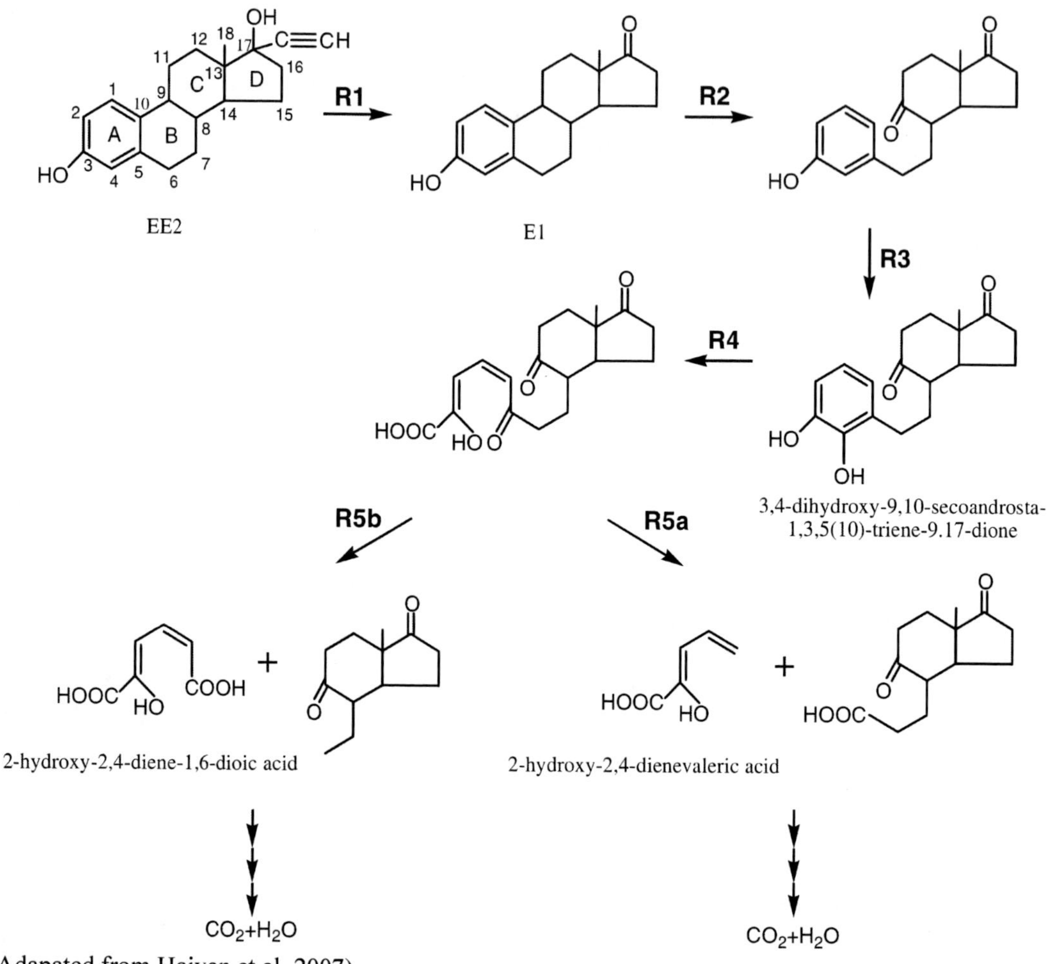

(Adapated from Haiyan et al. 2007)

Figure 3. Catabolic pathway of 17α-ethinylestradiol degradation by *Sphingobacterium* sp. JCR5.

Recent work by Kjær et al (2007) showed that biodegradation of steroid estrogens in soils is not promising as reported in earlier studies and therefore the threat to the aquatic environment of steroid estrogens is posed by root zone leaching of estrogens from manure-treated fields (Yu et al. 2004; Ying et al. 2005). A subject that has been hitherto overlooked in the recent past in studies involving biodegradation and sorption of steroid estrogens in soil. The steroid hormones 17β-estradiol (E2) and its degradation product estrone (E1) are of particular environmental concern as both are abundant in slurry from pregnant and cycling dairy and pigs and both are potential endocrine disruptors (lowest observable effect level (LOEL) 14 and 3.3 ng/L, respectively). The one year study in Denmark examined the transport of E1 and E2 from manure to tile drainage systems at two field sites on structured, loamy soil. The estrogens leached from the root zone to tile drainage water in concentrations exceeding the LOEL for as long as 3 months after application, with the maximum recorded concentration of E1 and E2 is 68.1 and 2.5 ng/L, respectively. Transport of estrogens from the soil to the aquatic environment was governed by pronounced macropore flow and consequent rapid movement of the estrogens to the tile drains. These findings suggest that the application of manure to structured soils poses a potential contamination risk to the aquatic environment with estrogen, particularly when manure is applied to areas where the majority of stream water originates from drainage water.

Degradation in Aquifer Material/groundwater

Biodegradation studies of steroid hormones in aquifer material and groundwater are limited. Although sorption behavior of these hormones for soils and sediments has been reported to be strong (e.g., Lee et al. 2003; Casey et al. 2003; Sarmah et al. 2007), colloid-facilitated transport, or flow through cracks or macropores during conditions of high rain event may allow the transport of these compounds to deeper reaches of soil profile (Das et al. 2004; Sarmah et al. 2006a), and may contaminate groundwaters (Peterson et al. 2000). Therefore studying their degradation behavior in aquifer material and groundwater is important. Ying et al. (2003) reported the half-life of estradiol and ethinylestradiol in aquifer material to be 2 and 81 days under aerobic condition, though they both persisted longer under anaerobic treatment. In contrast, in a recent study conducted on these hormones using slurries prepared with aquifer sediment and groundwater, Sarmah and Northcott (2007) reported rapid degradation of both natural and synthetic hormones with >90% of both compounds disappearing within the first few days of incubation. Degradation of the compounds did not follow first-order kinetics as reported for these hormones in a number of earlier studies. Rapid degradation under anaerobic treatment in Sarmah and Northcott's (2007) study could be attributed to the presence of electron acceptors such as NO_3^-, SO_4^{2-}, and Fe^{3+} in the media tested. Many aquifers have extensive anoxic zones in which denitrification, Fe (III) reduction, sulphate reduction, or methanogenesis predominate (Kazumi et al. 1997).

Veterinary Antibiotics (VAs)

Veterinary antibiotics (VAs) are widely used in many countries to treat various kinds of diseases and to protect the health of animals (Figure 4). As most of the antibiotics are water-soluble, as much as 90% of one dose can be excreted in urine and up to 75% in animal faeces as metabolite or in the parent form (Halling-Sorensen et al. 1998). Due to the application of manure as supplement to organic fertilizers on agricultural land, VAs enter the environment and their degradation products may form to some extent (e.g., Boxall et al. 2003; Sarmah et al. 2006b; Hu and Coats, 2007). These compounds can adsorb onto soil, leach into groundwater, or be transported to surface water by overland-flow or run-off. If the antibiotics are not efficiently degraded, there is potential for these residues to help maintain or develop antibiotic resistant microbial populations (Witte, 1998). Thus frequent application of manure on the same location may result in the continuous exposure of soil microbes to antibiotic residues and give rise to antibiotic resistant populations of bacteria. Whether VAs pose a substantial public health threat remains a matter of debate; however, some adverse effects of certain VAs on some terrestrial and aquatic species have been documented and place VAs within the priority groups of emerging contaminants in many developed countries (USA,

Europe, and Canada). Understanding the fate of antibiotics in various environmental matrices can help develop proper risk assessment guidelines for regulatory decision- making process.

Like any other organic contaminants, antibiotics are also degraded through biotic or abiotic processes, and in some cases conjugated metabolites may revert to the parent compound. Although the bioavailability and biodegradation of VAs is often thought to be simple, it has been reported that this is not necessarily true (Henschel et al. 1997; Halling-Sørensen et al. 1998). In the following sections, the biological degradation of some of the common VAs in various environmental media is discussed, along with their potential risks to the environment.

Tylosin

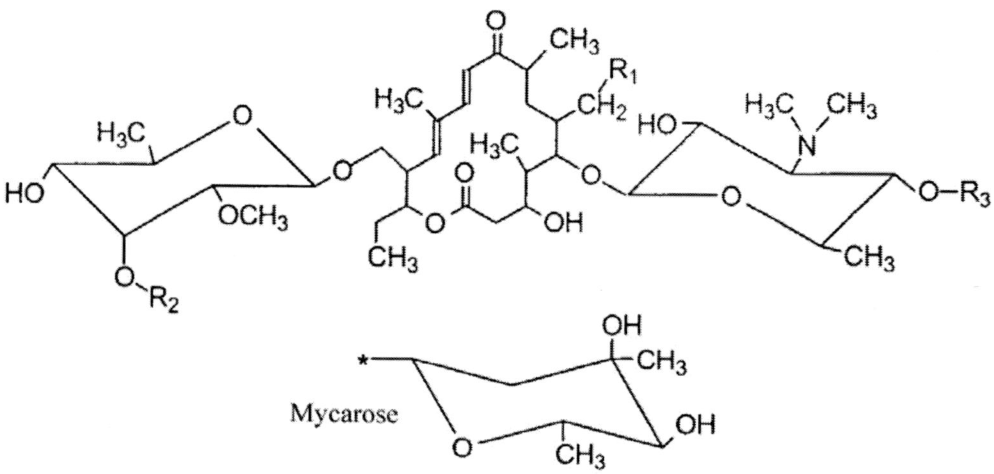

	Tylsoin- A	Tylosin- B	Tylosin- C	Tylosin- D
R_1	-CHO	-CHO	-CHO	$-CH_2OH$
R_2	$-CH_3$	$-CH_3$	-H	$-CH_3$
R_3	-mycarose	-H	-mycarose	-mycarose

Tetracyclines

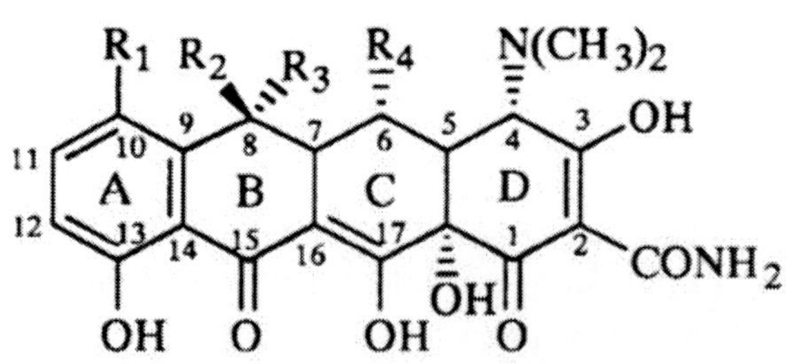

	R_1	R_2	R_3	R_4
Tetracycline (TC)	H	CH_3	OH	H
Chlortetracycline (CTC)	Cl	CH_3	OH	H
Oxytetracycline (OTC)	H	CH_3	OH	OH

Figure 4. (Continued on next page).

Sulfonamides

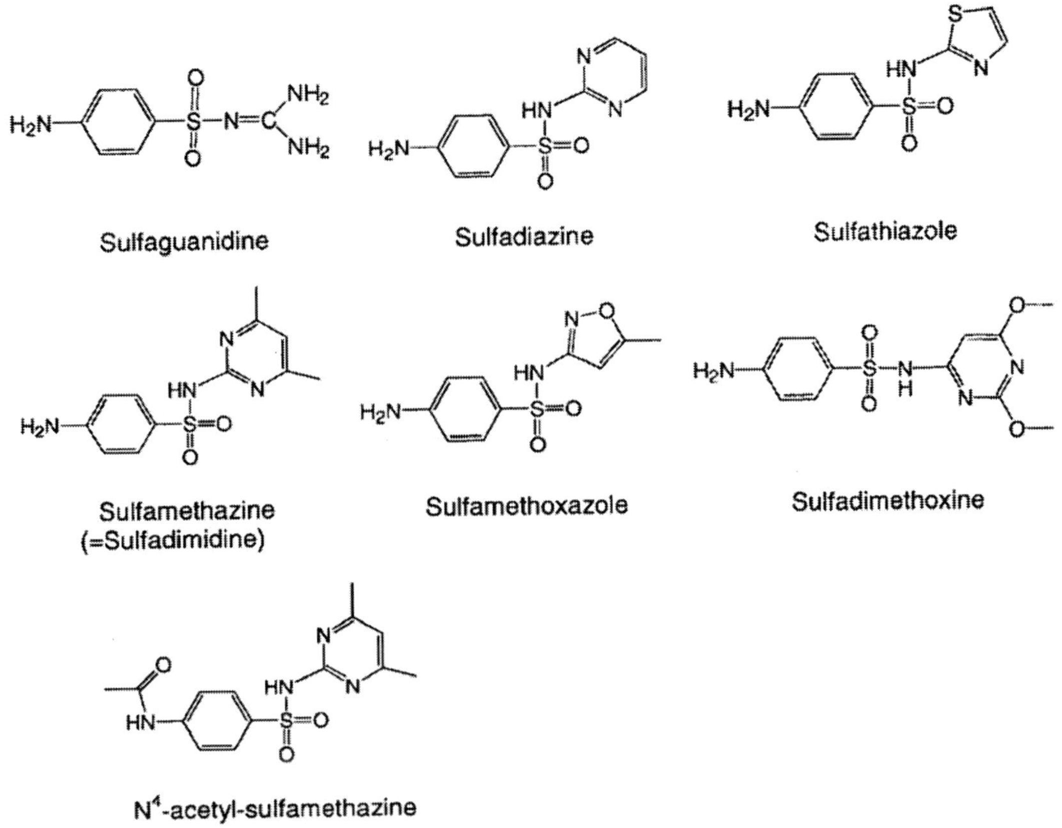

(Source: Sarmah et al. 2006)

Figure 4. Molecular structures of some commonly used veterinary antibiotics.

Degradation in Soil

Gilbertson et al. (1990) studied ceftiofur sodium, a wide-sprectrum cephalosporin antibiotic, in three soils (pH range 6.9–8.0), as well as in buffers of pH 5, 7 and 9. Their study showed that ceftiofur sodium (^{14}C) degraded to microbiologically inactive metabolites, with half-lives ($T_{1/2}$) of > 49, 22 and 41 days in three soils collected from California (pH 8.02), Florida (pH 6.96), and Wisconsin (pH 7.37). Gavalchin and Katz (1994), studied degradation of a range of VAs (bacitracin, penicillin, streptomycin, tylosin, bambermycins, erythromycin and chlortetracycline) as a function of temperature in a sandy surface soil (pH 6.0–6.3) from New Jersey, US, mixed with chicken faeces. Their study showed that persistence of these fecal-borne antibiotics varied according to their chemical structure and incubation temperature. Persistence under field conditions is also likely to be affected by the interplay of several factors such as temperature, humidity, rainfall, and the nature of soil properties as demonstrated by Donoho (1984) who reported the degradation of monensin (growth promoter used in pigs) was faster under field conditions than observed under laboratory study.

Kreuzig and Holtge (2005) investigated the fate of ^{14}C-labeled sufadiazine in manured soil under controlled laboratory test systems following the OECD guidelines. The authors used an initial concentration of 500 μg/kg of sulfadiazine and incubated at 20°C for over a period of 102 days. Their results showed that only 1% of the initially applied radiotracer was mineralized to ^{14}C-carbon dioxide and 82% were transferred to nonextractable residues within the total incubation period of 102 days. Schlüsener and Bester (2006) studied the persistence of erythromycin, oleandomycin, tylosin, and two growth promoters (tiamulin and salinomycin) used in animal husbandry using a typical German sandy loam soil, and an initial concentration of 2000 μg/kg. The first-order half-lives were 20, 27, 8, 16, and 5 days respectively for erythromycin, oleandomycin, tylosin, tiamulin and salinomycin when incubated in the dark at 20°C with moisture content of approximately 12% (w/w). Although the half-life for tiamulin in soil was found to be 16 days (Table 2) by Schlüsener and Bester (2006), an earlier study by Schlüsener et al. (2003) showed that 6 months after storing manure containing tiamulin, this did not show any sign of degradation, and thus manure utilized as fertilizer may contaminate the agricultural soil when it is used as fertilizer (Schlüsener et al. 2006).

Recently, Sassman et al. (2007) studied the aerobic degradation of tylosin A (TA) with an initial concentration of 20 mg/kg using two US soils (sandy soil, and a silty clay loam) over a 3-month (23°C) incubation period. The authors reported that after an initial lag phase of about 3 weeks, more than half of the applied TA was degraded within 3 days (between days 20 and 23 for sandy soil; and 21 and 24 for silty clay soil). As the degradation product of TA can retain significant antimicrobial activity (Loke et al. 2000; Sassman et al. 2007), the authors subsequently identified three products as: TA-acid, TB, and TB-acid, as a result of oxidation of the aldehyde group.

Because of their overall toxicity profiles, monensin and lasalocid (polyether ionophore antibiotic) fall among the recently classified high priority veterinary medicine for detailed risk assessment (Cappleton et al. 2006). These antibiotics are used in the beef and poultry production for the prevention of coccidial infections and for growth promotion in many parts of the world. In a laboratory incubation experiment (23°C), the degradation half-life for both monensin and lasalocid in soils was found to be < 4 days (Table 2) although rate of degradation appeared to be faster for monensin than lasalocid (Sassman and Lee, 2007). These studies confirm that veterinary antibiotics degrade rapidly in agricultural soils when experiments were performed under controlled laboratory conditions; however, under a medium containing these compounds or when manure is applied as fertilser to agricultural land, the compounds can behave differently and a multitude of factors, which are discussed below, can determine their fate.

Degradation in Manure/Slurry

Biodegradation studies of veterinary antibiotics reported in the literature also involved incubation experiments with soil mixed with either animal faeces or urine to which a known quantity of antibiotic was added and the inactivation observed through periodic sampling analysis either by LC-MS (liquid chromatography-mass spectrometry), GC-MS (gas chromatography-mass spectrometry) or microbiological assay (Halling-Sørensen et al. 1998; Sarmah et al. 2006b). While the laboratory incubation studies with soil are easy to perform as many of the parameters can be controlled, the real challenge would be to conduct biodegradation tests under realistic situations such as in the manure tank or field where the factors that influence the degradation processes are difficult to control.

More than a decade ago, Gilbertson et al. (1990) studied the degradation of ceftiofur sodium in the urine and faces of cattle and demonstrated the effect of sterilizing the faeces of cattle by the marked decline in the rate of degradation of the compounds compared with non-sterilized samples, thus indicating the important role of microorganisms in the degradation of the antibiotic. Although the metabolite could not be identified, the authors also observed similar results for pig faeces. Of late, a number of attempts have been made to study the biodegradability of a range of VAs in the laboratory using manure and/or slurry mixtures (Loke et al., 2000; Ingerslev and Sørensen, 2001; Kolz et al., 2005; Teeter and Meyerhoff, 2003), for instance, the use of methanogenic manure containing test system to study tylosin A degradation (Loke et al., 2000). Loke et al. (2000) reported the half-lives of tylosin A to be < 2 days under methanogenic conditions, which increased with the addition of more manure particles. However, the authors failed to support the concomitant decrease in the concentration of tylosin A with an increase in more manure to the system, was due to the result of either sorption, or abiotic or biotic degradation. It is conceivable that in a manure tank there would be a much greater concentration of colloid and particulate matter than in the laboratory test systems. This may increase the fraction of antibiotic that is sorbed and hence influence the overall rate of degradation.

There is a dearth of published information on the formation of tylosin metabolite during degradation studies involving manure or slurry. From a laboratory microcosm study of tylosin in aqueous manure soil systems, Oliveira et al. (2002) observed that degradation was rapid during the first 10 days and slowed to a steady rate through the formation and accumulation of metabolites in subsequent sampling events. The authors reported an initial half-life of 2.4 days for tylosin based on a 10-day pseudo first-order kinetics data (Oliveira et al. 2002). This half-life of tylosin was similar to that reported earlier by Loke et al. (2000). A study by Kolz et al. (2005) similarly found that tylosin degradation in manure-lagoon slurries (incubated at 22°C) exhibited biphasic kinetics with 90% disappearance occurring within < 5 days. The authors also observed the formation of tylosin degradates (tylosin B and D, dihydrodesmycosin and an unknown product) in anaerobic condition during their 8-month incubation study. The formation of degradates after an incubation period of 8 months suggested that degradation in lagoon slurries is not complete and there is a likelihood of the residues entering nearby agricultural fields (Kolz et al. 2005). Unlike oxygen limiting condition, degradation of tylosin in cattle, chicken and swine excreta has been found to be rapid under aerobic condition, with first-order half-lives at 20°C varying from 6.2, < 7.6, and 7.6 days respectively in cattle, chicken and swine excreta (Teeter and Meyerhoff, 2003).

Ingerslev and Halling-Sørensen (2001) simulated the biodegradability of 3 VAs in soil-manure slurries under aerobic laboratory conditions using aniline as the benchmark chemical and found that degradation half-lives for the compounds (4.1–8.1 days for tylosin, 5.9–8.8 days for olaquindox, and 9.7–26.9 days for metronidazole) did not seem to be influenced by the varied nature of soils and was not concentration dependent in the test system. Given the complex nature of real-world situations, where soil pH, redox conditions, temperature, prevailing soil water conditions, wetting and drying cycles, can vary, as can the size and type of bacterial populations, biodegradability of VAs may likely to be different in the field from what has been observed under controlled conditions in the laboratory. For example, under field condition, Halling-Sørensen et al. (2005) found the average degradation half-lives of chlortetracycline and tylosin A to vary between 25–34 days and 49–67 days in two Danish sandy soils, respectively. These half-lives in field soils were substantially higher than the reported values for these compounds when the experiment was conducted in the laboratory using soil microcosm study by Sassman et al. (2007).

Degradation in Surface Waters and Sediments

Information on biodegradation of VAs used for livestock purposes in both surface waters and sediments (freshwater and marine) is lacking. However, a large body of data exists on this aspect for VAs used specifically for aquaculture, which has been covered in a recent paper by Boxall et al. (2004). Some common VAs that are used in both animal and fish farms include OTC, sarafloxacin, sufadiazine, sulfamethoxine, and oxonilic acid. Biodegradation studies conducted on these compounds showed significant variation in the reported half-lives, and were often difficult to compare with one another for a single compound due to differences in experimental protocol and adopted laboratory conditions (e.g., Samuelsen et al. 1994; Pouliquen et al. 1992; Hektoen et al. 1995; Lai et al. 1995). For example, from a laboratory incubation study of OTC in marine sediment, no degradation was observed after 6 months of incubation period (Samuelsen et al., 1994). In contrast, in an earlier laboratory study by Samuelsen (1989), OTC was found to have a half-life of 30–64 days in sediment from a fish farm. Many other instances show the degree of variation observed in the degradation rate of VAs in these matrices.

Ingerslev et al. (2001) studied the primary aerobic and anaerobic biodegradability at intermediate concentrations (50-5000 μg/L) of the antibiotics olaquindox (OLA), metronidazole (MET), tylosin (TYL) and oxytetracycline (OTC) in a simple shake flask system simulating the conditions in surface waters. Assuming first-order degradation kinetics, ranges of half-lifes (Table 2) for aerobic degradation of the four antibiotics studied were obtained: 4-8 days (OLA), 9.5-40 days (TYL), 14-104 days (MET) and 42-46 days (OTC). OLA and OTC were degraded with no initial lag phase whereas lag phases from 2 to 34 days (MET) and 31 to 40 days (TYL) were observed for other substances. The biodegradation behavior was influenced neither by the concentrations of antibiotics nor by the time of the year and location for sampling of surface water. Addition of 1 g/L of sediment or 3 mg/L of activated sludge from wastewater treatment increased the biodegradation potential which is believed to be the result of increased bacterial concentration in the test solution. Biodegradation was significantly slower in tests conducted in absence of oxygen. Assessments of the toxic properties of antibiotics by studying the influence on the biodegradation rates of C-14-aniline at different concentrations of antibiotics showed that no tests were conducted at toxic concentrations.

Alkylphenol Polyethoxylates (APEs)

APEs are high-volume chemicals that have been used for more than 40 years as detergents, emulsifiers, wetting agents, and dispersing agents in many products and sectors. These sectors include textile and pulp and paper processing, paints, resins and protective coatings, oil and gas recovery, steel manufacturing, pest control products, power generation, and a wide range of consumer products, including cosmetics and cleaning products. APEs are a class of non-ionic surfactants, manufactured by reacting with an alkylphenol, such as nonylphenol (NP) and octylphenol (OP), with ethylene oxide to produce an ethoxylate chain. Nonylphenol ethoxylates (NPEO) account for approximately 80% of the world APE market, and octylphenol ethoxylates (OPEO) the remaining 20% (White et al. 1994).

The environmental safety of alkylphenol polyethoxylates (APEs) and their metabolites is a subject of concern. For example, NPs have increasingly gained international attention because of their potential to mimic the action of natural hormones in vertebrates (Lintelmann et al. 2003). Both NPs and NPEOs are discharged into the environment, mainly through incomplete removal in wastewater treatment facilities (Ying et al. 2002; Sarmah et al. 2005); and NPs are more resistant to biodegradation than their parent counterparts and therefore have been found in WWTP effluents and rivers in many parts of the world at concentrations up to μg/L level (Kolpin et al. 2002; Stachel et al. 2003). Owing to their strong hydrophobicity (log $K_{ow} \sim 3.9$-5.99), the metabolites of APEs show a tendency toward bioaccumulation in organisms (Heemken et al. 2001; Lintelmann et al. 2003), and consequently NPs represent a serious environmental and potential human health risk. The assessment of biodegradative processes affecting the environmental fate of these compounds is thus of utmost importance, and numerous studies have been carried out in recent times that are discussed below.

Degradation in Environmental Matrices

In general, biological degradation of NPEO results in a series of transformations that shorten the ethoxylate chain (Figure 5). Under aerobic conditions, NPEO degrades to NPEO with shorter chained ethoxylate groups or to NPEO with carboxylated ethoxylate and/or carbon chains. Concurrently, the nine unit hydrocarbon chain can also be shortened to form other alkylphenolic substances. Under oxygen limiting conditions such as those associated with sludge digestion in STPs, NPEO tends to be formed as the final product, and the shorter the ethoxylate chain, the more hydrophobic, persistent, and toxic the substance becomes (Porter and Hayden 2002).

Several possible degradation steps for NPEO under aerobic conditions have been demonstrated in recent times (e.g., Maki et al. 1996; Di Corcia et al. 1998; Manzano et al. 1999; Jonkers et al. 2001). Manzano et al. (1999) pointed out that historically it was believed that NPEO had the ethoxylate chain shortened to one or two ethoxylate groups, and that the ethoxylate chains were subsequently carboxylated. Following this understanding of the mechanism of NPEO transformation in the past; studies of NPEO and its breakdown products focused on measuring the concentrations of short-chained NPEO and short-chained carboxylated NPEO in environmental samples; however, carboxylated NPEO with long ethoxylate chains were not quantified in field samples as it was assumed they did not occur (Porter and Hayden, 2002). In contrast, Jonkers et al. (2001) demonstrated that NPEO with long ethoxylate chains actually degrades first to NP with carboxylated ethoxylate chains, forming long chained carboxylated NPEO, followed by the degradation of the ethoxylate chains. Therefore, it can be surmised from Jonkers et al.'s (2001) study that given NPEO with long ethoxylate chains may first degrade to carboxylated NPEO with long chains, previous measures of NPEO breakdown products in natural waters (Maki et al. 1996; Di Corcia et al.

1998; Manzano et al. 1999) may have been drastically underestimated. In their study, Jonkers et al. (2001) found that while > 99% of the NPEO was degraded by day 4, metabolites with carboxylated ethoxylate chains were still present in the reactors about a month from the start of the experiment, demonstrating that NPEO is not easily biodegradable, as some past studies claimed.

Under aerobic conditions, NPEO degrades to NPEO with shorter chained ethoxylate groups or to NPEO with carboxylated ethoxylate and/or carbon chains. According to Maki et al. (1996), NPEO with a carboxylated ethoxylate chain of length one was the final breakdown product of NPEO as obserbed under aerobic conditions in rivers. An earlier study by the same authors (Maki et al. 1994) also determined that under aerobic conditions the intermediate breakdown products of NPEO were NPEO with a two unit ethoxylate chain and NPEO with a carboxylated two unit ethoxylate chain. In contrast, DiCorcia et al. (1998) found that the ethoxylate chain of NPEO was first shortened and then carboxylated. The order of the breakdown or carboxylation of the ethoxylate chain remains a subject of debate due to observerd anomalies in the past studies.

Although data on the degradation of NPEOs and NP/OPs in natural systems and WWTPs are limited, several studies have been completed using controlled laboratory and field experiments (e.g., Ekelund et al. 1993; Ahel et al. 1994; Staples et al. 1999, 2001; Heinis et al. 1999; Khim et al. 1999; Hesselsoe et al. 2001). These studies have provided a range of half-lives depending on the kind of test systems and various conditions employed during the studies (Table 3). For example, half-lives of NP in seawater and in aerobic seawater and sediment (11°C) have been reported as 58 and 35 days respectively (Ekelund et al. 1993). Alkylphenol degradation in sediment has been reported to range from slow to practically non-existent in anaerobic environments (Johnson & Sumpter 2001). The potential for NP to persist in anaerobic environments is illustrated by a study demonstrating that the amount of NP remaining in sediment up to 15 months after addition was similar to that present after just 2 days (Heinis et al. 1999).

Given the conditions used within an experimental protocol, be it laboratory or field based, the half-lives of alkylphenols have been found to vary widely ranging from ≤ 1.2 days in water column, 28–104 days in sediment, and 8–13 days on macrophytes (Liber et al. 1999). These differences in the reported half-lives can be attributed to properties of respective media (e.g. pH, temperature, redox potential) and the type of specific degraders present in the media tested (Sarmah and Northcott, 2007). In a sludge-treated field study, Macromini et al. (1989) demonstrated that while aromatic surfactants (NP, NPEO1-2) were rapidly dissipated within the first month they still retained a residual concentration 10 months after application. Conversely, Topp and Starratt (2000) showed that 4-NP dissipated rapidly in Canadian agricultural soils, with an estimated half-life of 5–17 days. Under aerobic soil conditions, Hesselsoe et al. (2001) found the half-life of NP to be much lower (3–6 days). The same study demonstrated NP did not biodegrade after a period of 3 months under anaerobic conditions, and oxygen is the determining/limiting factor for the biodegradation of NP in sludge amended soils. The study showed it would take more than a year to reach aerobic conditions within a 2-cm sized sludge aggregate, and that until such conditions were achieved, NP would persist.

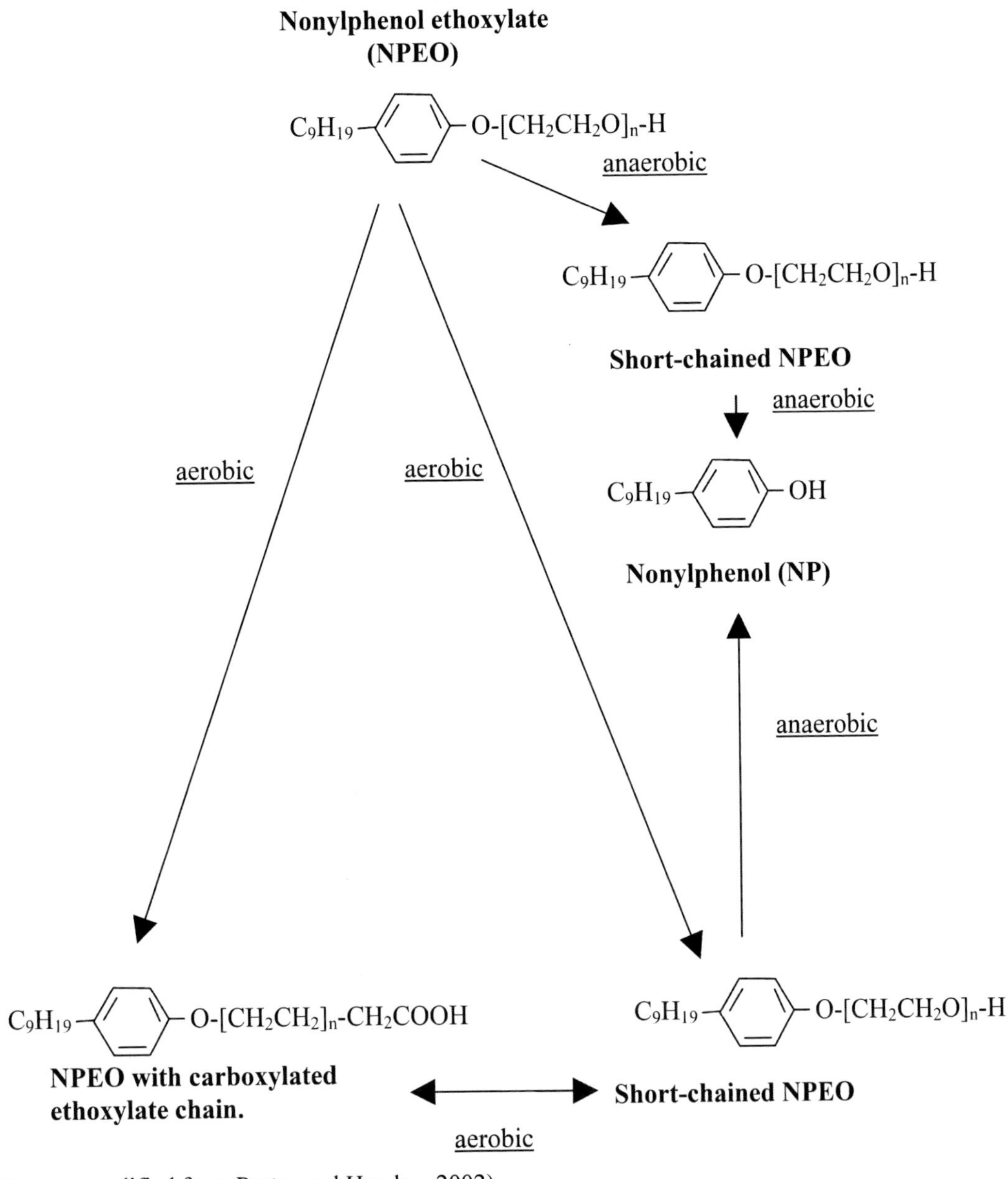

(Source: modified from Porter and Hayden, 2002)

Figure 5. Biological degradation pathways for Nonylphenol Ethoxylates (NPEO) in the environment under aerobic and anaerobic conditions.

Table 3. Available literature Half-life (HL) values on degradation of nonylphenols (NPs) and related products in various environmental matrices. Where HL values are not available, % mineralization of the compound is given with days in bracket

Compound (s)	Matrices	Concentration	Temp (°C)	HL (days)	References
NP	seawater	11 µg/L	11	58	Ekelund et al. (1993)
NP	seawater + sediment	11 µg/L	11	35	Ekelund et al. (1993)
NP	activated sludge	8.33 mg/L	28	80-90 (110)	Tanghe et al (1998)
NPEO	river water	5 mg/L	7-25	68-96 (30)	Manzano et al. (1999)
NP	OECD test solution	31 mg/L	22	20	Staples et al. (1999)
NP	soil + sludge	25 mg/kg	20-25	14.4 (28)	Gejlsbjerg et al. (2001)
4-*n*-NP	sludge + soil	-	20	3-6	Hesselsøe et al. (2001)
4-*n*-NP	aquifer material	1 µg/g	20	7	Ying et al. (2003)
4-*n*-OP	sea water	5 µg/L	20	5.8	Ying and Kookana (2003)
4-*t*-OP	marine sediment	1 µg/g	20	>20	Ying and Kookana (2003)
NP	composted biosloid	450 mg/kg	25-65	80 (15)	Xia and Pillar (2003)
NP	river sediment	2 µg/g	30	46.2-69.3	Chang et al. (2004)
NP1EO	river sediment	2 µg/g	30	49.5-77	Chang et al. (2004)
NP	sewage sludge	1-25 mg/kg	30	1.5-9.5	Chang et al. (2005)
NP1EO	sewage sludge	1-25 mg/kg	30	2.7-12.8	Chang et al. (2005)
NP	river sediment	2 µg/g	30	13.6-99	Yuan et al. (2004)
NP1EO	river sediment	2 µg/g	30	69.3-115.5	Yuan et al. (2004)
NP	water column*	-	-	≤ 1.2	Liber et al. (1999)
NP	pond sediment*	-	-	28-104	Liber et al. (1999)
NP	soil	10-100 mg/kg	30	5.6-12.8	Chang et al. (2007)
NP	soil	1-250 mg/kg	30	2.28-4.5	Topp and Starratt (2000)
4-*n*-NP	soil	1 mg/kg	20	4.6	Ying and Kookana (2005)
4-n-NP	soil[#]	80-240 mg/kg	30	7.2-19	Hseu (2006)
4-n-NP	river water + sediment	100 µg/L	16-18	96 (4)	Sarmah and Northcott (2007)
4-n-NP	aquifer material + groundwater	100 µg/L	16-18	> 95 (4)	Sarmah and Northcott (2007)

* littoral ecosystem study.

[#] treated with biosolids.

Table 4. Microorganisms capable of biodegrading or metabolizing BPA

Microor-ganisms	Strains	References
Bacteria	*Psudomonas paucimobilis* FJ-4	Ike et al. (2000)
	Pseudomonas sp.	Kang and Kondo (2002a)
	Pseudomonas putida	Kang and Kondo (2002a)
	Streptomyces sp.	Kang et al. (2004)
	Sphingomonas sp. strain AO1	Sasaki et al. (2005)
Fungi	*Pleurotus ostreatus* O-48	Hirano et al. (2000)
	Phanerochaete chrysosporium ME-446	Tsutsumi et al. (2001)
	Trametes versicolor IFO-7043	Tsutsumi et al. (2001)
	Trametes villosa	Fukuda et al. (2001), Uchida et al. (2001)
	Phanerochaete chrysosporum ME-446	Suzuki et al. (2003)
	Trametes versicolor IFO-6482	Suzuki et al. (2003)
	Aspergillus fumigatus	Yim et al. (2003)
	Fusarium sporotrichioides NFRI-1012	Chai et al. (2005)
	Fusarium moniliforme 2-2	Chai et al. (2005)
	Aspergillus terreus MT-13	Chai et al. (2005)
	Emericella nidulans MT-98	Chai et al. (2005)
	Stereum hirsutum	Lee et al. (2005)
	Heterobasidium insulare	Lee et al. (2005)
Planktons	Planktons *Chlorella fusca* var. *Vacuolata*	Hirooka et al. (2003)
	Nannochloropsis sp.	Ishihara and Nakajima (2003)
	Chlorella gracilis	Ishihara and Nakajima (2003)

Source: Kang et al. 2006.

Table 5. Available literature Half-life (HL) values on degradation of Bisphenol-A (BPA) in various environmental matrices. Where HL values are not available, % mineralization of the compound is given with days in bracket

Matrices	Concentration	Temp (°C)	HL (days)	References
water*	3 mg/L	22-15	2.5-4	Dorn et al. (1987)
OECD test solution	-	-	<1 (28)	Stone and Watkinson (1983)
OECD test solution[#]	7-25 mg/L	22-27	74.7-93.5 (28)	West et al. (2001)
river water, estuarine water	5 mgL	20	< 2	Klečka et al. (2001)
river water	0.2 mg/L	4	10-40 (20)	Kang and Kondo (2002)
river water	0.2 mg/L	20	4-7	Kang and Kondo (2002)
river water	0.2 mg/L	30	2-6	Kang and Kondo (2002)
river water	1 mg/L	30	2-3	Kang and Kondo (2002)
aquifer material	1 μg/g	20	unchanged over 70	Ying et al. (2003)
seawater	5 μg/L	20	d	Ying and Kookana (2003)
marine sediment	5 μg/L	20	> 98 (56)	Ying and Kookana (2003)
soil	60 μg/kg	20	14.5	Fent et al. (2003)
river water	1000 ng/ml	25-35	< 3	Kang and Kondo (2005)
seawater	1000 ng/ml	4-35	3-4	Kang and Kondo (2005)
soil	1 mg/kg	20	~20-30 (60)	Ying and Kookana (2005)
river water + sediment	100 μg/L	16-18	7	Sarmah and Northcott (2007)
aquifer material +	100 μg/L	16-18	> 98 (4)	Sarmah and Northcott (2007)
groundwater	100 μg/L	16-18	> 99 (4)	Sarmah and Northcott (2007)

* ship channel water, bayou water, treated process effluent.

[#] diluted activated sludge.

Table 6. Recent literature Half-life (HL) values (days or hours) on degradation of Phthalates in various environmental matrices. Where HL values are not available, % mineralization of the compound is given with days in bracket. (note: For data prior to 1997, refer to Staples et al. 1997)

Compound (s)	Matrices	Concentration	Temp (°C)	HL (days or h)	References
Di-ethyl hexyl phthalates (DEHP)	Sludge + quartz	4.1 nmol/g	20	39	Roslev et al. (1998)
Di-ethyl hexyl phthalates (DEHP)	Sludge slurry	4.1 nmol/g	20	58	Roslev et al. (1998)
Di-ethyl hexyl phthalates (DEHP)	Sludge + soil	4.1 nmol/g	20	58	Roslev et al. (1998)
Phthalic acid (PA)	Sludge + soil	4.1 nmol/g	20	2	Roslev et al. (1998)
Dimethyl phthalate (DMP)	STP sludge	10 mg/L	37	23.9 h	Jianlong et al. (2000)
Di-*n*-butyl phthalate (DnBP)	STP sludge	10 mg/L	37	32.1 h	Jianlong et al. (2000)
Di-*n*-octyl phthalate (DnOP)	STP sludge	10 mg/L	37	513.4 h	Jianlong et al. (2000)
Diethyl phthalate (DEP)	Soil	> 1 mg/g	20	0.75	Cartwright et al. (2000)
Diethyl phthalate (DEP)	soil	> 1 mg/g	20	10 (70)	Cartwright et al. (2000)
Diethyl phthalate (DEP)	River sediment	5 μg/g	30	0.5-8.8	Yuan et al. (2002)
Dipropyl phthalate (DPP)	River sediment	5 μg/g	30	0.5-11.4	Yuan et al. (2002)
Di-*n*-butyl phthalate (DnBP)	River sediment	5 μg/g	30	0.5-10.1	Yuan et al. (2002)
Diphenyl phthalate (DPhP)	River sediment	5 μg/g	30	0.5-6.4	Yuan et al. (2002)
(Butylbenzyl phthalate) BBP	River sediment	5 μg/g	30	0.5-10.5	Yuan et al. (2002)
Dihexyl phthalate (DHP)	River sediment	5 μg/g	30	3.1-19.6	Yuan et al. (2002)
Dicyclohexyl phthalate (DCP)	River sediment	5 μg/g	30	2.3-37.4	Yuan et al. (2002)
Di-ethyl hexyl phthalates (DEHP)	River sediment	5 μg/g	30	7.3-27.5	Yuan et al. (2002)
Di-ethyl hexyl phthalates (DEHP)	Sludge (lagoon, activated)	-	-	45.4-28.9	Amir et al. (2005)
Dimethyl phthalate (DMP)	Activated sludge	25-400 mg/L	30	6.27-10.4 h	Zeng et al. (2004)
Diethyl phthalate (DEP)	Activated sludge	25-400 mg/L	30	7.8-12.5 h	Zeng et al. (2004)
Di-*n*-butyl phthalate (DnBP)	Activated sludge	25-400 mg/L	30	10.3-15.4 h	Zeng et al. (2004)
Di-iso-butyl phthalate (DIBP)	Activated sludge	25-400 mg/L	30	15.0-23.3 h	Zeng et al. (2004)
Di-*n*-octyl phthalate (DnOP)	Activated sludge	25-400 mg/L	30	6.9-13	Zeng et al. (2004)
Di-ethyl hexyl phthalates (DEHP)	Activated sludge	25-400 mg/L	30	10.0-17.0	Zeng et al. (2004)
Di-ethyl hexyl phthalates (DEHP)	Activated sludge	-	20-60	2.0-144	Roslev et al. (2007)
Dimethyl phthalate (DMP)	Aqueous solution	10-300 mg/L	25	1.31	Hu and Wan (2006)
Diethyl phthalate (DEP)	Aqueous solution	10-300 mg/L	25	1.35	Hu and Wan (2006)
Di-*n*-butyl phthalate (DnBP)	Aqueous solution	10-300 mg/L	25	1.42	Hu and Wan (2006)
Di-*n*-octyl phthalate (DnOP)	Aqueous solution	10-300 mg/L	25	1.64	Hu and Wan (2006)
Di-ethyl hexyl phthalates (DEHP)	Aqueous solution	10-300 mg/L	25	2.2	Hu and Wan (2006)
Dimethyl phthalate (DMP)	Sediment slurry	100 μg/mL	30	22 (60)	Cheung et al. (2007)
Dimethyl isophthalate (DMIP)	Sediment slurry	100 μg/mL	30	65 (60)	Cheung et al. (2007)
Dimethyl terephthalate (DMTP)	Sediment slurry	100 μg/mL	30	72 (60)	Cheung et al. (2007)
Di-ethyl hexyl phthalates (DEHP)	wastewater	100-2000 mg/L	30	1.6-2.4	Chen et al. (2007)

Understanding NPEO degradation in anoxic environment is important because these conditions are common during wastewater transport and treatment process (Porter and Hayden, 2002). For example, Brunner et al. (1988) observed no formation of NP during aerobic sludge digestion, but significant NP accumulation during anaerobic sludge digestion. Under anaerobic conditions, NP becomes a significant and important final product of NPEO biodegradation by sequential shortening of the ethoxylate chain of NPEO until NP is formed as a final end product. Ejlertsson et al. (1999) found that although under anaerobic conditions NPEO was degraded to shorter chained NPEO and then to NP, further degradation of NP did not occur.

While NPEO is considered an inherently degradable compound the degree to which it is ultimately mineralised is highly dependant upon the specific conditions and combination of treatments in a WWTP. An average mass balance completed for NPEO and their metabolites in 11 Swiss WWTPs revealed that < 40% of the influent was completely degraded, with 20% adsorbed to sludge, and 40–45% remaining in the secondary effluent (Ahel et al. 1994). The nonylphenolic compounds in digested sludge were composed of 95% NP and 5% NP1EO and NP2EO, partly due to the lipophilic nature of NP and partly due to further degradation of NPEO to NP, during anaerobic digestion. These results demonstrate how biodegradable chemicals continue to concentrate in WWTP effluents and biosolids.

Factors Influencing NP Degradation

A number of factors are responsible for degradation of NP in various environmental matrices such as the temperature, pH, moisture content, and oxygen status of the medium containing NP. The effect of temperature on NP degradation was demonstrated by Topp and Starratt (2000), with higher rate of mineralization for NP with increases in temperature. However, the authors observed that NP mineralization reached maximum at optimum moisture content (15–16%), and was rather inhibited when moisture content was at its highest (24–55%) or lowest (1–3.6%). Hydrosolubility of NP is increased at alkaline pH values when NP is deprotonated (pKa~10–12) and thus becomes less sorbed to organic materials and potentially more bioavailable (Ivashechkin et al. 2004). It has been demonstrated that a pH value of 9 had a negative effect on NP degradation by *S. cloacae,* and degradation was slow, however, at pH range of 6–8 showed complete mineralization of 200 μg/kg of NP (Fuji et al. 2003). Oxygen appears to be one of the more critical parameters in NP degradation, with the majority of studies conducted in the past demonstrating the influence of oxygen on degradation of NP than under anoxic condition.

The structural feature within the molecule can also influence NP degradation. For example, the position and the length of alkyl chain can play a major role in NP degradation, with the length of the alkylphenol chain appearing seems to be the decisive criterion for the degradation pathway and the micoroganisms involved (Corvini et al. 2006). Most NP degrading bacteria belong to the *sphingomonads* and closely related genera, while NP metabolism is not restricted to defined fungal taxa (Corvini et al. 2006). For extensive information regarding the microbial degradation of NP and other alkylphenols, readers are referred to a recent mini-review presented by Corvini et al. (2006), where various factors influencing degradation of NP are critically analysed and presented with examples from wide range of literature.

Bisphenol-A (BPA)

Bisphenol A (BPA, 4,4'-isopropylidenediphenol) is a widely used intermediate in the production of polycarbonate plastic and a major component of epoxy resin used to prepare linings in food cans and dental sealants (Staples et al. 1998). The worldwide annual consumption of BPA is vast and increased from 0.73 billion kg in 1995 to 0.82 billion kg in 1998. Around 63% of BPA is used to manufacture polycarbonate plastic resins, 27% for epoxy resins, and the remainder for miscellaneous products including flame retardants (building block for tetrabromobisphenol-A) and polyester and other resins (Lee and Peart 1998).

BPA consists of two phenol rings connected by a methyl bridge, with two methyl functional groups attached to the bridge (Figure 1). BPA has a water solubility of 120 mg/L with a pK_a value of 9.6–10.2, thus making it more soluble at alkaline pH values (Howard, 1989). The safety of BPA has become a controversial issue because it not only possesses estrogenic endocrine disrupting effects (Krishnan et al. 1993; Brotons et al. 1995), but is also a suspected carcinogen (Ashby & Tennant 1988; Suarez et al. 2000).

Biodegradation is expected to be the dominant process for removal of BPA from the aquatic environment, and biodegradability has been studied in a number of laboratory test systems using both acclimated and nonacclimated microorganisms (Staples et al. 1998; Klečka et al. 2001). BPA can be biodegraded by microorganisms in the environment and can be also metabolized by enzymes existing in plants and animals, and many bacteria, fungi, certain plankotons and some enzymes capable of degrading or metabolizing BPA in various environmental matrices have been identified (Table 4).

Although the pathway(s) by which microorganisms metabolize BPA have not been elucidated, given that BPA is structurally analogous to both phenols and alkyl benzenes it can be expected that BPA is metabolized by similar pathways like phenols and alkylbenzes (Spivack et al. 1994). Using a gram-negative aerobic bacterium strain MV1, Spivack et al. (1994) presented a novel major pathway for metabolism of BPA (Figure 6). According to Spivack et al. (1994), although phenolic rings in BPA are available for attack, the initial oxidation is found to be on an unactivated aliphatic methyl group. The majority of BPA is metabolized by strain MV1 through oxidative rearrangement to form triol II, followed by dehydration to stilbene III and oxidative cleavage to 4-hydroxyacetophenone and 4-hydroxybenzaldehyde (Figure 6). Finally, the 4-hydroxybenzaldehyde is subsequently oxidized to 4-hydroxybenzoic acid, a compound shown to support the growth of strain MV1.

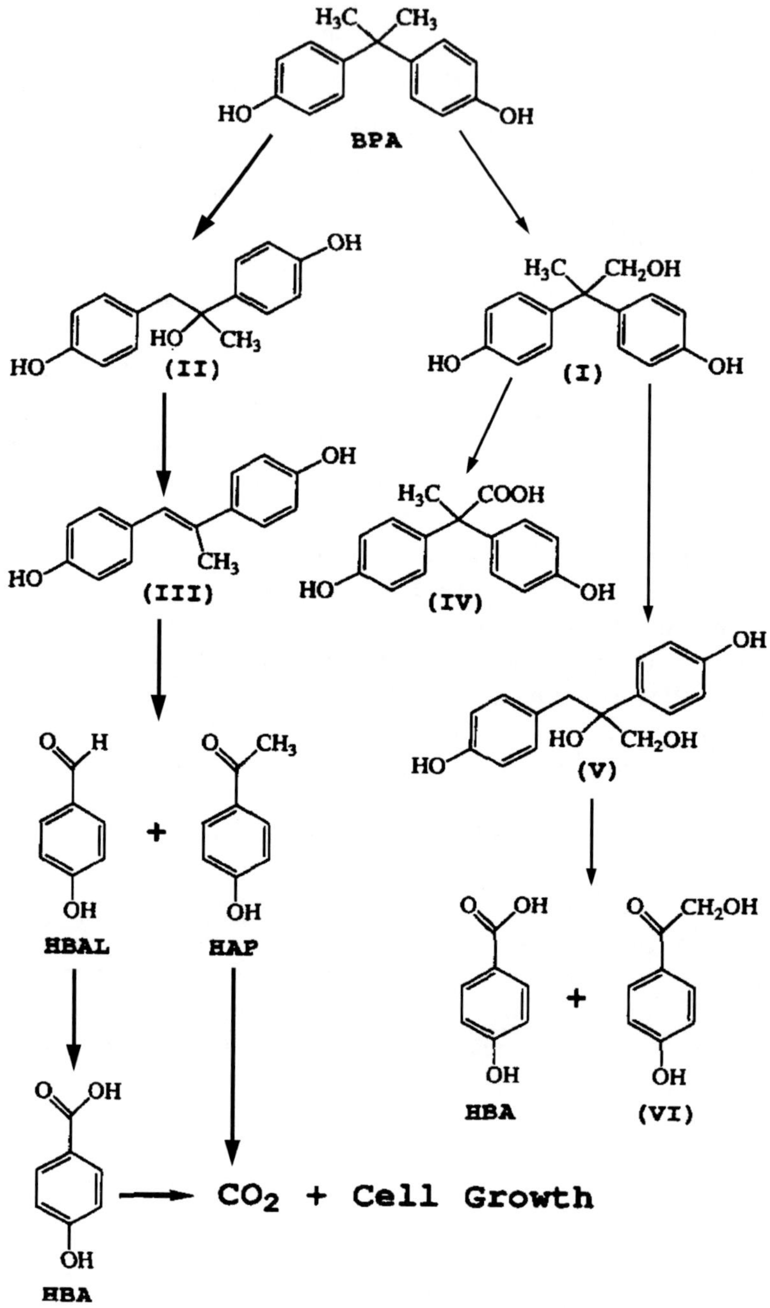

Figure 6. Metabolic pathway for BPA biodegradation by strain MV1. *I,* 2,2-bis(4-hydroxyphenyl)-l-propanol*Z;I,* 1,2-bis(4-hydroxyphenyb2-propanol; ZZZ, 4,4'-dihydroxy-a-methylstilbene*W*; ,2 ,2-bis(4-hydroxypheny1) propanoic acid; V, 2,3-bis(4-hydroxyphenyl)-l,2-propanediol; *VZ,* 4-hydroxyphenacyl alcohol; *BPA,* bisphenol A, *HBAL,* 4-hydroxybenzaldehyde;*H BA,* 4-hydroxybenzoic acid*H*; *A P* ,4 – hydroxyacetophenone (Source: modified from Spivack et al. 1994).

Degradation in Environmental Matrices

Aerobic biodegradation is the dominant loss process for BPA in all media except the atmosphere, where it is likely to be susceptible to rapid reaction with hydroxyl radicals (Staples et al. 1998). Dorn et al. (1997) studied degradation of BPA in natural waters using water collected from a BPA manufacturing facility's effluent, its receiving stream, and the downstream Houston Ship Channel into which the stream flowed. The water samples were spiked with BPA concentration of 3 mg/L and incubated at 22–25°C for 8 days. Their results showed that > 90% loss of BPA had occurred within 3–5 days, thus suggesting that BPA rapidly degraded in the media tested (Table 5). In contrast, other studies showed < 1% (Stone and Watkinson, 1983) and > 75% (West et al. 2001) mineralisation of BPA in OECD (Organisation for Economic Cooperation and Development) test systems with or without activated sludge. River die-away studies with BPA using a wide variety of surface waters have shown that BPA is degraded with average half-life (HL) of approximately 1 day (Kleča et al. 2001), irrespective of geographic location, sampling site (upstream or downstream of wastewater outfalls), sediment addition, or chemical concentration (0.05–5500 μg/L). Similar biodegradability of BPA in river water has been also reported by other researchers in recent times (Kang and Kondo, 2002; Kang and Kondo, 2005). For instance, Kang and Kondo (2002) observed that 10 out of 11 isolated bacteria from river water in China had BPA biodegradability range of 18–91% with the *Streptomyces* sp. strain having > 90% (for 10 days) biodegradability (Kang et al. 2004).

BPA degradation studies conducted using sea water and marine sediment also showed rapid degradation (Table 5). More recently, Sarmah et al. (2007) reported from a laboratory incubation study of BPA in slurries prepared with river water and sediment, and aquifer material and groundwater from a nearby dairy growing regions of New Zealand that > 95% of the applied BPA (100 μg/L) disappeared within just 4 days without exhibiting any lag phase. Studies conducted so far on BPA degradation in water in the US, Europe, Australia and New Zealand revealed that rapid and almost complete biodegradation occurs for this compound and is consistent with the testing carried out according to OECD procedure. This provides further evidence that BPA is unlikely to persist in aquatic environment and microorganisms capable of degrading the BPA molecule are ubiquitous in the aquatic environment (Klečka et al. 2001).

Biodegradation studies of BPA in agricultural soil are limited. Trace concentrations of BPA can reach the soil via application of sewage sludge from wastewater treatment systems that receive wastewaters containing BPA, or from leachate from uncontrolled landfills (Fent et al. 2003). From a laboratory study using ^{14}C-BPA, Fent et al. (2003) demonstrated rapid dissipation of BPA in soils representing a broad range of soil properties, with estimated HL of < 3 days at an incubated temperature of 20°C. However, Ying and Kookana (2003) observed slightly higher values of HL (7 days) in a south Australian soil. The predominant mechanism responsible for the loss of BPA in soil is thought to be aerobic biodegradation, although some of the BPA may be incorporated into the soil organic matter, rendering it irrecoverable by extensive extraction (Cousins et al. 2002). Degradation studies involving BPA and soil as medium showed the absence of a lag phase in the degradation kinetics, suggesting intrinsic ability of soil microbial biomass to rapidly transform the compound without any adaptation. This however, contrasted with the study by Ying and Kookana (2003), who found a lag phase of ~5 weeks for BPA degradation in sea water. In sea water, pH will be in the alkaline range, and given that BPA has a pK_a of 9.6–10.2, most of the molecules will be dissociated, rendering increasing solubility for the compound, and making it less bioavailable for the microorganisms. Differences and the breadth of microbial community may also be diverse in marine environment and thus adaptation by the microorganisms can take several days. Kang and Kondo (2005) suggested that BPA degradation in seawater may be caused by chemical hydrolysis rather than biological degradation, while organisms such as bacteria and flagellates can have an important effect on the chemical degradation of BPA.

PHTHALATES

Phthalates or phthalate esters are incorporated into plastics as plasticisers. They are predominantly incorporated into polyvinyl chloride (PVC) products to impart flexibility and, to a lesser degree, in paints, lacquers, and cosmetics (Cosmetic Ingredient Review 1992; Harris et al. 1997). Di-(2-ethyl-hexyl)-phthalate (DEHP) is the most common of the phthalate esters and has been in use since 1949. Annual global production of DEHP is estimated at 3–4 million tonnes (Stringer & Johnston 2001). Of all produced DEHP, 95% is used as plasticizer in PVC films and sheets. Other applications are as plasticizer in industrial paints and adhesives and as a dielectric fluid in condensers (Wams 1987; CEPA 1994a). Di-n-butylphthalate (DnBP) is used mainly as a plasticiser in polyvinyl emulsions used in adhesives and lacquer coatings, and paper coatings. DnBP is also used in cosmetics as a perfume solvent and fixing agent, a suspension agent in aerosols, an antifoaming agent, a skin emollient and a plasticizer in nail polish and hair spray (CEPA 1994b). Di-n-octyl phthalate is principally used as a plasticiser in PVC used to produce gloves, flooring and flexible plastic/epoxy sheets (CEPA 1993).

The phthalates are compounds with low water solubility/high fat solubility (lipophilic) and have low vapour pressure. Unless the side chains are very short, as in di-methyl and diethylphthalate, the polar carboxyl group contributes less than the physical properties of the phthalates. They behave as a fluid in temperature range between –50°C and 340°C (e.g., DEHP). Owing to their low water solubility, the phthalates hydrolyse relatively slowly, but the actual rate varies according to solubility and temperature. The properties of phthalates, such as solubility with different polymers, volatility, and effect on the polymer at different temperatures and so on, are modified with the side chains. While these are usually similar, they can also be different. They can be straight or branched aliphates, but cycloaliphatic and aromatic side groups also occur. Because of the physical rather than chemical incorporation of phthalates in the polymeric matrix, they readily leach from manufactured products and they are ubiquitous environmental contaminants. Release of phthalates occurs during production and via leaching and volatilization from plastic products during their use and/or after disposal (Staples et al. 1997). For detailed information on the fate of phthalate ester in the environment, readers are referred to Staples' (1997) excellent review article.

Degradation in Environmental Matrices

The metabolic breakdown by microorganisms is considered to be one of the major routes for degradation of phthalates. Some organisms selectively hydrolyse only one ester bond to give the monoalkyl phthalate plus an alcohol; the alcohol is then used for growth (Jianlong et al. 1996). It is believed that there are other organisms that are capable of complete degradation of either the monoalkyl or dialkyl phthalate. Much earlier, following pure culture studies, a general pathway of degradation for phthalate was proposed (Engelhardt et al. 1975; Keyser et al. 1976) as:

dialkyl phthalate→monoalkyl phthalate→phthalic acid→protocatechuic acid→ring cleavage and mineralization

Recently, Amir et al. (2005) summarized various catabolic pathways of phthalic acid ester biodegradation (Figure 7) based on a number of previous studies reported earlier (Wang et al. 2004; Amir et al. 2005 and ref. there in). Amir et al. (2005) concluded that the microbial metabolism seems to begin by the decomposition of alkyl side-chains before aromatic ring-cleavage, implying different catabolic pathways that include β-oxidation, demethylation or transesterification (Figure 7).

Numerous studies have showed that phthalate esters are degraded by a wide range of bacteria and actinomycetes under aerobic and anaerobic conditions in soil, sewage sludge, natural waters, and wastewater, which have been reviewed in detail by Staples et al. (1997). Soil systems may contain elevated DEHP concentrations due to the use of sewage sludge as a supplement to the fertiliser, and subsequently, some studies have been undertaken of biodegradation of phthalates in soils amended with or without sludge (e.g., Inman et al. 1984; Shanker et al. 1985; Cartwright et al. 2000; Roslev et al. 1998; Madsen et al. 1999). For instance, using flooded soil, Shanker et al. (1985) showed that only 30% of the initially applied phthalate was degraded within 1 month; however, under aerobic soil conditions, 90% was degraded. The half-lives of DEHP in contaminated Danish soil with an initial concentration of 21900 mg/kg, varied between 33 and 85 days, suggesting rapid dissipation (Maag and Løkke, 1990). Standardized aerobic biodegradation tests with sludge inocula have shown that phthalates undergo $\geq$ 50% ultimate degradation within 28 days (Staples et al. 1997).

It has been reported through observed and modelled data that > 40% of DEHP persisted in sludge-amended soil up to 1 year after sludge application (Madsen et al. 1999). In the same study, the level remaining in soil was even higher (70%) under anaerobic conditions. Elsewhere, in a Danish study, Roslev et al. (1998) observed bi-phasic degradation kinetics for DEHP and phthalic acid (PA) in sludge-amended soil, with estimated HL in the first stage varying between 2 and 58 days for PA and DEHP, respectively. However, in the second stage, HL for both the compounds were 2–7 fold higher, suggesting limited contribution of the soil microorganisms to the degradation of DEHP in sludge amended soil were probably due to immobilization of DEHP and its metabolites into the soil matrix (Roslev et al. 1998). Greater persistence of DEHP was also observed by Cartwright et al. (2000) with only 10% degradation of DEHP by day 70 (20°C) in a sandy clay loam soil from the UK. In contrast, owing to its lower K_{ow} coefficient value (2.38) than DEHP (K_{ow}~>7), DEP will act as membrane disruptive and therefore exhibited rapid biodegradation with HL of 0.75 d (20°C) in the same soil. Although HL of less than 1 week (Inman et al. 1984; Shanker et al. 1985) has been reported for dibutyl phthalate (DBP), using a mathematical model, Jensen et al. (2001) found a HL of 30 days for DBP in a Danish sandy soil near Askov, Denmark.

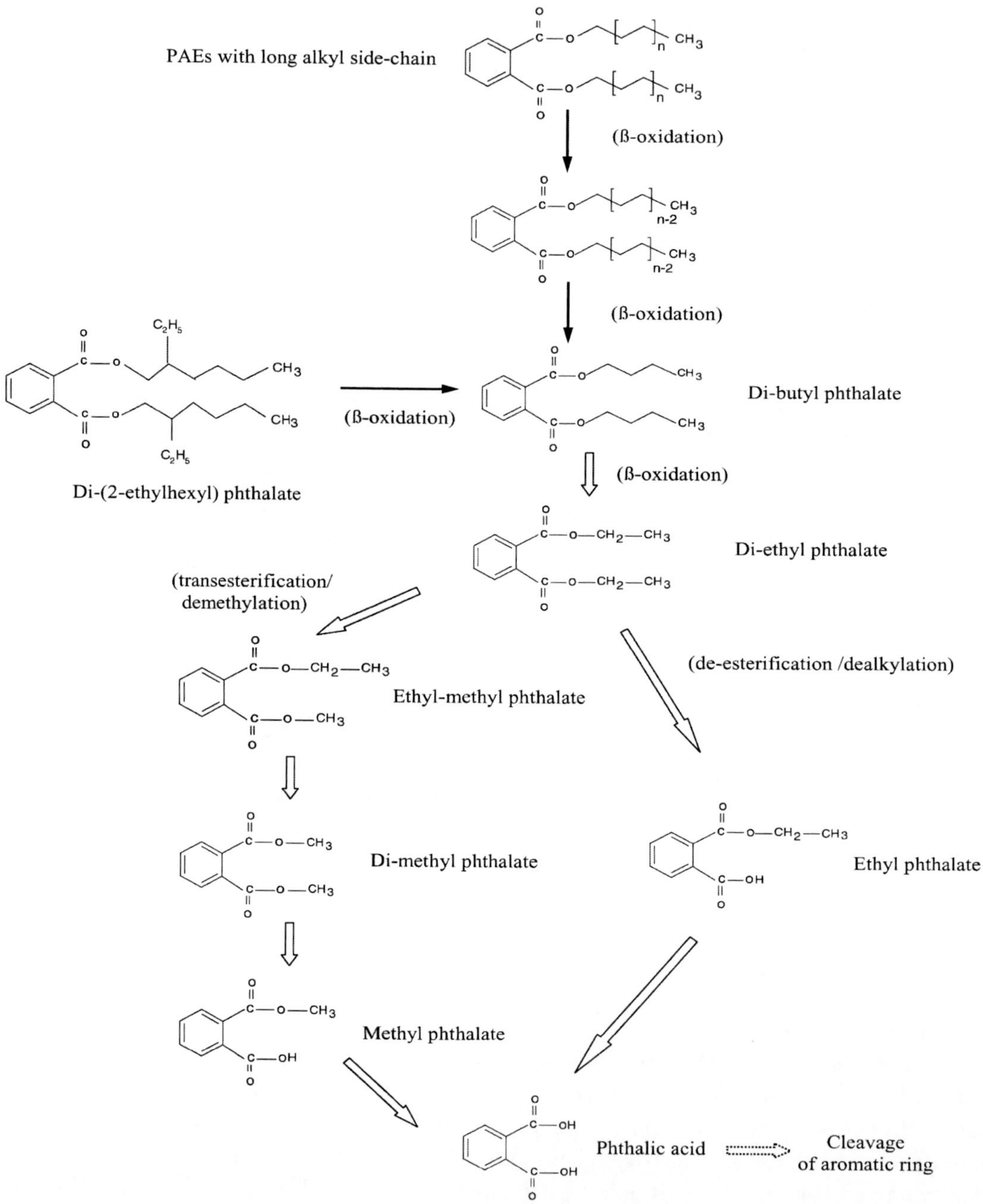

(Adapted from Amir et al. 2005)

Figure 7. Major biodegradation pathways of phthalic acid ester.

The fate of DEHP was recently investigated recently by Amir et al. (2005) during composting of both lagooning sludge (LS) and activated sludge (AS); the results showed that DEHP half-life under AS was shorter than that was obserbed under LS (Table 6). This was attributed to the presence of a wide range of indigeneous microbial communities in the activated sludge whose activity continues during composting. However, it is important to note that in the study by Amir et al. (2005), high temperature (70°C in AS vs ~52°c in LS) may have played a significant part in the overall DEHP degradation. Under anaerobic, oligotrophic and cold conditions, the half-life is expected to increase. For example, Jianlong et al. (2000) studied the microbial degradation of phthalic acid esters under anaerobic digestion of sludge collected from the primary anaerobic digester, a local wastewater treatment plant in Beijing, China. The authors observed that the biodegradation rate and the biodegradability of dimethyl phthalate (DMP), DnBP and di-n-octyl phthalate (DnOP) under aerobic conditions appeared to be related to the length of the alkyl-side chains. The authors reported that > 90% of DMP and DnBP with the short alkyl-side chain phthalates can be degraded with estimated HL of 24 and 32 hrs respectively, at an incubated temperature of 37°C, but HL for longer chain DnOP was more than 3 weeks (Table 6).

Zeng et al. (2004) isolated a novel bacterium (*Pseudomonas fluoresences* FS1) from activated sludge while studying biodegradation kinetics of phthalate esters. This bacterium has the ability to utilize phthalate ester as the sole source of carbon and energy under aerobic conditions. The authors estimated HL of 6 phthalate esters to vary from hours to days in a concentration range of 25–400 mg/L. The degradation of each phthalate ester was dependent on the length of the alkyl side chains and the alkyl branch chain on each compound and followed an order: DMP > DEP > DnBP > DIBP > DnOP > DEHP (Table 6). Hu and Wan (2006) reported similar characteristics for several phthalate esters biodegradation in aqueous solution using low and high concentrations, and estimated first-order half-life was found to vary from 1.3 to 2.2 days (Table 6).

These studies indicate that biodegradation is expected to be the dominant loss mechanism for phthalates in surface waters, sludge, soils and sediments under aerobic condition. However, under anaerobic conditions, degradation has been observed to be very slow to almost non-existent (Zioguo et al. 1989). Microorganisms responsible for anaerobic degradation of phthalate esters in environmental media exists (e.g., Shelton et al. 1984; Wang et al. 2000; Liang et al. 2007); however, biodegradation under oxygen limiting condition is most likely to occur only when microbes are unable to find any alternate energy or carbon source and depend on sulfate, nitrate or iron as electron acceptors. For example, Cheung et al. (2007) observed that DMP, dimethyl isophthalate (DMI), and dimethyl terephthalate (DMT) could not be mineralized over an extended period of 6 months, but transformation occurred to their respective monomethyl phthalate and/or phthalic acid. The authors isolated facultative anaerobes *Thaurea* sp., *Xanthobater* sp. and *Agrobacterium* sp. for DMP, DMI and DMT, respectively.

CONCLUSION

This overview shows that the degradation of steroid hormones is rapid in environmental matrices such as river water, sediment and terrestrial soil; however, rate of degradation is dependent on the prevailing oxygen status of the media tested. While the majority of the studies showed aerobic condition favored degradation of these hormones than anaerobic condition, given the breadth and adaptive ability of micro-organisms, even under anaerobic conditions steroid hormones could be degraded (Sarmah and Northcott, 2007). Under any given condition, whether aerobic or anaerobic, testosterone seemed to degrade more rapidly than 17β-estradiol. Mechanistically, the ethynyl group in the 17-position of the ethyhylestradiol prevents the formation of ketone (which has been observed for estradiol) and sterically hinders access to the hydroxyl group in the 17-position, thus making the compound more recalcitrant than the natural hormone, estradiol (Bolt, 1979). Naturally excreted hormones are continually added to our environment by direct excretal input from grazing animals, and through application of manure to land and sewage disposal. Therefore, even though they may degrade rapidly, accumulation of total hormonal loading in the ecosystem can actually magnify given that climatic conditions do not remain the same throughout the year, and that seasonal fluctuation in temperature and rainfall may actually alter their degradation rate. Given their short-half lives, steroid hormones are not likely to persist in the environment, however, potency of these hormones are of immediate concern due to their estrogenic or androgenic effects on aquatic species and therefore more emphasis should be placed on the biological potency of the chemicals rather than their concentrations in the environment.

Veterinary antibiotics undergo different rates of degradation and their degradation is dependent on the type of environmental medium, moisture content, and the incubated temperature or the prevailing pedo-climatic and/or weather conditions when the experiment is conducted under field conditions. The range of half-lives reported for antibiotics is broad and often due to conditions employed within each of the studies differ considerably, data obtained are specific to the conditions adopted in the study. Therefore, half-life values need to be treated with caution if the data are to be used for risk assessment purpose as the conditions for which data are generated may not necessarily represent in which the risk assessment for the compound is carried out.

Studies conducted in the past showed that while biodegradation of NPEO seems to be quick, complete mineralization does not take place, and if the conditions are anaerobic, metabolite such as NP can persist for a long time. Understanding the rates of conversion for these compounds in natural systems (river water, estuary) and anaerobically designed WWTPs should be considered in future research. Unlike industrial surfactants such as NPEO, available studies suggest that BPA degrades rapidly in environmental matrices such as water, soil, sediment and marine environment. The rate of degradation is dependent on the pH of the media, temperature, and the type of media used in the experiments. The ready biodegradability test results reported in the literature and the laboratory controlled studies in samples collected from river water, sediments, soil and sludge also suggest that persistence of BPA in the environment is low; however, the estrogenicity or the toxicity of BPA even at trace concentrations needs a closer look.

Because of their mutagenic, carcinogenic and toxic effects in certain type of organisms, biodegradation of phthalates has increasingly become a research topic. Among various phthalates, DEHP is one of the most commonly used phthalates, with an annual global production of approximately 10^6 tonnes, and biodegradation is the only sink for DEHP in aquatic and terrestrial systems (sewage, soils, sediments, and surface waters). While the biodegradation of phthalate esters appeared to be faster under aerobic conditions, significant amounts of DEHP and its metabolites with poor biodegradability may remain in sludge-amended soil for a long period despite the presence of microorganisms in the soil. Limited studies suggest that composting could be an important detoxification process in the removal of phthalates, mainly in the form of DEHP from sludge that is used as fertiliser on agricultural land. Overall, a range of HL values for various phthalates have been reported in the literature under aerobic conditions, while biodegradation under oxygen limiting conditions was found to be almost negligible.

The overview presented above reveals that these ECs are mostly biodegraded under conditions conducive to aerobic microorgnisms. However, to achieve a better insight into their ecotoxicological effect and overall impact on the ecosystem, these compounds should be studied under an anoxic natural environment. More efforts should be made to conduct long-term field trials so that realistic data can be obtained that can be also used for risk assessment purposes and for developing regulatory guidelines for policy makers at local and national levels.

References

Ahel, M.; Giger, W.; Schaffner, C. *Water Res.* 1994, 28, 1143–1152.

Alexander, M. 1994. *Biodegradation and Bioremediation—TD192.5.A43*, Academic Press, San Diego, California.

Amir, S.; Hafidi, M.; Merlina, G.; Hamdi, H.; Jouraiphy, A.; Gharous, M. E.; Revel, *J. C. Process Biochem.* 2005, 40, 2183–2190.

Andersen, H.; Siegrist, H.; Halling-Sorensen, B.; Ternes, T. A. *Environ sci Technol.* 2003, 37, 4021–4026

Ashby, J.; Tennant, R. W. *Mutat Res.* 1988, 204 (1), 17–115.

Baronti, C., Curini, R., D'Ascenzo, G., Di Corcia, A., Gentili, A, Samperi, R. *Environ. Sci. Technol.* 2000, 34, 5059–5066.

Berg, F. D.; Kuss, E. *Arch. Gynecol. Obstet.* 1992, 251, 17–27.

Bolt, H. M. *Pharmacol Ther.* 1979, 4, 155–181.

Boxall, A. B .A.; Kolpin, D. W.; Halling-Sørensen, B.; Tolls, *J. Environ Sci Technol.* 2003, 37, 286–294A.

Boxall, A. B .A.; Fogg L. A.; Blackwell, P.; Kay, P.; Pemberton, E. J.; Croxford, *A. Rev Environ Contam Toxicol.* 2004, 180, 1–91.

Brack, W.; Klamer, H. J. C.; de Ada, M. L.; Barcelo, D. Effect-directed analysis of key toxicants in European river basins - A reivew. *Environ. Sci. Poll. Res.* 2007, 14, 30 - 38.

Brotons, J. A.; Olea-Serrano, M. F.; Villalobos, M.; Pedraza, V.; Olea, N. *Environ Health Perspect.* 1995, 103 (6), 608–612.

Brunner, P.; Capri, S.; Macromini, A.; Giger, W. *Water Res.* 1988, 22, 1465–1472.

Cartwright, C. D.; Thompson, I. P.; Burns, R. G. *Environ Toxicol Chem.* 2000, 19, 1253–1261.

Casey, F. X. M.; Larsen, g. L.; Hakk, H.; Šimůnek, J. 2003, 37, 2400–2409.

CEPA (Canadian Environmental Protection Act). 1993. Priority substances list assessment report for Di-n-octyl phthalate. In: *Environment Canada and Health Canada,* Ottawa, Canada.

CEPA (Canadian Environmental Protection Act) (1994a) Priority substances list assessment report – Bis(2-ethylhexyl) phthalate. In: *Environment Canada and Health Canada,* Ottawa, Canada.

CEPA (Canadian Environmental Protection Act) (1994b) Priority substances list assessment report – Dibutyl phthalate. In: *Environment Canada and Health Canada*, Ottawa, Canada.

Chai, W.; Handa, Y.; Suzuki, M.; Saito, M.; Kato, N.; Horiuchi, C. A. *Appl Biochem Biotechnol.* 2005, 120, 175–182.

Chang, B. V.; Yu, C. H.; Yuan, S. Y. *Chemosphere,* 2004, 55, 493–500.

Chang, B. V.; Chiang, F.; Yuan, S. Y. *Chemosphere.* 2005, 60, 1652–1659.

Chang, B. V.; Chiang, F.; Yuan, S. Y. *Chemosphere.* 2007, 66, 1857–1862

Chen, J-A. Li, X.; Li, J.; Cao, J.; Qiu, Z.; Zhao, Q.; Xu, C.; Shu, W. *Appl Microbiol Biotechnol.* 2007, 74, 676–682.

Cheung, J. K. H.; Lam, R. K. W.; Shi, M. Y.; Gu, J.-D. *Sci Total Environ.* 2007. doi:10.1016/j.scitotenv.2007.03.030. (in press).

Colucci, M. S.; Bork, H.; Topp, E. *J Environ Qual.* 2001, 30, 2070–2076.

Colucci, M. S.; Topp, E. *J Environ Qual.* 2001, 30, 2077–2080.

Corvini, P. F. X.; Schäffer, A.; Schlosser, D. *Appl Microbiol Biotechnol.* 2006, 72, 223–243.

Cosmetic Ingredient Review, Final report on the safety assessment of butyl benzyl phthalate. *J Am Coll Toxicol.* 1992, pp. 11.

Cousins, I. T.; Staples, C. A.; Klecka, G. M.; Mackay, D. *A Human Ecol Risk Assess.* 2002, 8, 1107–1135.

Czajka, C. P.; Londry, K. L. *Sci Total Environ.* 2006, 367, 932–941.

Das, B. S.; Lee, L. S.; Rao, P. S. C.; Hultgren, R. P. *Environ Sci Technol.* 2004, 38, 1460–1470.

Daughton, C. G; Ternes, T. A. *Environ Health Perspect.* 1999, 107, 907–938.

De Liguoro, M.; Cibin, V.; Capolongo, F.; Halling-Sørensen, B.; Montesissa, C. *Chemosphere.* 2003, 52, 203–212.

DiCorcia, A.; Constantino, A.; Crescenzi, C.; Marinoni, E.; Samperi, R. *Environ. Sci. Technol.* 1998, 32, 2401–2409.

Donoho, A. L. *J. Anim Sci* 1984, 58, 1528–1539.

Dorn, P. B.; Chou, C. S.; Gentempo, J. J. 1987, 16, 1501–1507.

Ekelund, R.; Granmo, A.; Magnusson, K.; Berggren, M. *Environ Pollut.* 1993, 79, 59–61.

Engelhardt, G.; Wallnofer, P. R.; Hutzinger, O. *Bull Environ Contam Toxicol.* 1975, 13, 342–347.

Fan, Z.; Casey, F. X. M.; Hakk, H.; Larsen, G. *Chemosphere.* 2007, 67, 886–895.

Fent, G.; hein, w. J.; moendel, M. J.; Kubiak, R. *Chemosphere.* 2003, 51, 735–746.

Fotsis, T.; Jarvenpaa, P., Adlercreutz, H., *J. Steroid Biochem.* 1980, 12, 503–508.

Fuji, K.; Kikuchi, S.; Satomi, M.; Ushio-Sata, N.; Morita, N. *Appl Environ Microbiol.* 2002, 68, 2057–2060.

Fuji, K.; Yamamoto, T.; Tanaka, T.; Hirakawa, T.; Kikuchi, S. *J Ind Microbiol Biotech.* 2003, 30, 531–535.

Fukuda, T.; Uchida, H.; Takashima, Y.; Uwajima, T.; Kawabata, T.; Suzuki, M. *Biochem Biophys Res Commun.* 2001, 284, 704–706.

Gavalchin, J.; Katz, S. E. *J. AOAC Int.* 1994, 177, 481–485.

Gejlsbjerg, B.; Klinge, C.; Samsøe-petersen, L.; Madsen, T. *Environ Toxicol Chem.* 2001, 20, 2709–2716.

Gilberstson, T. J., Hornish, R. E., Jaglan, P. S., Koshy, K. T., Nappier, J. L., Stahl, G. L., Cazers, A. R., Nappier, J. M., Kubicek, M. F., Hoffman, G. A., Hamlow, P. J. *J. Agric. Food Chem.* 1990, 38, 890–894.

Gonsalves, D., Tucker, D. P. H. *Arch. Environ. Contam. Toxicol.* 1977, 6, 515–523.

Gotvajn, A. Z.; Zagorc-Koncan, J. *Chemosphere.* 1999, 38, 1339–1346.

Halling-Sørensen B.; Nielsen, S. N.; Lanzky, P. F.; Ingerslev, F.; Lutzhft, H. C. H.; Jorgensen, S. E. *Chemosphere.* 1998, 36, 357–393.

Halling-Sørensen, B.; Jacobsen, A. M.; Jensen, J.; Sengelov, G.; Vaclavik, E.; Ingerslev, F. *Environ. Toxicol. Chem.* 2005, 24, 802–810.

Hanselman, T. A.; Graetz, D. A.; Wilkie, A. C. *Environ Sci Technol.* 2003, 37, 5471–5478.

Hansen, P. D.; Dizer, H.; Hock, B.; Marx, A.; Sherry, J.; Memaster, M. *Trends Anal Chem.* 1998, 17, 448–451.

Harris, C. A.; Henttu, P.; Parker, M. G.; Sumpter, J. P. *Environ. Health Perspect.* 1997, 105, 802–810.

Haiyan, R.; Shulan, J.; Ahmad, N.; Dao, W.; Chengwu, C. *Chemosphere.* 2007, 66, 340–346.

Hektoen, H.; Berge, J. A.; Hormazabal, V.; Ynestad, M. *Aquaculture.* 1995, 133, 175–184.

Heemken, O. P.; Reincke, H.; Stachel, B.; Theobald, N. *Chemosphere.* 2001, 45, 249–259.

Heinis, L. J.; Knuth, M. L.; Liber, K.; Sheedy, B. R.; Tunell, R. L.; Ankley, G. L. *Environ Toxicol Chem.* 1999, 18, 363–375.

Henschel, K. P.; Wenzel, A.; Didrich, M.; Fliender, A. 1997. *Reg Toxicol Pharmacol.* 1997, 18, 220–225.

Hesselsøe, M.; Jensen, D.; Skals, K.; Olesen, T.; Møldrup, P.; Roslev, P.; Mortensen, G. K.; Henriksen, K. *Environ Sci Technol.* 2001, 35, 3695–3700.

Hirano, T.; Honda,Y.; Watanabe, T.; Kuwahara, M. *Biosci Biotechnol Biochem.* 2000, 64, 1958–1962.

Hirooka, T.; Akiyama, Y.; Tsuji, N.; Nakamura, T.; Nagase, H.; Hirata, K.; Miyamoto, K. J. *Eiosci Bioeng.* 2003, 95, 200–203.

Howard, P. H. *Handbook of Environmental Fate and Exposure Data.* Lewis, Chelsea, MI, 1989; vol. I.

Hseu, Z-Y. *Chemosphere.* 2006, 64, 1769–1776.

Hu, D.; Coats, J. *R. Env Toxicol Chem.* 2007, 26, 884-889.

Hu, X.; Wan, J. *J Synthetic Lubrication.* 2006, 23, 71–80.

Ike, M.; Jin, C. S.; Fujita, M. *Water Sci Technol.* 42, 31–38.

Ishihara, K.; Nakajima, N. *J. Mol. Catal.* 2003, B 23, 419–424.

Ingerslev, F.; Halling-Sørensen, B. *Environ Toxicol Chem.* 2000, 19, 2467–2473.

Ingerslev, F.; Toräng, L.; Loke, M. L.; Halling-Sørensen, B.; Nyholm, N. *Chemosphere.* 2001, 44, 865-872.

Ingerslev, F.; Halling-Sørensen, B. *Ecotoxicol. Environ. Safety.* 2001, 48, 311–320.

Ingerslev, F.; Torang, L.; Loke, M. L.; Halling-Sørensen, B.; Nyholm, N. *Chemosphere* 2001, 44, 865–872

Inman, I. C.; Strachan, S. D.; Sommers, L. E.; Nelson, D. W. *J Environ Sci Health Ser. B.* 1984, 19, 245–257.

Ivashechkin, P.; Corvini, P.; Dohmann, M. *Water Sci Technol.* 2004, 50, 141–147.

Jacobsen, A. M.; Lorenzen, A.; Chapman, R.; Topp, E. *J Environ Qual.* 2005, 34, 861–871.

Jensen, J.; van Langevelde, J.; Pritzl, G.; Krogh, P. H. *Environ Toxicol Chem.* 2001, 20, 1085–1081.

Jianlong, W.; Ping, L.; Quan, Y. *Environ Int.* 1996, 22, 737–741.

Jianlong, W.; Lujun, C.; Hanchang, S.; Yi, Q. *Chemosphere.* 2000, 41, 1245–1248.

Jobling, S.; Nolan, M.; Tyler, C. R.; Brighty, G.; Sumpter, J. P. *Environ Sci Technol.* 1998, 32, 2498–2506.

Johnson, A.; Sumpter, J. *Environ. Sci. Technol.* 2001, 35, 4697–4703.

Jonkers, N.; Knepper, T.; Voogt, P. *Environ. Sci. Technol.* 2001, 35, 335–340.

Jürgens, M. D.; Holthaus, K. I. E.; Johnson, A. C.; Smith, J. J. L.; Hetheridge, M.; Williams, R. J.; *Env Toxicol Chem.* 2005, 21, 480–488.

Kang, J. H.; Kondo, F. *Arch Environ Contam Toxicol.* 2002, 43, 265–269.

Kang, J. H.; Ri, N.; Kondo, F. *Lett App. Microbiol.* 2004, 39, 178–180.

Kang, J. H.; Kondo, F. 2005, 60, 1288–1292.

Kang, J. H.; Katayam, Y.; Kondo, F. *Toxicology.* 2006, 217, 81–90.

Kay, P.; Blackwell, P. A.; Boxall, A. B. A. Environ. Toxicol. Chem. 2004, 23, 1136–1144.

Keyser, P.; Basaya, G. P.; Eaton, R. W.; Ribbons, D. W. Environ Health Perspect. 1976, 8, 159–166.

Khim, J. S.; Villenteuve, D. L.; Kannan, K.; Lee, K. T.; Snyder, S. A.; Koh, C. H.; Giesy, J. P. *Environ. Toxicol. Chem.* 1999, 18, 2424–2432.

Kjær, J.; Olsen, P.; Bach, K.; Barlebo, H. C. Ingerslev F, Hansen M, Halling-Sørensen B. *Environ. Sci. Technol.* 2007, 41, 3911–3917;

Kreuzig, R.; Holtge, S. *Environ. Toxicol. Chem.* 2005, 24, 771–776.

Kazumi, J.; Caldwell, M. e.; Suflita, J. M.; Lovely, D. R.; Young, L. Y. *Environ Sci Technol.* 1997, 31, 813–894.

Klečka, G. M.; Gonsior, S. J.; West, R. J.; Goodwin, P. A.; Markham, D. A. *Environ Toxicol Chem.* 2001, 20, 2725–2735.

Kolz A. C.; Moorman, T. B.; Ong, S. K.; Scoggin, K. D.; Douglas, E. A. *Water Environ Res.* 2005, 77, 55–56.

Krishnan, A. V.; Starhis, P.; Permuth, S. F.; Tokes, L.; Feldman, D. Endocrinology. 1993, 132 (16), 2279–2286.

Kolpin, D. W.; Furlong, E. T.; Meyer, M. T.; Thurman, E. M.; Zaugg, S. D.; Barber, L. B.; Buxton, H. T. *Environ Sci Technol.* 2002, 36: 1202–1211.

Kühne, M.; Ihnen, D.; Möller, G.; Agthe, O. *J Vet Med Ser.* A. 2000, 47, 379–384.

Kurisu, F.; Tanaka, M.; Matsuo, T.; Yagi, S. *IWA Conference,* Sydney 2002.

Lai, H.T.; Liu, S. M.; Chien, Y. H. *J. Environ. Sci. Health A.* 1995, 30, 1897–1923.

Layton, A. C.; Gregory, B. W.; Seward, J. R.; Schultz, T. W.; Sayler, G. S. *Environ Sci. Technol.* 2000, 34, 3925–3931.

Lee, L. S.; Strock, T. J.; Sarmah, A. K.; Rao, P. S. C. *Environ. Sci. Technol.* 2003, 37, 4098–4105.

Lee, H. B.; Liu, D. *Water Air Soil Pollut.* 2002, 134, 353–368.

Lee, H. B.; Peart, T. *Water Qual Res J Canada.* 1998, 33, 389–402.

Lee, S. M.; Koo, B. W.; Choi, J. W.; Chai, D. H.; An, B. S.; Jeung, E. B.; Choi, I. G. *Biol Pharm Bull.* 2005, 28, 201–207.

Liang, D-W.; Zhang, T.; Fang, H. H. P. 2007. *Appl Microbiol Biotechnol.* 2007, 74, 221–229.

Liber, k.; Knuth, M. L.; Stay, F. S. *Environ Toxicol Chem.* 1999, 18, 357–362.

Lintelmann, J.; Katayama, A.; Kurihara, N.; Shore, L.; Wenzel, A. *Pure Appl. Chem.* 2003, 75, 631–681.

Loke, M. L.; Ingerslev, F.; Halling-Sørensen, B.; Tjørnelund, *J. Chemosphere.* 2000, 40, 759–765.

Lorenzen, A.; Chapman, R.; Hendel, J. G.; Topp, E. *J Environ Qual.* 2005, 34, 854–860.

Lukas, S. D.; Jones, D. L. 2006. *Soil Biol Biochem.* 2006, 38, 2803–2815.

Maag, J.; Løkke, H. Landfarming of DEHP contaminated soil. In Arendt F, Hinsenveld M, vanden Brink W. J. eds, *Contaminated Soil '90 Kluwer,* Amsterdam, The Netherlands, pp 975–982.

Macromini, A.; capel, P. D.; Lichtensteiger, T. H.; Brunner, P. H.; Giger, W. *J Environ Qual.* 1989, 18, 523–528.

Madsen, P. L.; Thyme, J. B.; Henriksen, K.; Møldrup, P.; Roslev, P. *Environ Sci Technol.* 1999, 33, 2601–2606.

Maki, H.; Masuda, N.; Fujiwara, Y.; Ike, M.; Fujita, M. *Appl. Environ. Microbiol.* 1994, 60, 2265–2271.

Maki, H.; Fujita, M.; Fujiwara, Y. *Bull. Environ. Contam. Toxicol.* 1996, 57, 881–887.

Manzanao, M. A.; Perales, J. A.; Sales, D.; Quiroga, J. *M. Water Res.* 1999, 33, 2593–2600.

Marengo, J. R.; Kok, R. A.; Obrien, K.; Velagaleti, R. R.; Stamm, J. M. 1997. *Environ Toxicol Chem.* 16, 462–471.

Metcalfe, C.; Metcalfe, T.; Kiparissis, Y.; Koenig, B. *Environ Toxicol. Chem* 2001, 20, 297–308.

Oliveira, M. F.; Sarmah, A. K.; Lee, L. S.; Rao, P. S. C. *Presented at the Soil Science Society of America National Meeting,* Indianapolis, Indiana, November 10–14, 2002.

Perlman, W. H. 1948. In The Hormones, G. Pincus and K. V Thimann (Eds.), Academic Press. NY, 1948, Vol. I, pp 351-405.

Peterson, E. W.; Davis, R. K.; Orndorff, H. A. *J. Environ. Qual.* 2000, 29, 826–834.

Porter, A J.; Hayden, N. J. (2002). *Nonylphenol in the Environment: A Critical Review* [http://www.emba.uvm.edu/~nhayden/npreview.pdf] accessed 21 may, 2006.

Pouliquen, H., Le Bris, H., Pinault, L. *Marine Ecol Prog Serv.* 1992, 89, 93–98.

Prudy RM, Maclusky NJ, Naftolin, F. 1985. In: *Estrogens in the Environment,* J. McLachlan (Ed) Amsterdam, Elsevier. pp 145–165.

Raman, D. R., Layton, A. C.; Moody, L. B.; Easter, J. P.; Sayler, G. S.; Burns, R. T.; Mullen, M. D. *Transac Amer Soc Agril Eng.* 2001, 44(6), 1881–1888.

Raman DR., Williams EL, Layton AC, Burns RT, Easter JP, Mullen MD, Sayler GS 2004. *Environ. Sci. Technol.* 38: 3567–3573.

Roslev, P.; Madsen, P. L.; Thyme, J. B. Henriksen, K. *Appl Environ Microbiol.* 1998, 64, 4711–4719.

Roslev, P.; Vorkamp, K.; Aarup, J.; Frederiksen, K.; Nielsen, P. H. *Water Res.* 2007, 41, 969–976.

Sarmah, A. K. *Water Wastes* NZ. 2003, 128, 36–38.

Sarmah AK, Northcott GL 2004. Animal waste as a potential source for steroid hormones: a New Zealand perspective. Programme and abstract book, *SETAC Europe 14th Annual Meeting,* Prague, Czech Republic, 18–22 April 2004.

Sarmah AK, Northcott GL 2005. Sorption of steroid hormones and a primary transformation product in representative NZ soils. Programme and abstract book, *SETAC Europe 15th Annual Meeting,* Lille, France, 22–26 May 2005.

Sarmah, A. K.; Northcott, G. L.; Tremblay, L. A. *Landcare Research Contract Report:* LC0506/037, November 2005. p 67.

Sarmah, A. K.; Northcott, G. L.; Leusch, F. D. L. ; Tremblay, L. A. *Sci. Total. Environ.* 2006a, 355, 135–144.

Sarmah, A. K.; Meyer, M. T.; Boxall, A. B. A. *Chemosphere.* 2006b, 65, 725–759.

Sarmah, A. K.; Northcott, G. L. *Environ Toxicol Chem.* 2007 (in press).

Sarmah, A. K.; Northcott, G. L. Scherr, F. F. *Environ Int..* 2007 (in press).

Samuelsen, O. B. *Aquaculture.* 1989, 83, 7–16.

Samuelsen, O. B.; Lunestad, B. T.; Ervik, A.; Fjelde, S. *Aquaculture.* 1994, 126, 283–290.

Sassman, S. A.; Sarmah, A. K.; Lee, L. S. *Environ Toxicol Chem.* 2007, 26 (8), in press.

Sassman, S. A.; Lee, L. S. *Environ Toxicol Chem.* 2007, 26 (8), in press.

Sasaki, M.; Maki, J. I.; Oshiman, K. I.; Matsumura, Y.; Tsuchido, T. *Biodegradation.* 2005, 16, 449–459.

Shanker, R.; Ramakrishna, C.; Seth, P. K. *Environ pollut Ser.* B. 1985, 39, 1–7.

Shelton, D. R.; Boy, S. A.; Tiedje, J. M. *Environ Sci Technol.* 1984, 18, 93–97.

Schlüsener, M. P.; Bester, K. Spiteller, M. *Anal Bioanal. Chem.* 2003, 375, 942–947.

Schlüsener, M. P.; Bester, K. *Environ pollut.* 2006, 143, 565–571.

Schlüsener, M. P.; von Arb, M.; Bester, K. *Arch Environ Contam Toxicol.* 2006, 51, 21–928.

Shi, J. H.; Suzuki, Y.; Lee, B. D.; Nakai, S.; Hosomi, M. 2002. *Water Sci Technol.* 2002, 45, 175–179.

Shi, J. H.; Fujisawa, S.; Nakai, S.; Hosomi, M. *Water Res.* 2004, 38, 2323–2330.

Spivack, J.; Leib, T. K.; Lobos, J. H. 1994. *J Biol Chem.* 1994. 269, 7323–7329

Stachel, B.; Ehrhorn, U.; Heemken, O.-P.; Lepom, P.; Reincke, H.; Sawal, G.; Theobald, N. *Environ Pollut.* 2003, 124, 497–507.

Staples, C. A.; Peterson, D. R.; Parkerton, T. F.; Adams, W. J.; *Chemosphere* 1997, 35, 667–749.

Staples, C. A.; Dorn, P. B.; Klecka, G. M.; O'Block, S. T.; Harris, L. R.; *Chemosphere.* 1998, 36, 2149–2173.

Staples, C.A.; Williams, J. B.; Blessing, R. L.; Varineau, P. T. *Chemosphere.* 1999, 38, 2029–2039.

Staples, C.A.; Naylor, C. G.; Williams, J. B.; Gledhill, W. E. *Environ Toxicol Chem.* 2001, 20, 2450–2455.

Stone, C. M.; Watkinson, R. J. *Sittingbourne Research Centre (SBRC).* Sittingbourne, UK: Shell Research Limited: 1983.

Stringer, R.; Johnston, P. Kluwer Academic Publishers, The Netherlands. 2001.

Stuer-Lauridsen F.; Kjølholt, J.; Høibye, L.; Hinge-Christensen, S.; Ingerslev, F.; Hansen, M.; Andersen, Krogh K.; Andersen H. R.; Halling-Sørensen, B.; Hansen, N.; Køppen, B.; Bjerregaard, P.; Frost B. 2005 *Survey of Estrogenic Activity in the Danish Aquatic Environment Environmental Project* Nr. 977 Danish EPA

Stumpe, B.; Marschner, B. *Sci Total Environ.* 2007, 374, 282–291.

Suarez, S.; Sueiro, R. A.; Garrido, *J. Mutat Res.* 2000, 470 (2), 221–228.

Tanghe, T.; devriese, G.; Verstraete, W. *Water Res.* 1998. 32, 2889–2896.

Ternes, T. A.; Stumpf, M.; Mueller, J.; Haberer, K.; Wilken, R. D.; Servos, M. *Sci. Total Environ.* 1999a, 225, 81–90.

Ternes T. A.; Kreckel, P.; Mueller, J. 1999b. Sci. *Total Environ.* 225: 91–99.

Teeter, J. S.; Meyerhoff, R. D. *Environ Res*. 2003, 93, 45–51.

Topp, E.; Starratt, A. *Environ Toxicol Chem.* 2000, 19, 313–318.

Tsutsumi, Y.; Haneda, T.; Nishida, T. *Chemosphere.* 2001, 42, 271–276.

Uchida, H.; Fukuda, T.; Miyamoto, H.; Kawabata, T.; Suzuki, M.; Uwajima, T. *Biochem biophys Res Commun.* 2001. 287, 355–358.

Vader, J. S.; van Ginkel, C. G.; Sperling, F. M. G. M.; de Jong, J.; de Boer, W.; de Graaf,; J. S. *Chemosphere.* 2000, 41, 1230–1243.

Van Gool, S. *Possible environmental effects of antibiotic residues in animal manure.* Tijdschrift voor Diergeneeskunde: The Netherlands, 1999, pp 8-10 (in Dutch, English summary).

Wams, T. *J. Sci Total Environ.* 1987, 66, 1–6.

Wang, J.; Chen, L.; Shi, H.; Qian, Y. *Chemosphere.* 2000, 41, 1245–1248.

Weerasinghe, C. A.; Towner, D. *Environ Toxicol Chem.* 1997, 16, 1873–1876.

Weber, S.; Leuschner, P.; Kämpfer, P.; dott, W.; Hollender, J. 2005. *Appl Microbiol Biotechnol.* 2005, 67, 106–112.

West, R. J.; Goodwin, P. A.; Klecka, G. M. Bull. *Environ Contam Toxicol.* 2001, 67, 106–112.

White, R.; Jobling, S.; Hoare, S. A.; Sumpter, J. P.; Parker, M. G. *Endocrinology.* 1994, 135, 175–182.

Williams, R. J.; Jurgen, M. D.; Johnson, A. C. *Water Res.* 1999, 33, 1663–1671.

Winckler, C.; Grafe, A. *Abschätzung des Stoffeintrags in Böden durch Tierarzneimittel und pharmakologisch wirksame Futterzusatzstoffe.* UBA-Texte 44/00, Berlin, 2000.

Witte, W. *Science*. 1998, 279, 996–997.

Xia, K.; Pillar, G. D. Anthropogenic organic chemicals in biosolids from selected wastewater treatment plants in Georgia and South Carolina. P. 806-809. In K. Hatcher (ed.) proc. *2003 Georgia Water Res. Conf.*, Athens, GA. 23–24 Apr. 2003. Univ. of Georgia Press. Athens.

Ying, G. G.; Kookana, R. S.; Dillon, P. *Water Res.* 2003, 37, 3785–3791.

Ying, G.; Kookana, R. S. *Environ Sci Technol.* 2003, 37, 1256–1260.

Ying, G. G.; Kookana, R. S. *Environ Toxicol Chem.* 2005, 24, 2640–2645.

Ying, G. G.; Williams, B.; Kookana, R. S. *Environ. Int.* 2002, 28, 215–226.

Yoshimoto, T.; Nagai, F.; Fujimoto, J.; Watanabe, K.; Mizukoshi, H.; Makino, T.; Kimura, K.; Saino, H.; Sawada, H.; Omura, H. *Appl Microbiol Biotechnol.* 2004, 70, 5283–5289.

Yuan S, Y.; Yu, C.; Chang, B. V. *Environ Pollut*. 2004, 37: 3785–3791.

Zeng, F.; Cui, K.; Li, X.; Fu, J.; Sheng, G. *Process Biochem*. 2004, 39, 1125–1129.

Ziogou, K.; Kirk, P. W. W.; Lester, J. N. *Water Res.* 1989, 23, 743–748.

In: Leading-Edge Environmental Biodegradation
Editor: L. E. Pawley, pp. 95-129
ISBN: 978-1-60021-903-0

Chapter 3

BIOREMEDIATION OF DIESELFUEL-CONTAMINATED SOILS IN THE ATACAMA DESERT, CHILE

***A. Godoy-Faúndez,*[a,e] *B. Antizar-Ladislao,*[b] *I. Franke-Whittle,*[c] *J.L. Turrión-Gómez*[d] *and L. Perez-Roepke*[e]**

[a] Pontificia Universidad Católica de Chile, Dept. Ingeniería Química y Bioprocesos, Av. Vicuña Mackenna 4860, Santiago, Chile.

[b] Universidad de Cantabria, Dept. Ciencias y Técnicas del Agua y del Medio Ambiente, 39005 Santander, Spain.

[c] Universität Innsbruck, Institut für Mikrobiologie, Technikerstrasse 25, A-6020 Innsbruck, Austria.

[d] Universidad de Salamanca, CIALE, Dept. Microbiología y Genética, Av. Campo Charro, lab-301, 37007 Salamanca, Spain.

[e].Universidad Andrés Bello, Av. República 220, Santiago, Chile.

ABSTRACT

Since the 1980s, composting has been used as an alternative for bioremediation in open sites, with the microbial communities involved being a focal point of study. The use of this innovative bioremediation approach has been necessary due to recent changes in waste management regulations around the world, which focus towards more integrated, practical, sustainable and economic schemes than landfilling.

Currently, not much literature on the feasibility of bioremediation technologies to soils contaminated with mixtures of organic compounds exists. Even less information on the application of composting as a bioremediation technology is available. This chapter will include a case study on the successful application of in-vessel composting of a fuel-contaminated (diesel and gasoline) desert-mining soil/sawdust in the Atacama Desert (Chile). The physical and chemical factors that influence the distribution, abundance, richness diversity and evenness of microbial communities during composting of a fuel-contaminated soil during bioremediation will be discussed.

This chapter will also provide a review of the research conducted to date using molecular tools for the biomonitoring of changes in microbial communities as an indicator of remediation and restoration of oil-contaminated soils. In particular, the investigation of microbial community structure changes based on molecular tools including terminal restriction fragment length polymorphism (T-RFLP), BIOLOG EcoPlate^{TM} and the COMPOCHIP microarray will be presented. The advantages and disadvantages of certain molecular tools for environmental management will also be discussed. Future needs for bio-monitoring of microbial community structure changes during bioremediation of oil-contaminated soils will be identified, and a discussion of the technological management, regulatory and sustainable aspects of using composting as a bioremediation technology will be included.

I. Introduction

For both commercial and human health reasons, it is important to identify and clean up sites that have become heavily contaminated by accidental oil spills around the world. Thus, a variety of different physical and chemical treatments have been applied in the treatment of contaminated soils (Lundstedt *et al.* 2006, Negri *et al.* 2004). Over the last twenty years, biological treatments, such as bioremediation, an inexpensive and environmentally-friendly approach, have shown promising results in the field (Antizar-Ladislao *et al.* 2004, Semple *et al.* 2001). The bioremediation of organic compounds results in the mineralization or transformation of harmful waste materials to more stable and harmless forms through the actions of microbial communities that use such compounds as carbon and energy sources. The success of the process is dependent on the biodegradability of the contaminant and the existing environmental conditions (physical-chemical factors). Furthermore, physical and chemical factors influence the distribution, abundance, richness, diversity and evenness of microorganisms.

Both *in-situ* and *ex-situ* bioremediation techniques exist. Among ex-situ techniques, closed systems (in-vessel systems or bioreactors) have been reported to be highly successful, where the control and optimization of environmental conditions increase the speed of the biological degradation process and the effectiveness of removal (Antizar-Ladislao *et al.* 2006a). Thus, the development and optimization of new treatment technologies using closed systems applied to contaminated soils are nowadays the focus of numerous studies.

However, there are still a limited number of published studies on the applicability of bioremediation techniques using an in-vessel composting system of contaminated soils with mixtures of organic compounds, and the majority of these studies have been conducted on soils contaminated with mono-compounds at laboratory scale and rarely applied to a real-case scenario. Additionally, most of previous studies have used bioremediation systems as a black box, whereby the microbial communities have not been investigated in detail. Thus, future efforts should include studies concerning changes in the diversity of microbial communities, based on advanced molecular tools allowing a greater understanding of bioremediation systems and potentially allowing for a high-quality predictor for the course of the bioremediation process (Franke-Whittle *et al.*, 2005).

In this chapter, we will present a case study of the application of in-vessel composting, as a bioremediation treatment technology of a fuel-contaminated desert-mining soil/sawdust in the Atacama Desert (Chile). This chapter will present the fate of contaminants and changes in the microbial communities during in-vessel composting and will also discuss the relationship of the microbial community changes with physical and chemical parameters.

In summary, this chapter will determine optimal operation conditions during in-vessel composting of contaminated desert soils and contaminated sawdust, and will include discussion of technological, regulatory and sustainable aspects of treatment technologies to decontaminated soils in Chile.

II. Bioremediation of Soils Contaminated with Fuel Oils

A. Desert Soils Contaminated with Fuel Oils

Deserts are arid regions, and thus desert soils are normally exposed to evaporation rates potentially twice as great as the precipitation. The world's deserts are divided into four categories: subtropical, cool coastal deserts, cold winter deserts and polar regions. Desert soils contaminated with fuel oils, were collected in the Atacama desert, Chile. The Atacama Desert is a cool coastal desert, characterized by cool temperatures because of offshore ocean currents, and is also the world driest desert after Antarctica.

Diesel oil or fuel is a complex mixture consisting basically of paraffinic, olefinic and aromatic hydrocarbons and, in smaller amounts, molecules containing sulphur, nitrogen, metals, oxygen, etc. Diesel oils are composed of molecules with 8-40 carbon atoms and are usually heavier, more viscous and less volatile than gasoline.

Desert soils may become contaminated by fuel due to various reasons, mainly because of accidental spills. Petroleum contamination is recognized as a significant threat to desert soils, however documented research on the environmental consequences of terrestrial spills in deserts is still scarce. In fact, up to date, there are only a few reports on the characterization and treatment of fuel contaminated desert soils. One particular case is located is that of desert soil being contaminated with oil in Kuwait (Arabian Desert) following Iraqi invasion in the early 1990s. This case has been the subject of a study and application of treatment technologies (Balba 1998, Al-Daher 2001). Another study was conducted in the Antartic polar desert, where the effects of a small temperature increase on the removal of crude oil and diesel fuel contamination under various sub-Antarctic conditions was investigated (Delille *et al.* 2004).

Due to the limited number of studies available on desert soils, their characterization once pollution has occurred, and posterior feasibility (treatment) studies, the aim of this chapter is to present the results obtained in a study of characterization and treatment of fuel-contaminated desert-soils and sawdust in the Atacama Desert, Chile. Additionally, necessary information will be provided for those readers who are not familiar with this particular topic.

B. Bioremediation Approaches Commonly Used

Soils contaminated with mixtures of organic compounds may be treated using a wide range of available techniques (Table 1), although they are often treated by solid-phase technologies (CLU-IN 2007). In general, different approaches have been applied for the bioremediation of contaminated soils, and can be categorised as in-situ or ex-situ, depending directly on the location of the treatment in reference to the location of the contamination. In-situ treatments do not require additional costs of excavations and confinement; nevertheless they present important disadvantages, for example a high variability and thus difficulty to achieve a steady state and adequate decontamination end point as well as a difficult control process and operational verification. Ex-situ treatments mainly involve the containment of the contaminated soil or transport of it to a hazardous waste landfill (HWL) for further treatment or some type of containment system. This sentence is too long and complex. Ex-situ treatments normally require shorter times for maximum decontamination, facilitated by the optimization of operational parameters, normally at lab scale and pilot plant scale. Additionally, ex-situ treatments offer the advantage of adding external agents which may enhance the biodegradation of the target contaminants, i.e., oxygen if the treatment is more efficient under aerobic conditions, nutrients and inocula to stimulate resident microbial communities, surfactants to facilitate contaminant bioavailability, etc. Two major decision criteria to select in situ or ex situ treatment are cost, including initial investment and operational cost, and time necessary to clean-up the contaminated site. These will be dependent on the type of contaminant or mixtures of contaminants, extension of contamination, type of soil, environmental risk, human exposure and others. In general, in-situ treatments are cheaper but present uncertainty regarding the final results, whereas ex-situ treatments are more expensive but offer a better control of operational parameters and process (USEPA 2002).

A widely used in-situ bioremediation approach is bio-venting, where aerobic biodegradation is enhanced by adding air or oxygen, water, nutrients, surfactants and/or microorganisms in order to achieve a final maximum removal of the target contaminant or mixture of contaminants (Dejonghe *et al.* 2001, USEPA 2002). Phytoremediation, which is defined as the process by which plants are used to remove, degrade, contain or render harmless environmental contaminants (Cunningham *et al.* 1996) is an attractive alternative to disruptive and expensive engineering remediation methods (Macek *et al.* 2000, Kechavarzi *et al.* 2007).

Among most widely used ex-situ bioremediation approaches to treat fuel-contaminated soils are biopiles, landfarming and composting. The use of biopiles has been used particularly in the petroleum industry (Guerin 2000), where contaminated soils are aerated and mixed with pretreated soils, nutrients and/or microorganisms as inocula to enhance contaminant biodegradation. Landfarming is another widely used ex-situ treatment that involves soil excavation and relocation in a remote site in thin layers. With the use of agriculture machinery the contaminated soil is ploughed facilitating a better contact between the contaminated soil, nutrients and resident microbial communities as well as a periodic aeration (Guerin 1999). The contaminated soil can be also irrigated to facilitate appropriate moisture content. A successful ex-situ bioremediation treatment currently under investigation for a systematic and efficient application to contaminated soils is in-vessel composting. In this treatment, contaminated soils is excavated and mixed with organic amendments (e.g. green waste, sludge) which provides an appropriate carbon to nitrogen to phosphorous balance and may also stimulate residential microbial communities (Antizar-Ladislao *et al.* 2004).

Table 1. Treatment technologies applied to fuel contaminated soils

Technology	Development status [1]	Use rating	Cleanup time	Technology function
In situ biological treatment				
Bioventing	Full	Wide	NA	Destruct
Enhanced biodegradation	Full	Wide	NA	Destruct
Phytoremediation	Pilot	Limited	More than 3 years	Destruct
In situ physical / chemical treatment				
Chemical oxidation	Full	Limited	NA	Destruct
Electrokinetic separation	Full	Limited	NA	Destruct
Fracturing	Full	Limited	NA	Extract
Soil flushing	Pilot	Limited	NA	Extract
Soil Vapor Extraction	Full	Wide	NA	Extract
Solidification / Stabilization	Full	Limited	NA	Extract/ Destruct
In situ thermal treatment				
Thermally enhanced SVE	Full	Limited	> 3 years	Extract
Ex situ biological treatment (assuming excavation)				
Biopiles	Full	Wide	0.5-1 year	Destruct
Composting	Full	Wide	0.5-1 year	Destruct
Landfarming	Full	Wide	> 1 year	Destruct
Slurry phase biotreatment	Full	Limited	0.5-1 year	Destruct
Ex situ physical / chemical treatment (assuming excavation)				
Chemical extraction	Full	Limited	0.5-1 year	Extract/ Destruct
Chemical reduction / oxidation	Full	Limited	0.5-1 year	Destruct
Dehalogenation	Full	Limited	NA	Destruct
Separation	Full	Limited	> 1 year	Extract
Soil Washing	Full	Limited	0.5-1 year	Extract
Solidification / stabilization	Full	Limited	> 1 year	Immob.
Ex situ thermal treatment (assuming excavation)				
Hot gas decontamination	Full	Limited	< 0.5 year	Destruct
Incineration	Full	Limited	< 0.5 year	Destruct
Pyrolysis	Full	Limited	< 0.5 year	Destruct
Thermal desorption	Full	Wide	< 0.5 year	Extract
Other treatment				
Excavation and off-site disposal	NA	Wide	< 0.5 year	Extract/ Immob.

[1] Full scale: technology has been used in real site remediation.

NA = Not Available.

However, little literature exits on the bioremediation of mixtures of commercial products typically found in real consecutive accidental spills. The rate of biodegradation of such contaminated wastes would depend on biological and physicochemical factors, such us low numbers of microbes, insufficient oxygen or nutrient availability, temperature and water availability (Antizar-Ladislao *et al.* 2005, 2006a). The rate of biodegradation would also depend on several physicochemical properties related to the contaminant (i.e. hydrophobicity, volatility, polarity), solid matrix (i.e. organic matter, porosity) (Mader *et. al* 1997), and its bioavailability (Antizar-Ladislao *et al.* 2006b, Semple *et al.* 2003). Additionally, biodegradation of fuel oils following real consecutive accidental spills is a complex process that depends on the nature and amount of contaminants (aliphatic and aromatic hydrocarbon mixtures) present (Mrayyan and Battikhi 2005).

a. Composting as a Sustainable Bioremediation Approach

Composting is an aerobic process whereby organic materials are biologically decomposed. Heat is produced during the process, and conventional composting processes typically comprise four major stages. The four stages (mesophilic, thermophilic, cooling and maturation) vary in temperature and in the composition of the microbial communities (Bolta *et al.* 2003, Liang *et al.* 2003, VanderGheynst and Lei 2003).

Composting systems are generally divided into three categories: windrow, static pile, and in-vessel. In the windrow approach, the solid waste mixtures (soil and amendment) are composted in long rows and aerated by convective air movement and diffusion. The mixtures are periodically mechanically turned to expose the organic matter to ambient oxygen. This approach to composting leads to a characteristic fluctuating temperature regime as core temperatures will fall upon turning due to the introduction of cooler air masses and the release of entrapped heated air, and will rise when maximum microbial activity resumes. Over time, peak temperatures fall as the level of substrates decline. Windrowing is a simple mechanical system with no effective control of either oxygen or temperature levels within the organic mass undergoing decomposition. The technique, though popular commercially due to its ease of implementation and relative low cost of installation, has been criticized microbiologically since considerable periods of time elapse when either oxygen, temperature or both will become limiting for microbial diversity and decomposition.

In the static pile approach, piles of solid waste mixture, often with bulking agents (wood chips, straw) as a matrix improver, are aerated by using a forced-aeration system, installed under the piles to maintain a minimum oxygen level throughout the compost mass. Aeration may be on a timed basis (Epstein 1997) leading to high core temperatures and a severely restricted microbial diversity or via temperature feedback control (Antizar-Ladislao *et al.* 2004). This latter approach is based on a greater understanding of the microbial ecology involved whereby the control of air flow is dictated by reference to an upper temperature limit above which fans blow continuously until heat loss through latent heat of vaporization of water reduces the core temperature to an acceptable operational level. In both the Epstein (1997) and the Finstein *et al.* (1986) static systems, there are substantial variations in temperature throughout the composting mass and the cooler edges are routinely covered with a thick layer of finished compost acting as a thermal blanket to ensure that these edges reach regulatory temperature levels. Although the Finstein *et al.* (1986) system allows for the control of maximum core temperature, and in theory supplies an excess of oxygen for decomposition, both static approaches suffer from a gradation of temperature (and oxygen) throughout the organic waste mass. In static systems, active composting will occur over a three to four week period depending on the nature of the substrate being processed. This active phase is often followed by a two to three month period of maturation.

In-vessel composting takes place in a partially or completely enclosed container in which environmental conditions can be controlled. Enclosed vessels more closely approximate a laboratory incubator where the organic mass and its associated microflora should be exposed to a more even temperature profile. Control of temperature in in-vessel systems is usually achieved through recycling of exhaust gases with intermittent mixing of fresh air to maintain an agreed temperature (Antizar-Ladislao *et al.* 2004, Fraser and Lau 2000). In theory, such systems should allow for excellent control of temperature within the vessel and considerably less variation of temperature through the composting mass. However, in-vessel systems have serious limitations in general composting due to limited throughput and high installation costs (Das and. Keener 1994). Because of this, they are often used as a form of pre-treatment "bioreactor" for up to 5 days prior to conventional composting,(Epstein 1997, Haug, 1993, Sasek *et al.* 2003).

The application of composting as a bioremediation technology to treat hazardous wastes has been shown to be effective in biodegrading polycyclic aromatic hydrocarbons (PAHs) (Antizar-Ladislao *et al.* 2004, 2006a, Canet *et al.* 2001), chlorophenols (Laine and Jorgensen 1997), polychlorinated biphenyls (Michel *et al.* 2001), explosives (Gunderson *et al.* 1997) and petroleum hydrocarbons (Al-Daher *et al.* 2001, Balba *et al.* 1998) at laboratory and/or field-scales. At present, few studies can be found on the applicability of composting to contaminated desert mining soils (subject of case study in Section III) with low organic matter content, high mineral content and high salinity (Mader *et al.* 1997, Rhykerd *et al.* 1995). Literature on the bioremediation of desert soils in Kuwait (Al-Daher *et al.* 2001, Balba *et al.* 1998), refers to open systems where *in-situ* landfarming, bioventing and composting treatment technologies were implemented (Al-Awadhi *et al.* 1996, Al-Daher *et al.* 1998) to treat more than 20 Mm^3 of oil accumulated in the desert soil following the Iraqi invasion of Kuwait in the early 1990s. Research with Kuwait desert soils indicates that biodegradation of total petroleum hydrocarbons above 60% can be reached following eight months of compost soil piles treatment.

b. Role of Fungal and Bacterial Microorganisms During Bioremediation

Fungi, although representing almost 75% of the microbial soil biomass and accounting for up to 100 meters of hyphae per gram of dried soil (Kästner and Mahlo 1996), have been less studied than bacteria in bioremediation processes. A review by Martín-Moreno *et al.* (2004) documents well the literature on the use of fungi in bioremediation and in the recovery of soils. Basidiomycetes, which cause white rot decay of wood have been employed in bioremediation for the degradation of a variety of environmental persistent pollutants such as PAH, and produce unspecific extracellular enzymes for the oxidation of aromatic rings (Kirk *et al.* 1992, Paszczynski and Crawford 1995, Bogan *et al.* 1996a, Bogan *et al.* 1996b).

White rot fungi in different combinations have been used in the biodegradation of PAHs in microcosms containing wheat straw and non-sterile coal-tar contaminated soil (Canet *et al.* 2001). Results suggested that the degradation of PAH was an indirect consequence of native microflora metabolism rather than an effect of the addition of the fungal species. Within this same perspective, litter-decomposing fungi have been studied for their ability to grow on different lignocellulosic substrates as well as for their ability to secrete lignolytic enzymes and degrade PAHs (Steffen *et al.* 2006). This work showed that the agaric basidiomycetes *Stropharia rugosoannulata* and *Stropharia coronilla* were the most efficient PAH degraders among the different species used. Other studies have been conducted using other fungi, such as the deuteromycete fungus *Cladosporium sphaerospermum*. This fungus, isolated from soil of an aged gas manufacturing plant, was able to degrade 23% PAHs, including high molecular weight PAHs in non sterile soils after four weeks, correlating well fungal establishment with laccase secretion and PAH degradation (Potin *et al.* 2004). The evaluation of the arbuscular mycorrhizal fungi (AMF) occurrence has been also tested in soil samples from a landfarming area used for fifteen years for petrochemical waste treatment (de Paula *et al.* 2006). Results showed that a significant growth of different plant species occurred when AMF were present. These AMF could be contributing to phytoremediation of soils containing not only PAHs but also heavy metals that are a human health risk (Gohre and Paszkowski 2006).

Similar approaches have been used to study PAH degradation, such as the inoculation of soils with basidiomycetous and ascomycetous fungi to induce the degradation of PAHs either derived from commercial or lignite tar (Grams *et al.* 1999). It was observed that 22 to 38% PAH could no longer be recovered from pasteurized soils inoculated with the mentioned fungi, three to 10 days after spiking. Additionally, *Irpex lacteus* and *Pleurotus ostreatus* used for the bioremediation of PAH-contaminated soils (Bhatt *et al.* 2002) was able to remove up to 67% of fluorine, 56% of phenanthrene and 49% of anthracene. The use of three species of white rot fungi (Novotny *et al.* 1999), including *Pleorotus ostratus* showed the highest PAH degradation activity reducing the levels of anthracene, pyrene and phenanthrene by 81-87%, 84-93% and 41-64% within two months, respectively. The presence of cadmium or mercury did not alter the ability of *Pleorotus ostratus* to degrade PAHs (Baltrian *et al.* 2000).

The removal of aromatic herbicides by fungi isolated from pesticide contaminated soils is reported in a study by Bordjiba *et al*, (2001). The most frequently found species in the contaminated soils were *Aspergillus fumigatus, A. niger, A. terreus, Absidia corymbifera* and *Rhizopus microsporus* var *microsporus*, species not sensitive to pesticides.

Trichoderma, an anamorphic Hypocreaceaea (class Ascomycetes) is commonly found in the environment, and especially in soils (Wessels 1997). It is a well known biocontrol microorganism of several plant diseases (Benítez *et al.* 2004, Harman *et al.* 2004, Montealegre *et al.* 2005, Pérez *et al.* 2002), and has also proved to be useful in soil bioremediation (Esposito and da Silva 1998), in colonization of corn rhizosphere in pesticide containing soils (Rigot and Matsumura 2002) and in the degradation of phenols (Karetnikova and Zhirkova 2005). *Trichoderma* would thus appear to be a candidate for the accomplishment of multiple roles in polluted soils: colonization of plant roots, bioremediation of soils and biocontrol of plant diseases.

Unlike bacteria, fungi do not depend on continuous water-filled pathways for their translocation and spread in unsaturated-water soils. This is because of the expression and exposure of hydrophobins (Wessels 1997), small hydrophobic proteins that allow fungi to breach water - air interfaces and to grow in the pores formed between soil aggregates, that may also be penetrated by filamentous fungi through the use of special-shaped hyphae that disperse the soil (Wösten *et al.* 1999). Negligible bacterial movement can be detected in soils where vectors such as flowing water (Wong and Griffin 1976), earthworms (Singer *et al.* 2001) or plant roots (Kuiper *et al.* 2001) are absent.

The coexistence of fungi and bacteria in soil, and their metabolic cooperation in different systems has been discussed (Kohlmeier *et al.* 2005). The extensive colonization of fungal hyphae by bacteria suggests that fungi play an important role in bacterial movement and growth, by the formation of growth-promoting exudates (Arora and Gupta 1993). On this basis, "bioavailability" of bacteria to pollutants can be achieved using fungi to mobilize bacteria, rather than to provide bioavailability by mixing bacteria and contaminants (Semple *et al.* 2004). The combination of the fungus *Penicillium janthinellum,* bacterium *Stenotrophomonas maltophilia* VUN 10,010 and bacterial consortium VUN 10,009, isolated from separate creosote- and manufactured-gas-plant-contaminated soils, allowed a significant degradation of pyrene, chrysene, benz[*a*]anthracene, benzo[*a*]pyrene and dibenz[*a,h*]anthracene as well as bacterial growth. This process scarcely occurred in the absence of the fungus (Boonchan *et al.* 2000). The co-culture of any of these four bacteria (*Pseudomonas aeruginosa*, *Ralstonia picketii, Pseudomonas* sp. and *Psedomonas cepacea)* with any of the four fungi (*Penicillium* sp., *Trichoderma viride, Alternaria tenuis* and *Aspergillus terrus*) grown on sugarcane bagasse, showed a synergism in phenanthrene removal that depended on the bacteria-fungus co-culture, as compared to the pollutant removal by fungi that ranged from 35% to 50%, and the removal by bacteria that only reached 20% (Chávez-Gómez *et al.* 2003).

More recently, the use of *Pythium ultimum*, a fast growing oomycete that exhibits hydrophilic mycelial surfaces and of *Rhexocercosporidium*, an ascomycete that exhibits hydrophobic mycelial surfaces, with a naphthalene and phenanthrene-degrading *Pseudomonas putida*, showed that an efficient translocation of the bacteria occurred in the presence of *Pythium ultimum*. The study also showed that effective phenanthrene degradation occurred only in the presence of the fungal mycelia (Wick *et al.* 2007). The development and spread of fungal hyphae into the soil would be useful to facilitate a bacterial degradation approach of pollutants that are present not only at the soil surfaces but also deep inside the soils. A hyphal network could provide a physical space that could be used by bacteria to penetrate into the soil, taking advantage of the mechanisms used by fungi to grow and colonize the soils.

From this point of view, biocontrol fungi such as those belonging to the *Trichoderma* genera, resistant to different type of pollutants, could be used both to colonize soils and mobilize biocontrol bacteria (Montealegre *et al.* 2005) preventing the development of phytopathogens, and providing a network for mobilization of pollutant degrading bacteria (Leben 1984).

More studies about the roles of fungi concurring during bioremediation processes are necessary, in order to better understand the synergistic relationships between bacterial and fungal communities.

C. Bioremediation Monitoring

a. Microbiological Tools

An important first step towards understanding the roles of various microorganisms including bacteria, actinomycetes and fungi in a particular environmental sample is a determination of the composition and relative abundances of different microorganisms present in the environment. Until the 1980's, the determination of microbial community structures and the identification of microorganisms in environmental samples was conducted using culture-based studies, and different microorganisms were identified based on their phenotype. However, modern molecular techniques estimate that less than 1% of microbial species can grow under standard laboratory conditions (Amann *et al.* 1995, Kaeberlein *et al.* 2002, Ward *et al.* 1992). Other approaches to analyze the composition of microbial communities exist and are capable of discriminating between different bacterial strains. One such approach is the immunological direct-detection method enzyme-linked immunosorbent assay (ELISA) (Watanabe *et al.* 1996). Although this immunological method has a high resolution, it is not useful for a wide identification of microorganisms. Chemical biomarkers such as phospholipid fatty acids (PLFA) can also be used in the characterization of microbial communities present in contaminated waste mixtures under bioremediation conditions (Antizar-Ladislao and Russell 2007).

Molecular techniques based on the comparative analysis of 16S ribosomal DNA sequences are being widely used in microbial ecology studies, and have allowed the discovery of many novel and non-culturable bacteria. Furthermore, molecular techniques may help to discriminate between most of the microorganisms present in an environmental sample by analyzing the microbial genotype. The 16S rRNA gene encodes a RNA molecule of the ribsosome and is the most commonly used DNA sequence for the detection and phylogenetic analysis of microorganisms (Santos *et al.* 2004). The gene is approximately 1500 bp in length, is highly conserved, being found in all *Bacteria* and *Archaea*, and contains both variable and conserved regions. The gene is not laterally transferable and is the best molecular chronometer known (Woese *et al.* 1987). Large and rapidly growing databases of 16S rRNA gene sequences also exist.

Amplification of the 16S rRNA gene, followed by cloning, screening of the clone library and sequence analysis is becoming a commonly used approach used for microbial community analysis. Due to the power of the polymerase chain reaction (PCR) to amplify very small amounts of DNA, bacteria present at very low levels in an environmental sample are now detectable. The exponential nature of the PCR amplification process is very sensitive to any disturbance of amplification efficiency, and this can easily result in major PCR bias (Felske *et al.* 1998). PCR bias can occur such that certain templates within a mixed community type will inherently be favored in amplification. The use of PCR to amplify target DNA however greatly increases the detection sensitivity of microorganisms in an environmental sample in comparison to non-PCR based methods. Also, the physical, chemical and biological steps involved in the molecular analysis of an environmental sample are another contributing bias in microbial community analysis (von Wintzingerode *et al.*, 1997). Despite these potential pitfalls in molecular ecology, the application of molecular tools has resulted in great advances into the knowledge of microbial diversity. The clone library approach has many advantages over other molecular methods, as it provides precise identification and quantification of the phylotypes present in samples. The cloning approach has been taken by many researchers to investigate the microbial diversity of numerous environmental samples including lab scale anaerobic digesters (Chouari *et al.* 2005), Antarctic desert soils (Smith *et al.* 2006) and deep-sea carbonate crusts (Heijs *et al.*2006).

Amplified 16S rDNA can also be subjected to DGGE, a method that can be used to assess the diversity of a microbial community through the separation of PCR-amplified 16S rDNA over a denaturing gradient. DGGE was originally invented to detect point mutations in medical studies (Fisher *et al.* 1979), but has been frequently used in recent years to investigate microbial communities in diverse environmental samples such as composts, and bioremediated environments (Green *et al.* 2004, Haruta *et al.* 2004). During DGGE, PCR products migrate through a polyacrylamide gel, encountering increasingly higher concentrations of chemical denaturant. Upon reaching a threshold denaturant concentration, the lower melting domains of the double-stranded PCR product begin to denature at which time migration slows dramatically. Different patterns of bands will be produced as a result of differing sequences of DNA (from different bacteria) denaturing at different denaturant concentrations. Each band theoretically represents an individual bacterium present in the community. Bands can be excised from gels, and the DNA extracted, PCR amplified, and sequenced. Thus, the identity of members of the microbial community of a particular sample can be determined.

Temperature gradient gel electrophoresis (TGGE) is a method similar to DGGE, but instead uses a temperature gradient for separation of different DNA molecules. A high concentration of chemical denaturant is included in the gel, with the concentration being constant over the entire gel, such that increasing temperature is required for full denaturation. TGGE has been used to assess the diversity of microbial communities in different environments, including zinc-polluted soils and mealy-bugs (Brim *et al.* 1999, Franke-Whittle *et al.* 2005).

Terminal restriction fragment length polymorphism (T-RFLP) is another well known method for comparison of rDNAs. The method involves the amplification of DNA using universal 16S rDNA primers and the subsequent digestion of the product with restriction enzymes. Due to the differing sequences of different bacteria, the occurrence and location of the restriction sites will not be the same, and as a result, various patterns will be produced upon digestion. Sequence databases exist and can be searched for taxon-specific restriction sites (Moyer *et al.*, 1996). Tiquia (2005) used T-RFLP's to study microbial community dynamics in manure composts and found that the bacterial and fungal community profiles changed over the composting process. Bacterial communities showed a higher diversity compared to fungal communities. The results of their study demonstrated that distinctive community patterns from manure composts could be rapidly generated using T-RFLP analysis and used to monitor the dynamics of microbial communities qualitatively and quantitatively.

Microscopic detection of specific microbial cells achieved by fluorescence *in situ* hybridisation (FISH) is a technique which can be used to detect and localize the presence or absence of specific microorganisms within a sample. It uses fluorescent probes which bind only to DNA in cells with which the probes show a high degree of sequence similarity. FISH, targeting the 16S rRNA gene, allows the detection and enumeration of specific phylogenetic groupings of microorganisms, without prior isolation. Jjemba *et al.* (2006) investigated soil bacteria that degrade PAHs in soils, using probes specific for the *Alpha-*, *Beta-* and *Gammaproteobacteria.* They reported that using the group specific probes, bacteria of the respective sub-classes could be detected and were distinguishable.

In summary, the application of molecular methods has resulted in great advances into the knowledge of microbial diversity, since molecular methods can be applied to various environments. However, in the course of experimentation, it should always be considered that the results obtained using these methods may include biases caused by the extraction efficiency of nucleic acids, PCR error, or other unknown factors. In particular, the application of such analytical methods to the biodegradation process is sometimes limited by the properties of compost, such as huge amounts of organic compounds and the presence of humic acids (Stach *et al.* 2001). These factors can make it difficult to purify nucleic acids. Other limitations include the difficulty of conducting quantitative studies with some environmental samples, the lack of discrimination between living and non-living microorganisms, the impossibility to discern between strains, the limited amount of known sequences, and the fact that most molecular methods are labour-intensive. Molecular methods often require the use of traditional microbiology which allows the isolation of microorganisms from specific environmental conditions. Then, their DNA, RNA, proteins and biochemical reactions can be investigated, and thus it is possible to isolate microorganisms involved in the bioremediation process during composting. Additionally, RNA extraction from soils and compost can be complicated and it is very difficult to discriminate between the majority of the microbial species present in an environmental sample. Other limitations include those arising from uncultivable microorganisms, which restrict the understanding of relationships of microbial communities with their surrounding environment (Kaeberlein *et al.* 2002). In summary, a polyphasic approach, including traditional cultivation and molecular methods, is desirable in order to understand the technical biases and limitations of each method, and the results obtained should be interpreted with extreme caution.

To completely understand the various processes that may occur during bioremediation of contaminated wastes using composting approaches, a more general and functional approach is required which would allow the investigation of microbial genes present in an environmental sample. Thus, new techniques have emerged, which search for genes responsible for contaminant biodegradation during bioremediation. These are:

Metagenomics. The direct sequencing of environmental samples has provided valuable insight into the lifestyles and metabolic capabilities of uncultured organisms occupying various environmental niches. Recent efforts include sequencing of individual large-insert bacterial artificial chromosome (BAC) clones as well as small-insert libraries made directly from environmental DNA (Tringe *et al.* 2005).

Proteomics. Mass spectrometry-based proteomic methods are applied to evaluate gene expression, identify key activities and examine partitioning of metabolic functions (Ram *et al.* 2005).

Tagged mutagenesis. A single mutant strain is marked by a tag signal and identified by parallel selection. Thus, genes expressed by microorganisms can be identified during exposure to conditions in their natural environments (Groh *et al.* 2005).

Environmental gene expression analysis. Hybridization based microarrays with soil cDNA allow for the quantitative identification of biodegradation genes, and thus a characterisation of activated genes (Rhee et al. 2004). Additionally, probes varying in taxonomic specificity can be used, allowing for rapid classification, and perhaps quantification of the organisms present within a sample (Bae *et al.* 2006). Other methodologies applied to the quantification of functional genes in soils are MPN-PCR, competitive-PCR and real time PCR (Sharma *et al.* 2007).

Substrate-induced gene expression screening (SIGEX). Relevant substrates generally induce catabolic-genes, and, in many cases, their regulatory elements (Uchiyama *et al.* 2005).

The use of individual microbes for complex environment tasks such as bioremediation of contaminated soils represents a great challenge for environmental biotechnology. Foremost, there is substantial discord between the laboratory conditions where the organism is manipulated and the *in situ* environment that is targeted by the microbial community. The application of molecular techniques such as those described in this section have allowed for a better understanding of the dynamics and composition of different microbial communities in the environment. Nevertheless, discovery of new gene pools involved in the biodegradation of pollutants and the development of new techniques such as enzymatic bioremediation is still a challenge which impedes a complete understanding of the microbiological processes involved during bioremediation of contaminated soils (Alcalde *et al.* 2006). Thus, newly advanced, robust and easy-to-use biomonitoring tools need to be developed.

b. Cutting-Edge Biomonitoring Tools

The recent introduction of nucleic acid microarrays for the identification of bacteria in microbial ecology studies provide a powerful tool for the parallel detection of 16S rRNA genes, and have proved useful in microbial studies of phylogenetically diverse microbial groups (Gushin *et al.* 1997, Loy *et al.* 2002, Small *et al.* 2001). Microarray technology was first described in the late 1990s, and since then the technology has become more and more commonly used in various fields of research. DNA microarrays are based on the relatively old technology of DNA hybridization between two complementary strands of nucleic acids. The core innovation of the microarray technique is the ability to attach nucleic acids to a solid matrix in a precise location to create a densely packed array (Maskos and Southern 1992, Southern *et al.* 1999). DNA microarrays exploit the preferential binding of complementary, single-stranded nucleic-acid sequences (Brazma *et al.* 2000). Arrays are composed of many discretely located probes spotted onto a solid surface, each probe being specific for a given strain, subspecies, species, genus, or higher taxon (Loy and Bodrossy 2005). Each spot represents a unique probe sequence. Spots are usually 100-200 μm in size and are located within 200-500 μm of each other (Call *et al.* 2003).

DNA array technology offers the possibility to analyze an entire array of microorganisms concerning their presence or absence in a particular sludge sample in a single experiment. The extracted DNA from samples to be hybridized is labelled with fluorescent dyes such as Cy3 and Cy5 (Brazma *et al,* 2000). Hybridization with the labelled samples is followed by washing steps of differing stringency and scanning using a high-resolution scanner (Call *et al.* 2003). The relative expression levels of the genes and abundance of a target in the sample and in control populations can be estimated from the fluorescence intensities and colours of each spot (Brazma *et al.* 2000).

Microarrays do however, also have their drawbacks. Various methodological deficiencies, such as the laborious protocols for the preparation of target molecules and insufficient detection limits constrict a broad application of microarrays in the analysis of microbial communities (Zhou, 2003, Peplies *et al.* 2004). Biases resulting from DNA extraction procedures as well as from PCR also exist (von Wintzingerode *et al.* 1997, Vianna *et al.* 2005). Weaker fluorescent signals can represent either a low concentration of a particular gene within the DNA extract or alternatively, a gene having sufficient but less than perfect similarity within the printed oligonucleotide (Dubois *et al.* 2004). Target secondary structure may also account for lower hybridization signals (Wang *et al.* 2002). Nonetheless, this technology has been applied and successfully used to study microbial communities in several environmental habitats such as a hypersaline cyanobacterial mat from a lake in Sinai, Egypt (Loy *et al.* 2002), ready-to-eat vegetable salads (Rudi *et al.* 2002), landfill methanotroph communities (Stralis-Pavese *et al.* 2003), soils (Small *et al.* 2001), and faecal samples (Wang *et al.* 2002). Diagnostic microarrays can also be used to investigate the effect of soil microbial diversity on soil fertility and the effect of agricultural practices on soil microbial communities (Bodrossy, 2003).

A microarray spotted with 369 different probes specific to microorganisms that have been previously reported in the degradation process of organic waste during composting, for plant, animal and human pathogens, and for plant disease suppressive bacteria was developed by Franke-Whittle *et al.* (2007). The COMPOCHIP 16S rRNA microarray was developed with the aim of being able to provide a direct characterization of the bacteria present in a particular compost sample. Composts contain a large and diverse community of microorganisms that are involved in the digestion of organic wastes (Franke-Whittle *et al.*, 2005). The COMPOCHIP array was applied to three different compost types (green compost, manure mix compost, and anaerobic digestate compost) of different maturity (2, 8 and 16 weeks), in order to determine if there were any differences in the microorganisms in the 3 compost types and maturity stages. Fig. 1 shows the COMPOCHIP microarray hybridized with DNA extracted from a 2 week green compost.

Figure 1. COMPOCHIP microarray image following hybridization of Cy5-labeled DNA extracted from 2 week green compost. Signals obtained for the print controls (Cy3-labeled), hybridization controls (central column of spots) and the different probes are evident.

Multivariate analysis showed that the bacterial composition of the 3 composts were different at the beginning of the composting process and became more similar upon maturation (Fig. 2). Certain probes were found to be more influential in discriminating between different composts. Interestingly, the 2 week manure mix compost was found to have the highest number of positive signals of all compost samples, and higher signals for many probes than found in any other sample. The diversity and intensity of signals then decreased significantly over the following weeks, such that not many signals were detected in the 16 week manure mix compost sample. For both the anaerobic digestate and green composts, many of the probes gave hybridization signals above the threshold in the 2 and 8 week samples. In fact, higher signal to noise ratio (SNR) values were obtained in the 8 week sample for many of the probes than in the 2 week samples. The 16 week samples were found to closely resemble the 16 week manure mix compost sample, not yielding many signals with SNRs above the threshold. This study showed the potential usefulness of the COMPOCHIP microarray in determining microbial communities in composts. This microarray could also be applied to other soil and compost samples in order to gain an impression of the microbes present in a single test. A microarray containing oligonucleotides targeting key microorganisms in various bioremediation processes could be designed and would allow a direct monitoring of the microbial communities present in a bioremediation process.

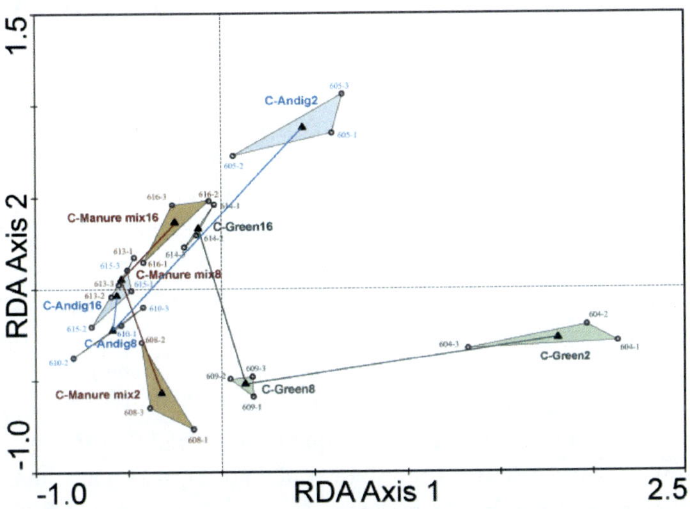

Figure 2. 2D-ordination-plot of three compost communities evolving over time analyzed by canonical analysis (redundancy analysis-RDA). The means for each sample are shown together with the three replicates.

Sole-carbon-source tests can also be used to monitor changes in microbial communities. Such tests are non-molecular, and reportedly reflect the composition of a microbial community. The method first described by Garland and Mills (1991), involves a commercially available microtiter plate (BIOLOG), which can be used to simultaneously test the utilization of 95 substrates as sole carbon sources. Utilization of a carbon source is indicated by colour development of a redox indicator dye, and changes in the overall patterns of carbon source utilization rates can be assessed by multivariate statistics (Baath *et al.*, 1998). The BIOLOG test was originally developed for the classification of bacterial isolates based on the ability of the isolates to oxidize the different carbon sources. However, this method has also been adapted and used to characterize the metabolic capacities of a microbial community. BIOLOG has been commonly used to detect differences between microbial communities in soil and the rhizosphere (Garland 1996, Pfender *et al.* 1996, Zak *et al.* 1994). Due to the necessity of faster analysis of microbial communities, the initial BIOLOG plate GN was modified and redesigned at the request of a group of microbial ecologists. The result was a new BIOLOG microplate, the EcoPlate^{TM}, which contains 31 of the most useful carbon sources found in soils and used by microorganisms. The carbon sources are present in triplicate, and the 31 different C-sources (amines, aminoacids, carbohydrates, carboxylic acids and polymers) allow community level physiological profiles (CLPPs) to be established based on the capacity of microorganisms to utilize these different C substrates. In addition, it is feasible to establish the average metabolic response of a sample by calculating the average well colour development. Despite the simplicity and low cost of using CLPPs for the characterization of different samples, relatively few reports of the application of this methodology in compost studies exist.

III. Bioremediation of Contaminated Desert Soils with Diesel Fuel by In-Vessel Composting in the Atacama Desert, Chile

Chile is a regional economic leader in South America. Its economy is ore-dependent and it is considered to be a mining country due to continuous investments in the mining industry. Chile is currently the worlds leading extractor of copper, and copper sales represent approximately 40% of its commercial balance and around 9% of its Gross Internal Product. Relations between the mining industry, environment and society have not been smooth due to continuous environmental impacts and confrontations between the industry and human communities. Topics of debate range from the type of exploitation (e.g. open pit mine) to cultural disturbances. Nevertheless, the Chilean population has accepted the mining industry as a very important development motor.

In Chile there are two principal mining companies, CODELCO (Copper Corporation) a Governmental Company, and Minera Escondida Ltd (MEL), a private investment of BHP-Billiton, the world's largest diversified resources company. MEL owns Chiles most important copper mine. The mine is located in the Atacama region (north of Chile; 170 km SE of Antofagasta; 3,100 m above sea level), and is responsible for the production of 8% of the world's copper. The mine has an annual income tax of approximately US$ 140 million and investments total more than US$ 4 billion. The construction of the mine in the Atacama region started in 1988, and the first batch of ore was processed in 1990. MEL's production of copper at the Atacama region represents 20% of total copper production in Chile, and includes a work-force close to 2,300 people, 1,900 permanent jobs through outsourcing, and 8,000 additional permanent jobs in diverse productive activities associated to the mining industry, especially in the North of Chile. The role of Minera Escondida Ltd. is vital to the Chilean economy, since its production accounts for 2.5% of the country's Gross Internal Product (Minera Escondida, 2007).

Despite the economic benefits that the mining industry offers to Chile, environmental concerns related to the production of hazardous wastes exist. It is well known that the mining industry has a negative environmental impact, due to continuous excavations, dust production, noise emission by heavy machinery, and disturbance of flora and fauna. A non-evident impact is that resulting from the daily operation of heavy machinery, where repetitive spills of fuel oils have occurred during the repair and maintenance of machinery. Soils and sawdust have frequently been used as cheap and readily available adsorbent materials to help in the clean up of the spilled fuel oils, which consist of complex mixtures of aliphatic and aromatic hydrocarbons. This common practice in the mining industry has caused a real case environmental problem of current concern, whereby large amounts of hazardous wastes, comprised mainly of desert mining soils and sawdust contaminated with fuel oils, have accumulated over time and need to be cleaned-up.

Our present aim is to minimize the environmental impact caused by mining activities in Chile, and in particular in the Atacama region, which may be extended to other regions around the world. Little literature exits on the degradation and fate of mixtures of commercial products typical in real consecutive accidental spills in desert mining soils, under bioremediation conditions. In-vessel composting bioremediation technology is presented as a potential treatment option, that can be optimized to produce results complying with current environmental laws, and that minimizes human and animal exposure to contaminated sites.

A. Physicochemical Characteristics of Desert Mining Soils in the Atacama Desert, Chile

Chile has principally a Mediterranean climate, however, in the North, the Atacama desert exists, one of the worlds most important deserts, comprising the worlds biggest mining cluster and industry of copper and other ores. The Atacama Desert is an extreme, arid and temperate desert that extends from 20°S to 30°S along the Pacific coast of South America. Soils in the Atacama desert are taxonomically characterized as either aridisol, inceptisol or entisol, and have a thin development of A horizon above thin B horizons. Atacama desert's soils often also have carbonate accumulation in a K horizon and are closer in composition to mother rock due to a continuous eolic erosion and a low pluvial activity typical of dry climates (USDA 1999). The low pluvial activity increases only in summers with the "Bolivian Winter" (Miller 1976). Additionally, soils are hyper-saline and this is one of the main reasons why these soils are dedicated principally to mining activities. The organic matter content is lower than 0.1% and concentrations of copper and sulphur are higher than the typically reported agronomic values (Thompson 1978). Minor and major essential elements are present at concentrations lower than that typical of agronomic soils.

B. Historical Contamination

The maintenance of machinery and trucks at Minera Escondida Ltd. has resulted in continual accidental spills of mixtures of commercial fuel oils (diesel and gasoline). Typically, these fuel oils have been 'cleaned up' by absorption with sawdust and native desert mining soils in order to contain the contaminants and minimize the detrimental environmental and human impact of these contaminants. The contaminated mixtures of soil and sawdust are contained and disposed of in HWL. The HWL is normally exposed to environmental conditions typical of the desert climate, with daily oscillations of temperature between 0°C during the night and above 40°C during the day, and the hazardous wastes are not exposed to direct-solar radiation. This common practice in the mining industry has caused a real case environmental problem of current concern, where large amounts of these hazardous wastes have accumulated over time and need to be disposed of.

C. Bioremediation Strategy

Due to the limited number of case studies available on the use of bioremediation technologies to treat contaminated desert soils and the need to decontaminate large amounts of hazardous wastes in the Atacama desert (Chile), this research sought to determine the feasibility of bioremediation by aerated in-vessel composting of mixtures of desert mining soil and sawdust contaminated with fuel oils as a bioremediation strategy to clean-up fuel contaminated wastes. The specific objectives of the project were to determine the effectiveness of different ratios of contaminated soil to contaminated sawdust (S:SD) contained and disposed of for two years in a HWL. Another objective was to find correlations between the percentage of contaminant removal and various physical and chemical parameters which may indicate the best conditions for maximum contaminant removal. A third objective was to investigate the microbial community changes during in-vessel composting treatment.

Our experimental design set-up consisted of the implementation of 30 laboratory-scale cylindrical aerated composting reactors operated continuously for 56 days with a moisture content close to 50% and maintained at mesophilic temperatures. Contaminated desert-soil and sawdust were treated at different ratios of contaminated soil to contaminated sawdust, S:SD 1:0, 3:1, 1:1, 1:3 and 0:1 (2,000 g total composting mixture, wet weight). Prior to filling in the composting reactors, contaminated wastes were separately passed through a 5 mm sieve followed by a 2 mm sieve. Atmospheric air warmed to 60 ^{0}C was introduced (pump of 1.1 KW and a maximum flow of 550 m^3 h^{-1}) to each reactor and circulated through an internal perforated piping to ensure sufficient oxygen concentration in the reactors (Fig. 3). Temperature was monitored with K type thermocouples, connected to a system OPTO 22 and stored in an ASCII file in a PC Pentium III. Recorded temperatures throughout the length of the treatment showed that the composting reactors reached mesophilic temperatures of 30 ± 2^{0}C at the surface of the substrate in contact with the head space and 40 ± 2^{0}C at the bottom of the substrate. Additionally, gas streams from the inlet and exhaust were continuously monitored for carbon dioxide production as evidence of aerobic biodegradation. Each reactor was homogenized on a daily basis to avoid stratification and oxygen content limitation. Sterilized distilled water (SDW) was added when needed to maintain 50% moisture content in all reactors during the treatment.

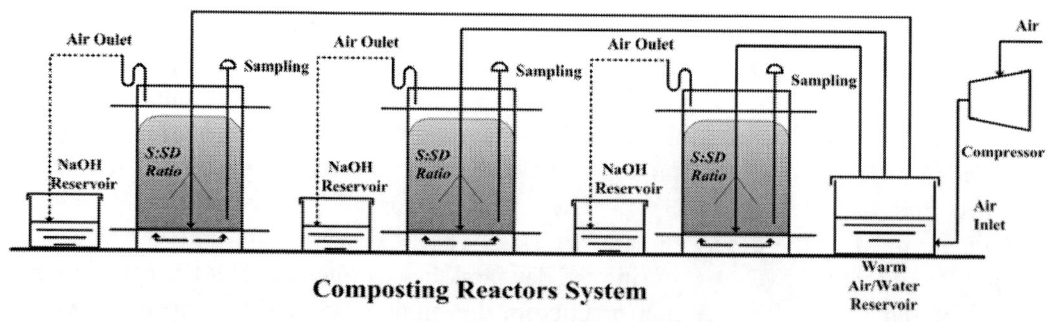

Figure 3. Schematic laboratory set up of in-vessel composting reactors.

Sampling was conducted in triplicate after 0, 14, 28, 42 and 56 days. The samples were collected using a core tube (70 cm long, 5 cm i.d.) in order to obtain stratified samples at different radial distances in each reactor. Each sample was separately manually homogenized with a stainless steel spatula in a sterile bag and stored in an amber glass flask for further analysis.

Our criterion was that the autochthonous microbial communities were acclimatized to cold and warm temperatures and to the presence of pollutants. The corresponding abiotic controls comprised the same S:SD ratios, but reactors were radiated with Gamma rays of a Cesium 137 source to 25 kGy doses three times using an Experimental Nuclear Reactor (Chilean Commission of Nuclear Energy, CCHEN).

Chemical and microbiological analyses were performed using standard procedures (Sadzawka *et al.* 2000, USEPA 1986). pH, E.C., moisture content, organic matter content, nitrogen content, oligoelements and heavy metals, organic contaminants, and cultivable heterotrophic bacteria were monitored. Additionally, changes in the microbial communities, number of bacterial cells and percentage of live cells were determined using fluorescence microscopy by directs counts and a fluorescence microplate reader by LIVE/DEAD BacLight Bacterial Viability kit (Molecular Probes). Terminal-restriction fragment length polymorphism (T-RFLP) and BIOLOG Ecoplate^{TM} tests were conducted, and diversity indexes calculated using the Shannon-Wiener index (H′). The average well colour development (AWCD) was determined for the BIOLOG Ecoplates^{TM}.

Finally statistical analyses were conducted. A two-factor ANOVA test (factor 1, treatment or S:SD ratio; factor 2, time) applied for each degradation curve to elucidate the effect of S:SD ratio treatment and time of the removal of fuel oil. A t-test was conducted to investigate possible differences between initial and final values of diesel fuel content in each treatment and final diesel fuel content following different treatments. Principal Components Analysis (PCA) and Linear correlations (r) within-groups of physical and chemical parameters and removal percentage in different reactors were also investigated to determinate correlations between treatments.

D. Changes in Physicochemical Conditions

Since pH is one of the key factors for microbial metabolism, its value in the composting of substrates was monitored every 14 days (Fig. 4). Before treatment, all reactors were found to have pH values of 7.0 -7.5 and no statistical differences were found between reactors. Once the treatment started, pH changes were observed in the reactors with significant lower pH at lower S:SD ratios. Statistical significant differences were found between initial and final times ($p<0.05$ by t-test) in the majority of the reactors. Only the reactors S:SD - 1:1 and S:SD - 3:1 presented no statistical differences among them. No pH changes were observed in the abiotic controls (data not shown).

Lower pH values in treatments of lower S:SD ratios could be explained by the formation of canalizations in the presence of sawdust. This may have facilitated the aeration of the contaminated mixture promoting aerobic biodegradation, and thus CO_2 production and anaerobic zones where fermentation processes may have occurred, acidifying the mixtures. On the other hand, in the reactor with contaminated desert soil only (S:SD 1:0), the pH tendency was to rise over time.

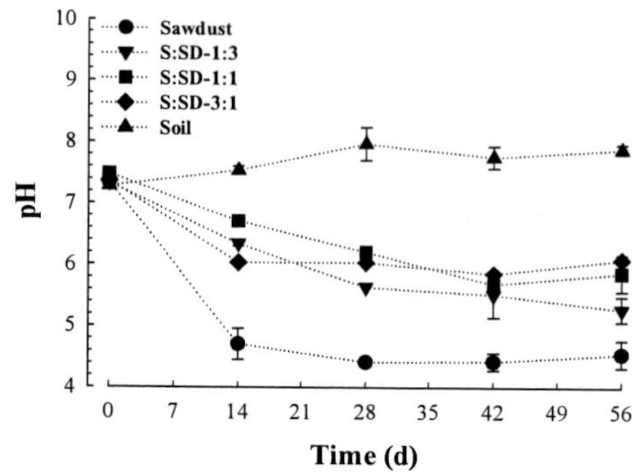

Figure 4. Temporal changes of pH.

Soils in the northern Chile Atacama desert are hyper-saline (electrical conductivity; EC > 10.0 mmhos cm^{-1}) and this is one of the main reasons for why these soils are dedicated principally to mining activities. Increasing levels of salinity in the reactors treating mixtures of contaminated desert soil and contaminated sawdust were observed over time (data not shown). This indicated a higher content of free ions in the composting mixtures over the duration of the experiment. In the reactors containing sawdust, the EC remained at a constant levels (≈12 mmhos cm^{-1}), however, in the reactor with S:SD - 1:0, the EC progressively decreased during the treatment to values closer to 7 mmhos cm^{-1}. Thus, no significant effect of salinity was observed in the composting treatments ($p>0.05$, correlation index and PCA) in this study.

Major and minor elements were monitored since they are essential for the catabolic activity of microorganisms. Thus, fuel oils degradation and biomass production depend directly of an accurate balance of these nutrients. Levels of Mn and available N, P, K were higher (range between 1,000 and 10,000 ppm) in the contaminated soil, sawdust and mixtures of contaminated soil and contaminated sawdust at various S:SD ratios as compared to normal values of B and Zn (30 - 180 ppm, respectively). Nevertheless, available N, P and K concentration values were low in all reactors as compared to reported agronomic values (data not shown). Initial N content was different in the reactors depending on the S:SD ratio, and the highest consumptions of total N was observed in the reactors with a S:SD 1:0, 1:3 and 3:1. The tendency of total N was to reach a steady state and similar concentration values ($p>0.05$) in the final points in all reactors (Fig. 5a). Total N did not change in the abiotic controls during treatment (data not shown). In the present study, it was assumed that indigenous soil microbial populations were adapted to the fuel oils present in the soil and to the environmental conditions. Values of available nutrients in this study were lower than those values typically reported in composting processes, which was not surprising because desert soils are mainly composed of mineral surfaces with low organic matter contents The levels of S and Cu were higher than those normally encountered in typical desert soils, and it was assumed that autochthonous microbial communities were adapted to the presence of high levels of S and Cu.

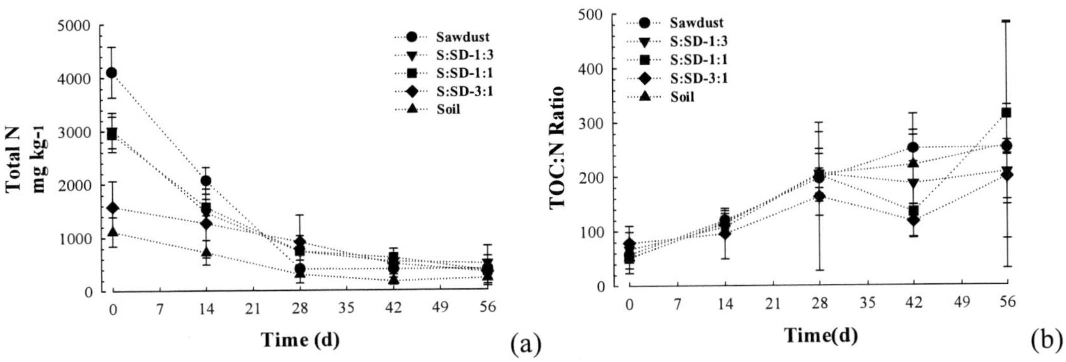

Figure 5. Temporal changes of: (a) N and (b) ratio TOC:N.

With respect to organic matter content, the composting reactors with mixtures of contaminated desert soil and contaminated sawdust contained a higher organic matter content compared to the composting reactors with only contaminated desert soil. This is most likely due to the capacity of sawdust to adsorb organic contaminants. Thus, composting reactors with a S:SD ratio of 3:1, 1:1 and 1:3 had organic matter contents of 24%, 30% and 49% respectively; the control reactor (S:SD 1:0) presented an organic matter content of 14%, a high value in comparison with typical non-contaminated desert soils (Atacama and Kuwait region, organic matter content < 1%). TOC/N ratios were within the range 50:1 and 100:1, higher ratios than those normally encountered in a composting process (Al-Daher *et al.* 1998). Temporal increases in TOC/N ratios similar in all reactors indicated that the temporal concentration of N decreased over time as compared to the temporal concentration of organic matter content (Fig. 5b). TOC/N ratios decreased from day 28 to day 42 in the reactors with a S:SD ratio of 3:1, 1:1 and then increased again. This temporal decrease needs further investigation. Temporal trends of total N and TOC/N ratios (Fig. 5a, b) indicate that N may have been a limiting nutrient in the composting reactors.

E. Changes in Diesel Fuel Concentration

In order to follow the biodegradation of fuel oils present in the contaminated soil and sawdust mixtures, total hydrocarbon concentration was monitored. Different initial concentrations of fuel oil were observed in the reactors with higher initial hydrocarbon concentrations present in sawdust (Fig. 6a). The initial concentration of fuel oils in the mixtures corroborated the higher adsorbing capacity of sawdust. Following 56 days of treatment, different biodegradation curves were obtained in all reactors, with most of the biodegradation occurring during the first 42 days of treatment (Fig. 6a). Biodegradation curves indicated a similar trend among composting reactors with S:SD ratios of 1:0 and 3:1 with smooth decreases reaching contaminant removal between 30% and 40% after 42 days of continuous treatment (Fig. 6b). Reactors with S:SD ratios 1:1 and 1:3 with higher levels of sawdust presented higher removal rates. The reactor with a S:SD ratio 1:0 presented the lowest removal rate (60%). The concentration of fuel oils remained unchanged in the abiotic controls (data not shown). These removal percentages are similar and concordant with those reported in a previous commercial diesel oil contamination study (Marchal *et al.* 2003).

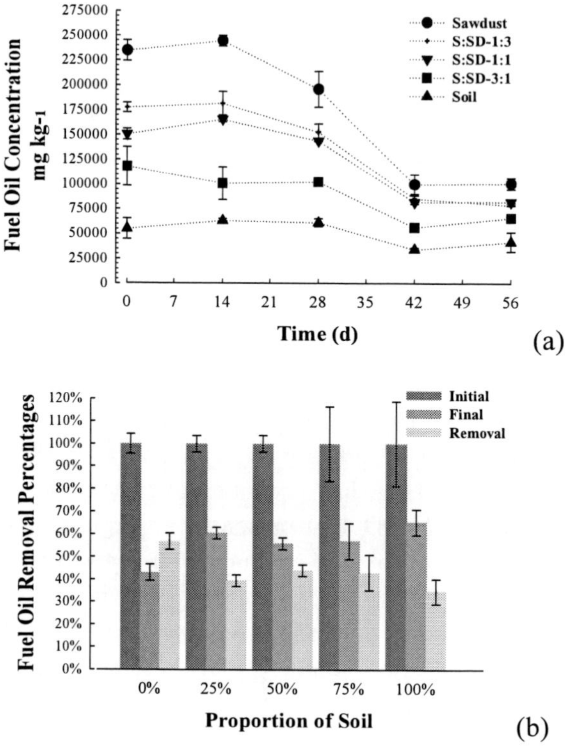

Figure 6. (a) Temporal changes in fuel oil concentration in experimental reactors, and (b) Fuel oil removal rates in the experimental reactors.

F. Changes in Microbial Communities

There have been few reported studies whereby molecular tools have been used to identify microorganisms or diversity of microbial communities in desert soils (Ros *et. al* 2006). Studies conducted in the Atacama Desert have until very recently focused on the investigation of the diversity of heterotrophic bacteria based on 16S rRNA gene sequences, in order to improve our current understanding of the diversity of bacteria living in extreme environments. Such environments are close to their limit of dryness to sustain microbial life, and thus they have been compared with soils from the surface of Mars (Mars-like soils) (Navarro-González *et al.* 2003), a focus of diverse studies by NASA (Clements, 2002).

These microorganisms are typically the first organisms to react to chemical and physical changes in the environment. For example, it has been observed that an increase in the moisture gradient along the Atacama region results in a greater bacterial diversity of heterotrophically culturable bacteria (Clement 2002) and it is then more likely to find principally autotrophic (litotrophic) heterotrophic microorganisms. Drees *et al.* (2006) investigated the bacterial community structure of the Atacama Desert and reported a predominance of *Gemmatimonadetes* and *Planctomycetes*, which is unusual when compared to typical soil populations. The unique phylogenetic distribution ofthe organisms identified in the study of Drees *et al.* (2006) and in studies of other arid soils would suggest that the hyperarid environment does select for specific groups of bacteria. Nevertheless, more research is still needed to obtain a better knowledge of the microbial community composition in the Atacama Desert, and especially of the presence of heterotrophic bacteria in mineral soils.

The study presented in this chapter has used traditional and cutting-edge biomonitoring tools to investigate microbial community structure changes during bioremediation of fuel-contaminated soils in the Atacama desert, where the organic degradation take place in a similar fashion to a natural Martian environment (Stocker and Bullock, 1997). Direct counts and the percentage of viable bacteria indicated that count numbers were high at the beginning of the treatment and that no significant changes were observed during the course of the treatment (over 10^6 cells gr^{-1}). Additionally, viable bacteria comprised over 60% of total number of cells (Fig. 7a, b). These preliminary results indicated that most probably resident microbial communities were adapted to the presence of organic contaminants following consecutive accidental spills and storage of the contaminated wastes in the LHM, prior to the composting treatment.

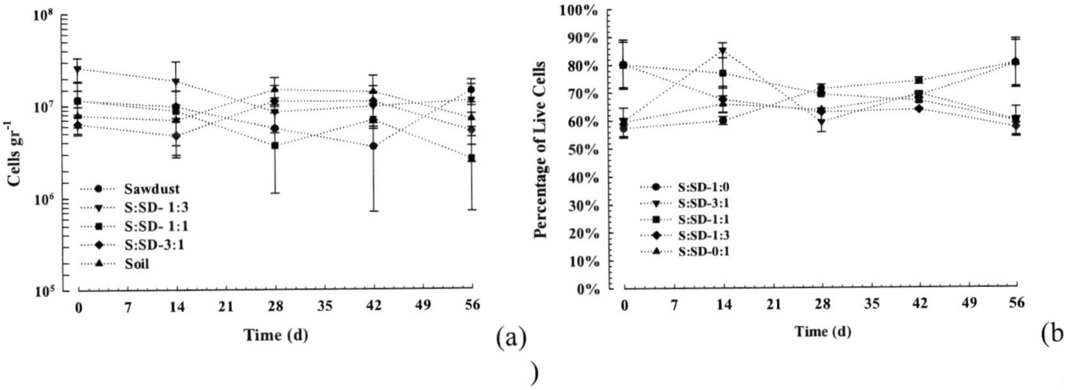

Figure 7. Temporal changes of: (a) Cell per gram of soil and (b) percentage of live cells in the experimental reactors.

When cutting-edge biomonitoring tools were applied to the Atacama desert soil samples, contradictory results were obtained, indicating that the changes in number of phylotypes were not in agreement with the changes in the metabolic activity (Fig. 8a, b). In the reactors where only soil was treated, an increase in diversity with a maximum reached after 42 days of treatment was seen, according to BIOLOG EcoPlate^{TM}. According to T-RFLP, maximum diversity was reached after 28 days of treatment and then decreased to a minimum after 42 days of treatment. The reactors where only sawdust was treated presented a trend in diversity similar to that of only soil according to BIOLOG EcoPlate^{TM}, but constant to a value close to 3 according to T-RFLP. Additionally, reactors where only sawdust was treated had, according to T-RFLP, higher diversity than all other samples.

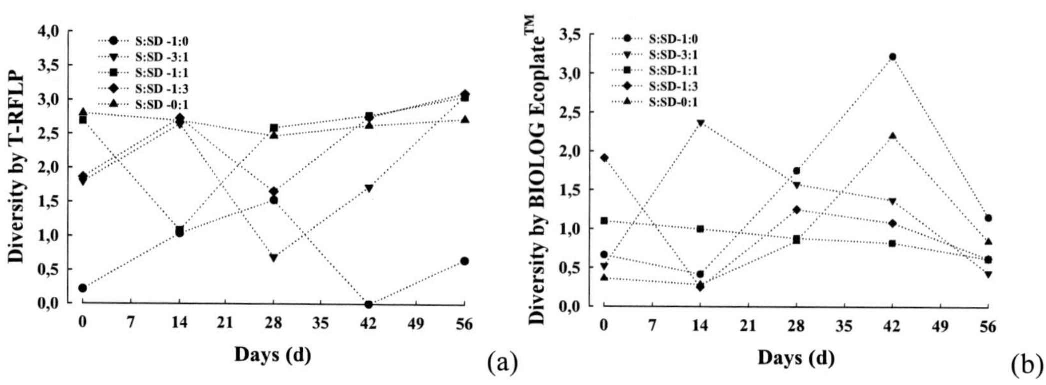

Figure 8. Temporal changes of diversity by (a) T-RFLP and (b) BIOLOG EcoPlate™ in the experimental reactors.

The AWCD indicated that the microbial activity in composting reactors increased with treatment duration (Fig. 9). According to these results, the reactors with a S:SD ratio 3:1 presented the highest metabolic activity during the length of the treatment and the reactors with a S:SD ratio 0:1 presented the lowest metabolic activity. AWCD results were in agreement with the biodegradation curves (Fig. 6a), and thus higher microbial activities resulted in higher fuel removal rates. Finally, higher microbial activities are observed towards the end of the treatment process and at higher sawdust contents in the composting mixture.

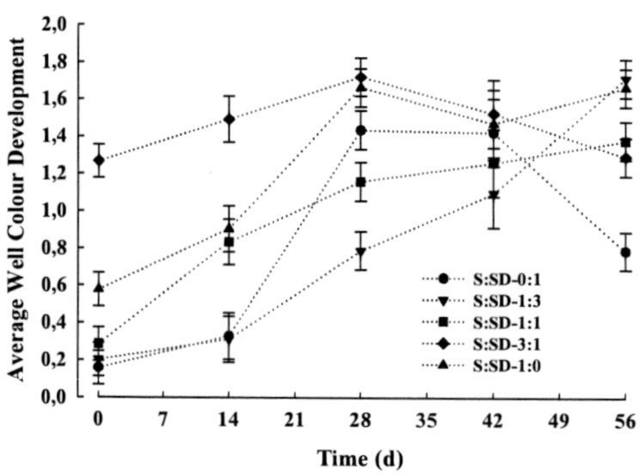

Figure 9. Temporal changes of AWCD in the experimental reactors.

G. Optimal Conditions for Maximum Removal of Fuel Diesel

Statistical analysis using the results of different treatments indicated that 2 factors explained 83.6% of the total variance. Factor Time explained 73.3% of the total variance, and factor Treatment (S:SD ratio) explained 10.3% of the total variance. These results were confirmed with PCA analysis considering all physicochemical factors, where the reactors are dispersed in two axes that explain over 70% of variance (Fig. 10). Comparison of the results of the abiotic and experimental reactors, indicated that all treatments presented significant differences, indicating that biodegradation was the main mechanism of contaminant removal from fuel oil contaminated soils and sawdusts. The mesophilic temperatures used in this experiment were lower than those required for volatilization (Loehr *et al.* 2001), which further supports the results obtained.

A linear correlation analysis including the various physicochemical parameters monitored in this study during treatment indicated that the removal of fuel oils was significantly influenced by pH, TOC/N ratio and total N. The reactor with a S:SD ratio 1:0 presented a significant correlation of fuel oils removal with TOC:N ratio which may indicate that N may have been a limiting factor for fuel oils biodegradation during treatment. The reactor with a S:SD 1:3 also presented a significant correlation of fuel oils removal with the total N present, which may have had some nutrient implications during treatment. The other reactors presented no significant correlations with the physicochemical parameters monitored. These correlations may indicate that the main mechanism of fuel oils removal during treatment is aerobic biodegradation with subsequent production of CO_2 resulting in lower pHs, as observed in this study

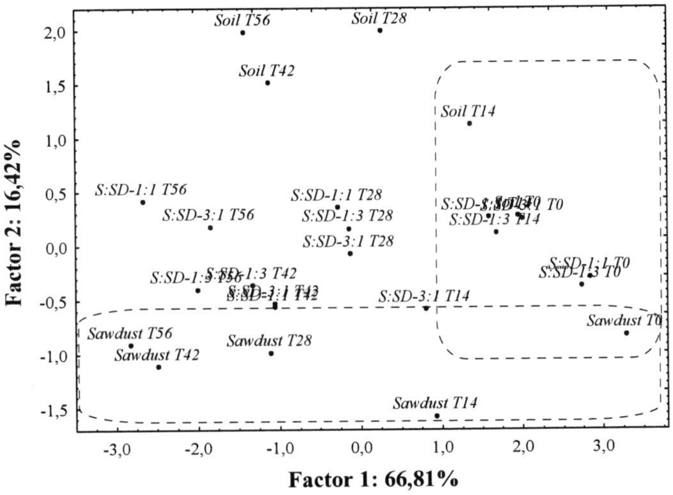

Figure 10. Effect of Factor 1 (time) and Factor 2 (treatment, S:SD) on fuel oils removal in all treatments investigated.

The major difference between treatments was observed for treatments with only soil (S:SD 1:0) and mixtures of contaminated soil and contaminated sawdust. No statistical difference was observed between treatments with a S:SD 1:1 and 1:3 and both were statistically different from treatments with a S:SD 0:1 and 3:1. Furthermore, the reactors with S:SD ratios 0:1 and 1:3 presented a significant difference between the initial and final fuel oil concentration ($p<0.005$). These results may suggest a lower biodegradation capability of residential microorganisms to biodegrade the fuel oils present in the contaminated soil and thus the importance of mixing both soil and sawdust, for a higher removal of fuel oils in soil. Additionally, diesel fuel initially present in the contaminated soil could be more recalcitrant than the fuel oil initially present in the contaminated sand. Treatment or S:SD ratio thus is the factor that more strongly affects the removal of fuel oil under in-vessel composting conditions of aged fuel oil contaminated mixtures of soil and sawdust. Soil to organic amendment ratios has proved to be necessary for in-vessel composting optimization (Antizar-Ladislao *et al.*

2006a). For an elevated operation yield (i.e. higher amount of contaminated mixture treated), a low S:SD ratio is advantageous, since a high S:SD ratio may show inhibitory effects on the residential microbial populations.

H. Regulatory and Sustainable Aspects of In-Vessel Composting as a Bioremediation Technology in Chile.

Environmental concern due to the presence of hazardous wastes is relatively new in Chile. Current Chilean legislation contemplates the application of bioremediation technologies for environmental decontamination (Ministry of Health 2004). Nevertheless, the existing legislation which is the only one that contemplates the integral management, manipulation and disposal of hazardous wastes, includes bioremediation technologies, such as composting as a valid alternative practice. At present there is not a coherent or clear guide to implement the existing normative. In this present year, 2007, a Ministry of Environment has been created in Chile with the mission of designing protocols to implement environmental policies. These protocols will include measures to treat contaminated wastes including contaminated soils, facilitating an appropriate application of the environmental policies among administrations, industry and citizens. Nowadays in Chile there is also a lack of technical expertise within the field of the application of bioremediation technologies. An undergoing constant discussion among administration, industry and citizens in Chile should lead to solutions to resolve environmental problems of current concern.

The feasibility study presented in this chapter provides evidence that the application of in-vessel composting as a bioremediation technology to treat hazardous wastes is a real treatment alternative to decontaminate wastes. The application of this bioremediation technology is encouraged by economic pressures since the cost of disposal of hazardous wastes in a HWL are high (more than US$100 per cubic meter of wastes) and by regulatory pressures over the mining industry, as they are enforced to present plans of closing and abandonment of the mine.

IV. Final Remarks

Whilst present research on the application of bioremediation technologies to fuel-contaminated desert soils is still scarce, the investigations presented in this chapter have proved that composting is a good sustainable environmental technology that may be used to remove organic contaminants from fuel-contaminated desert soils, sawdust and mixtures. This bioremediation technology has proved to be beneficial for the amelioration of contaminated soils by reducing the concentration of organic contaminants, which promotes soil sustainability and soil re-use in contrast to other treatment technologies, i.e., landfill or incineration approaches. Additionally, a number of cutting-edge biomonitoring tools can be used to assess the extent of physical and chemical environmental changes during the application of bioremediation technologies.

The coexistence of fungi and bacteria in soil, and their metabolic cooperation in different systems should be considered in future bioremediation technologies, mainly based on the fact that extensive colonization of fungal hyphae by bacteria may play an important role in bacterial movement and growth in soil substrates. Fungi thus increase the "bioavailability" of bacteria to pollutants. In addition, biocontrol fungi such as those belonging to the *Trichoderma* genera (resistant to different type of pollutants) could be used for multiple purposes: colonization of soils to provide a network for the mobilization of pollutant degrading bacteria used in bioremediation processes, and/or of biocontrol bacteria to prevent development of phytopathogens. Nevertheless, more studies are necessary to establish the role of concurring fungi during a bioremediation process, considering that bacterial communities and fungal microbiota can co-operate and participate synergistically.

The bioremediation of desert mining soils in the Atacama Desert was possible using aerated in-vessel composting under controlled conditions, where the heterotrophic microorganisms were capable of degrading organic compounds. This is in concordance with other authors who studied desert soils and Martian conditions. The removal curves obtained had a tendency towards stabilization over time but with different end points, showing the differential capacity of absorption of the different composts due to presence of sawdust. These differences in the final concentration can be explained by differences in the degree of absorption, microbial communities present, interactions between them, structural properties of the pollutants and controlled conditions of treatment. The differential end points may however also be due to N limitation, where the right balance of nutrients must be reached.

Two principal factors, time and the presence of sawdust, explain the performance of bioremediation. The sawdust could be allowing major levels of removal (absorption) and the time may perhaps show that adaptation of the microbial communities for bioremediation does not happen straight away. Biomonitoring tools used in this study proved to be useful and complementary. Nevertheless, it is suggested that the application of the COMPOCHIP microarray could result in a better resolution of microbial communities present in the samples. The presence of adapted microbial communities and accessibility to aged contaminants was essential for experimental treatment of the Atacama desert soils. In conclusion, this study has further raised the importance of support with experimental data to the Chilean Legislation and support of the application of cutting edge tools in the bioremediation of desert soils.

REFERENCES

Al-Awadhi N., Al-Daher R., El-Nawawy A., Balba M.T. (1996) Bioremediation of oil-contaminated soil in Kuwait .1. Landfarming to remediate oil-contaminated soil, *J. Soil Contamination* 5: 243-260.

Alcalde M., Ferrer M., Plou F.J., Ballesteros A. (2006) Environmental biocatalysis: from remediation with enzymes to novel green processes, *Trends Biotechnol.* 24:281-7.

Al-Daher R., Al-Awadhi N., El-Nawawy A. (1998) Bioremediation of damaged desert environment using the windrow soil pile system in Kuwait, *Environ. Int.* 24: 175-180.

Al-Daher R., Al-Awadhi N., Yateem A., Balba M.T., El-Nawawy A. (2001) Compost soil piles for treatment of oil-contaminated soil, *Soil Sed. Contamination* 10: 197-209.

Amann R.I., Ludwig W., Schleifer K.H. (1995) Phylogenetic identification and in situ detection of individual microbial cells without cultivation. *Microbiol. Rev.* 59: 143-69.

Antizar-Ladislao B., Lopez-Real J.M., Beck A.J. (2004) Bioremediation of polycyclic aromatic hydrocarbon (PAH)-contaminated waste using composting approaches, *Crit. Rev. Env. Sci. Tecnol.* 34: 249-289

Antizar-Ladislao B., Lopez-Real J., Beck A.J. (2005) Laboratory studies of the remediation of polycyclic aromatic hydrocarbon contaminated soil by in-vessel composting, *Waste Manag.* 25: 281-289.

Antizar-Ladislao B., Lopez-Real J., Beck A.J. (2006a) Degradation of polycyclic aromatic hydrocarbons (PAHs) in an aged coal-tar contaminated soil under in-vessel composting conditions, *Environ. Pollut.* 141: 459-468.

Antizar-Ladislao B., Lopez-Real J., Beck A.J. (2006b) Investigation of organic matter dynamics during in-vessel composting of an aged coal-tar contaminated soil using fluorescence excitation-emission spectroscopy, *Chemosphere* 64: 839-847.

Antizar-Ladislao B., Russell N.J. (2007) In-vessel composting as a sustainable bioremediation technology of contaminated soils and waste. In: *Environmental Research Trends*. Editor: Marvin A. Cato., Nova Science Publishers, Inc. New York.

Arora D. K., Gupta S. (1993) Effect of different environmental conditions on bacterial chemotaxis toward fungal spores, *Can. J. Microbiol.* 39: 922-931.

Baath E., Diaz-Ravina M., Frostegard A., Campbell C. (1998) Effect of metal-rich sludge amendments on the soil microbial community, *Appl. Environ. Microbiol.* 64: 238-245.

Bae J.W., Park Y.H. (2006) Homogeneous versus heterogeneous probes for microbial ecological microarrays, *Trends Biotechnol.* 24:318-323.

Bhatt M., Cajthaml T., Sasek V. (2002) Mycoremediation of PAH-contaminated soils, *Folia Microbiol (Praha)* 47: 255-258.

Balba M.T., Al-Daher R., Al-Awadhi N., Chino H., Tsuji H. (1998) Bioremediation of oil-contaminated desert soil: The Kuwaiti experience, *Environ. Int.* 24: 163-173.

Baltrian P., in Der Wiesche C., Gabriel J., Nerud F., Zadrazil F. (2000) Influence of cadmium and mercury on activities of ligninolytic enzymes and degradation of polycyclic aromatic hydrocarbons by *Pleurotus ostreatus* in soil, *Appl. Environ. Microbiol.* 66: 2471-2478.

Benítez T., Rincón A. M., Limón M. C., Codón A. C. (2004) Biocontrol mechanisms of *Trichoderma* strains, *Intl. Microbiol.* 7: 249-260.

Bodrossy L. (2003) Diagnostic oligonucleotide microarrays for microbiology. In: *A beginners guide to microarrays*. Pg 43-84. Editor: Blalock E. Kluwer Academic Publishers, Massachusetts USA.

Bogan B.W., Schoenike B., Lamar R.T., Cullen D. (1996a) Manganese peroxidase mRNA and enzyme activity levels during bioremediation of polycyclic aromatic hydrocarbon contaminated soil with *Phanerochaete chrysosporium*, *Appl. Environ. Microbiol.* 62: 2381-2386.

Bogan B.W., Schoenike B., Lamar R.T., Cullen D. (1996b) Expression of *lip* genes during growth in soil and oxidation of anthracene by *Phanerochaete chrysosporium*, *Appl. Environ. Microbiol.* 62: 3697-3703.

Bolta S.V., Mihelic R., Lobnik F., Lestan D. (2003) Microbial community structure during composting with and without mass inocula, *Compost Sci. Util.* 11: 6-15.

Boonchan S., Britz M.L., Stanley G.A. (2000) Degradation and mineralization of high-molecular-weight polycyclic aromatic hydrocarbons by defined fungal-bacterial cocultures, *Appl. Environ. Microbiol.* 66: 1007-1019.

Bordjiba O., Steiman R., Kadri M., Semadi A., Guiraud P. (2001) Removal of herbicides from liquid media by fungi isolated from contaminated soil, *J. Environ. Qual.* 30: 418-426.

Brazma A., Robinson A., Cameron G., Ashburner M. (2000) One-stop shop for microarray data. *Nature* 403: 699-700

Brim H., Heuer H., Krogerrecklenfort E., Mergeay M., Smalla K. (1999). Characterisation of the bacterial community of a zinc-polluted soil. *Can. J. Microbiol.* 45: 326-338.

Call D.R., Borucki M.K., Loge F.J. (2003) Detection of bacterial pathogens in environmental samples using DNA microarrays. *J. Microbiol. Meth.* 53: 235-243.

Canet R., Birnstingl J.G., Malcolm D.G., Lopez-Real J.M., Beck A.J. (2001) Biodegradation of polycyclic aromatic hydrocarbons (PAHs) by native microflora and combinations of white-rot fungi in a coal-tar contaminated soil, *Bioresour. Technol.* 76: 113-117.

Chávez-Gómez B., Quintero R., Esparza-García F., Mesta-Howard A. M., Zavala Díaz de la Cerna F. J., Hernández-Rodríguez C. H., Guillén T., Poggi-Varaldo H. M., Barrera-Cortés J., Rodríguez-Vásquez R. (2003) Removal of phenanthrene from soil by co-cultures of bacteria and fungi pregrown on sugarcane bagasse pith, *Bioresour. Technol.* 89: 177-183.

Chouari R., Le Paslier D., Daegelen P., Ginestet P., Weissenbach J., Sghir A. (2005) Novel predominant archaeal and bacterial groups revealed by molecular analysis of an anaerobic sludge digester. *Environ. Microbiol.* 7: 1104-1115.

Clements M. (2002) Studies of the Culturable Bacterial Diversity Along the North-south Moisture Gradient of the Atacama Desert. NASA. AMES. Second Astrobiology Science Conference NASA AMES Research Center - April 7 - 11, 2002. http://www.astrobiology.com/.

CLU-IN (2007) Technology Innovation Program, United States Environmental Protection Agency. http://www.clu-in.org/techfocus/

Cunningham S.D., Andersson T.A., Schwab A.P., Hsu F.C. (1996) Phytoremediation of soils contaminated with organic pollutants, *Adv. Agronomy* 56: 55-113.

Das K., Keener H.M. (1997) Numerical model for the dynamic simulation of a large scale composting system, *Transactions ASAE* 40: 1179-1189.

de Paula A.M., Soares A.R.F.S., Siquiera J.O. (2006) Biomassa, atividade microbiana e fungos micorrízicos em solo de "landfarming" de residuos petroquímicos, *R. Bras. Eng. Agric. Ambiental* 10: 448-455.

Dejonghe W., Boon N., Seghers D., Top E., Verstraete W. (2001). Bioaugmentation of soils by increasing microbial richness: missing links, *Environ. Microbiol.* 3: 649-657.

Delille D., Coulon F., Pelletier E. (2004) Effects of temperature warming during a bioremediation study of natural and nutrient-amended hydrocarbon-contaminated sub-Antarctic soils, *Cold Regions Sci. Technol.* 40: 61-70.

Drees K., Neilson J.V., Betancour J.L., Quade J., Henderson D.A., Pryor B.M., Maier R.M. (2006) Bacterial community structure in the hyper-arid core of the Atacama Desert, Chile, *Appl. Environ. Microbiol.* 72:7902-7018.

Dubois J.W., Hill S., England L.S., Edge T., Masson L., Trevors J.T., Brousseau R. (2004) The development of a DNA microarray-based assay for the characterization of commercially formulated microbial products, *J. Microbiol. Meth.* 58: 251-262.

Epstein E. (1997) *The Science Of Composting*, Technomic Publishing Company, Lancaster.

Esposito E., da Silva M. (1998) Systematics and environmental application of genus *Trichoderma*, *Crit. Rev. Microbiol.* 24: 89-98.

Felske A., Akkermans A.D.L., De Vos W.M. (1998). Quantitation of 16S rRNAs in complex bacterial communities by multiple competitive reverse transcription-PCR in temperature gradient gel electrophoresis. *Appl. Environ. Microbiol.*, 64: 4581-4587.

Finstein M.S., Miller F., Strom P.F. (1986) Waste treatment composting as a controlled system. In: H. Rehm, G. Reed (Eds.), *Biotechnology*, Verlag Chemie, Weinheim, Federal Republic of Germany, pp. 363-398.

Fisher S.G., Lerman L.S. (1979) Length-independent separation of DNA restriction fragments in two-dimensional gel electrophoresis, *Cell* 16: 191-200.

Franke-Whittle, I.H., Klammer S.H., Insam H. (2005) Design and application of an oligonucleotide microarray for the investigation of compost microbial communities, *J. Microbiol. Meth.* 62:37-56.

Franke-Whittle I.H., Knapp B.A., Fuchs J., Kaufmann R., Insam H. (2007) Changes in the bacterial community of different composts over time, investigated using the COMPOCHIP DNA microarray, (Submitted for publication)

Fraser B.S., Lau A.K. (2000) The effects of process control strategies on composting rate and odor emission, *Compost Sci. Util.* 8: 274-292.

Garland J.L., Mills A.L. (1991) Classification and characterisation of heterotrophic microbial communities on the basis of patterns of community level-sole-carbon-source-utilization, *Appl. Environ. Microbiol.* 57: 2351-2359.

Garland, J.L. (1996) Patterns of potential C source utilization by rhizosphere communities, *Soil Biol. Biochem.* 28: 223-230.

Godoy-Faúndez A., Antízar-Ladislao B., Reyes-Bozo L., Camaño A., Saéz-Navarrete C. (2007). Bioremediation of contaminated mixtures of desert mining soil and sawdust with fuel oils by aerated in-vessel composting in the Atacama Region (Chile), (Submitted for publication)

Gohre V., Paszkowski U. (2006) Contribution of the arbuscural mycorrhizal to heavy metal phytoremediation, *Planta* 223: 115-1122.

Grams G., Voigt K-D., Kirsche B. (1999) Degradation of polycyclic-aromatic hydrocarbons with three to seven aromatic rings by higher fungi in sterile and unsterile soils, *Biodegradation* 10: 51-62.

Green, S.J., Michel Jr. F.C., Hadar Y., Minz D. (2004) Similarity of bacterial communities in sawdust- and straw-amended cow manure composts, *FEMS Microbiol. Lett.* 233: 115-123.

Groh J.L., Luo Q., Ballard J.D., Krumholz L.R. (2005) A method adapting microarray technology for signature-tagged mutagenesis of Desulfovibrio desulfuricans G20 and Shewanella oneidensis MR-1 in anaerobic sediment survival experiments, *Appl. Environ. Microbiol.* 71:7064-7074.

Guerin T.F. (1999) Bioremediation of phenols and polycyclic aromatic hydrocarbons in creosote contaminated soil using ex-situ landtreatment, *J. Hazard. Materials*, 1999, 65: 305-315.

Guerin T.F. (2000) The differential removal of aged polycyclic aromatic hydrocarbons from soil during bioremediation, *Environ. Sci. Pollut. Res.*, 2000, 7: 19-26.

Gunderson C.A., Kostuk J.M., Gibbs M.H., Napolitano G.E., Wicker L.F., Richmond J.E., Stewart A.J. (1997) Multispecies toxicity assessment of compost produced in bioremediation of an explosives-contaminated sediment, *Environ. Toxicol. Chem.* 16: 2529-2537.

Guschin, D.Y., Mobarry, B.K., Proudnikov, D., Stahl, D.A., Rittmann, B.E., Mirzabekov, A.D. (1997) Oligonucleotide microchips as genosensors for determinative and environmental studies in microbiology. *Appl. Environ. Microbiol.* 63: 2397-2402.

Harman G.E., Howell C.R., Viterbo A., Chet I., Lorito M. (2004) *Trichoderma* species – opportunistic, avirulent plant symbionts, *Nature Reviews* 2: 43-56.

Haruta S., Kondo M., Nakamura K., Chanchitpricha C., Aiba H., Ishii M., Igarashi Y. (2004) Succession of a microbial community during stable operation of a semi-continuous garbage-decomposing system, *J. Biosci. Bioeng.* 98:20-27.

Haug R.T. (1993) *The Practical Handbook of Compost Engineering*, Lewis Publishers, CRC Press, Boca Raton, Florida.

Heijs S.K., Aloisi G., Bouloubassi I., Pancost R.D., Pierre C., Sinninghe Damsté J.S., Gottschal J.C., Van Elsas J.D., Forney L.J. (2006) Microbial community structure in three Deep-Sea carbonate crusts. *Microbial Ecol.* 52: 451-462.

Jjemba P.K., Kinkle B.K., Shann J.R. (2006) In-situ enumeration and probing of pyrene-degrading soil bacteria, *FEMS Microbiol. Ecol.* 55: 287-298.

Kaeberlein T., Lewis K., Epstein S.S. (2002) Isolating "uncultivable" microorganisms in pure culture in a simulated natural environment, *Science* 296:1127-1129.

Karetnikova E.A., Zhircova A.D. (2005) Degradation of phenols formed during lignin pyrolysis by microfungi of genera *Trichoderma* and *Penicillium*, *Izv. Akad. Nauk. Ser. Biol.* 5: 539-544.

Kästner M., Mahro B. (1996) Microbial degradation of polycyclic aromatic hydrocarbons in soils affected by the organic matrix of compost, *Appl. Microbiol. Biotechnol.* 44: 668 – 675.

Kechavarzi C., Pettersson K., Leeds-Harrison P., Ritchie L., Ledin S. (2007) Root establishment of perennial ryegrass (*L-perenne*) in diesel contaminated subsurface soil layers, *Environ. Pollut.* 145: 68-74.

Kirk T.K., Lamar R.T., Glaser J.A. (1992) The potential of white-rot fungi in bioremediation. In: S. Mongkolsuk, P.S. Lovett, J.E. Trempy (Eds.) *Biotechnology and environmental science-molecular approaches*. New York, Plenum Press. Pp. 131 – 138.

Kohlmeier S., Smits T.H.M., Ford R.M., Keel C., Harms H., Wick L.Y. (2005) Taking the fungal highway: mobilization of pollutant-degrading bacteria by fungi, *Environ. Sci. Technol.* 39: 4640-4646.

Kuiper I., Bloomberg G.V., Lungtenberg B.J.J. (2001) Selection of a plant-bacterium pair as a novel tool for rhizostimulation of polycyclic aromatic hydrocarbon-degrading, *Mol. Plant-Microbe Interact.* 14: 1197-1205.

Laine M.M., Jorgensen K.S. (1997) Effective and safe composting of chlorophenol-contaminated soil in pilot scale, *Environ. Sci. Technol.* 31: 371-378.

Leben C. (1984) Spread of plant pathogenic bacteria with fungal hyphae, *Phytopathology* 74: 983-986.

Liang C., Das K.C., McClendon R.W. (2003) The influence of temperature and moisture contents regimes on the aerobic microbial activity of a biosolids composting blend, *Bioresour. Technol.* 86: 131-137.

Loehr R.C., Smith A., Corsi R.L. (2001), VOC and SVOC emissions from slurry and solid phase bioremediation processes, *Pract. Periodical of Haz., Toxic, and Radioactive Waste Mgmt.* 5: 211-224.

Loy A., Lehner A., Lee N., Adamczyk J., Meier H., Ernst J., Schleifer K.H., Wagner. M. (2002) Oligonucleotide microarray for 16S rRNA gene-based detection of all recognized lineages of sulfate-reducing prokaryotes in the environment, *Appl. Environ. Microbiol.* 68: 5064-5081.

Loy A., Bodrossy L. (2005) Highly parallel microbial diagnostics using oligonucleotide microarrays, *Clinica Chimica Acta* 363: 106-119.

Lundstedt S., Persson Y., Oberg L. (2006) Transformation of PAHs during ethanol-Fenton treatment of an aged gasworks' soil, *Chemosphere* 65: 1288-1294.

Macek T., Mackova M., Kas J. (2000) Exploitation of plants for the removal of organics in environmental remediation, *Biotechnol. Adv.* 18: 23-34.

Mader B.T., UweGoss K., Eisenreich S.J. (1997) Sorption of nonionic, hydrophobic organic chemicals to mineral surfaces, *Environ. Sci. Technol.* 31: 1079-1086.

Marchal R., Penet S., Solano-Serena F., Vandecasteele J.P. (2003) Gasoline and diesel oil biodegradation, *Oil Gas Sci. Technol.* 58: 441-448.

Martín-Moreno C., González-Becerra A., Blanco-Santos M. J. (2004) Tratamiento biológico de suelos contaminados: contaminación con hidrocarburos. Aplicaciones de hongos en tratamientos de biorrecuperación, *Rev. Iberoam. Micol.* 21: 103-120.

Maskos U., Southern E.M. (1992) Parallel analysis of oligodeoxyribonucleotide (oligonucleotide) interactions. I. Analysis of factors influencing oligonucleotide duplex formation, *Nucleic Acid Res.* 20: 1675-1678.

Michel F.C., Quensen J., Reddu C.A. (2001) Bioremediation of a PCB-contaminated soil via composting, *Compost Sci Util.* 9: 274-284.

Mihial D., Viraraghavan T., Jin Y. (2006) Bioremediation of Petroleum-Contaminated Soil Using Composting, *Practice Periodical Hazard. Toxic Radioactive Waste Manag.* 10:108-115

Miller A. (1976) *Climates of Central and South America*, W. Schwerdtfeger, Ed. (Elsevier, Amsterdam), pp.113–145.

Minera Escondida (2007) *http://www.escondida.cl/Escondida/ingles/index.html.*

Ministry of Health (2004) Decreto Supremo No. 148: Reglamento Sanitario Sobre Manejo de Residuos Peligrosos (Sanitari normative on handling hazardous wastes), Santiago, República de Chile.

Montealegre J., Herrera R., Velásquez J. C., Silva P., Besoaín X., Pérez L. M. (2005) Biocontrol of root and crown rot in tomatoes under greenhouse conditions using Trichoderma harzianum and Paenebacillus lentimorbus. Additional effect of solarizaton, *Electronic Journal of Biotechnology* 8: Issue of December 15, 2005. *http://www.ejbiotechnology.info/content/vol8/issue3/full/7/*

Moyer C.L., Tiedje J.M., Dobbs F.C., Karl D.M. (1996) A computer-simulated restriction analysis of bacterial small-subunit rRNA genes: efficacy of selected tetrametric restriction enzymes for studies of microbial diversity in nature. *Appl. Environ. Microbiol.* 62: 2502-2507.

Mrayyan B., Battikhi M.N. (2005) Biodegradation of total organic carbons (TOC) in Jordanian petroleum sludge, *J. Hazard. Mater.* 120: 127-134.

Navarro-González R., Rainey F., Molina P., Bagaley D., Hollen B., de la Rosa J., Small A., Quinn R., Grunthaner F., Cáceres L., Gomez-Silva B., McKays C. (2003). Mars-Like Soils in the Atacama Desert, Chile, and the Dry Limit of Microbial Life, *Science* 302: 1018-1021.

Negri M., Manfredini A., Saponaro S., Sorlini C., Bonomo L., Valle A., Zanardini E. (2004) Solid phase treatment of an aged soil contaminated by polycyclic aromatic hydrocarbons, *J. Environ. Sci. Health A – Toxic/Hazard. Substances Env. Eng.* 39: 1-17.

Novotny C., Erbanova P., Sasek V., Kubatova A., Cajthaml T., Lang E., Krahl J., Zadrazil F. (1999) Extracellular oxidative enzyme production and PAH removal in soil by exploratory mycelium of white rot fungi, *Biodegradation* 10: 159-168.

Paszczynski A., Crawford R. L. (1995) Potential for bioremediation of xenobiotic compounds by white-rot fungus *Phanerochaete chrysosporium, Biotech. Prog.* 11: 368 – 379.

Peplies, J., Chun Kwan Lau S., Pernthaler J., Amann R., Glockner F. O. (2004) Application and validation of DNA microarrays for the 16S rRNA-based analysis of marine bacterioplankton. *Environ. Microbiol.* 6: 638-645.

Pérez L.M., Besoaín X., Reyes M., Pardo G., Montealegre J. (2002) The expression of extracellular fungal cell wall hydrolytic enzymes in different *Trichoderma harzianum* isolates correlates with their ability to control *Pyrenochaeta lycopersici*, *Biological Res.* 35: 401-410.

Pfender, W.F., Fieland V.P., Ganio L.M., Seidler R.J. (1996) Microbial community structure and activity in wheat straw after inoculation with biological control organisms, *Appl. Soil Ecol.* 3: 69-78.

Potin O., Veignie E., Rafin C. (2004) Biodegradation of polycyclic aromatic hydrocarbons (PAHs) by *Cladosporium sphaerospermum* isolated from an aged contaminated soil, *FEMS Microbiol. Ecol.* 51: 71-78.

Ram R.J., Verberkmoes N.C., Thelen M.P., Tyson G.W., Baker B.J., Blake R.C. 2nd, Shah M., Hettich R.L., Banfield J.F. (2005) Community proteomics of a natural microbial biofilm, *Science* 308:1915-1920.

Rhee S.K., Liu X., Wu L., Chong S.C., Wan X., Zhou J. (2004) Detection of genes involved in biodegradation and biotransformation in microbial communities by using 50-mer oligonucleotide microarrays, *Appl. Environ. Microbiol.* 70:4303-4317.

Rhykerd R.L., Weaver R.W., McInnes K.J. (1995) Influence of salinity on bioremediation of oil in soil, *Environ. Pollut.* 90: 127-130.

Rigot J., Matsumura F. (2002) Assessment of the rhizosphere competency and pentachlorophenol-metabolizing activity of pesticide-degrading strain of Trichoderma harzianum introduced into the root zone of corn seedlings, *J. Environ. Sci. Health B.* 37: 201-210.

Ros M., Goberna M., Moreno J.L., Hernandez T., García C., Insam H., Pascual J.A. (2006) Molecular and physiological bacterial diversity of a semi-arid soil contaminated with different levels of formulated atrazine, *Appl. Soil Ecol.* 34: 93-102.

Rudi K., Flateland S.L., Hanssen J.F., Bengtsson G., Nissen H. (2002) Development and evaluation of a 16S Ribosomal DNA array-based approach for describing complex microbial communities in ready-to-eat vegetable salads packed in a modified atmosphere, *Appl. Environ. Microbiol.* 68, 1146-1156.

Sadzawka A., Grez R., Mora M., Saavedra N., Carrasco M., Rojas C. (2000) *Métodos De Análisis Recomendados Para Los Suelos Chilenos*, Chilean Normative for soils. http://www.sag.cl.Chile.

Santos S.R., Ochman H. (2004) Identification and phylogenetic sorting of bacterial lineages with universally conserved genes and proteins, *Environ. Microbiol.* 6:754-759.

Sasek V., Bhatt M., Cajthaml T., Malachova K., Lednicka D. (2003) Compost-mediated removal of polycyclic aromatic hydrocarbons from contaminated soil, *Arc. Environ. Contam. Toxicol.* 44: 336-342.

Seymour R.M., Donahue D., Bourdon M., Evans J.R., Wentworth D. (2001) Intermittent aeration for in-vessel composting of crab processing waste, *Compost Sci Util.* 9: 98-106.

Semple K.T., Reid B.J., Fermor T.R. (2001) Impact of composting strategies on the treatment of soils contaminated with organic pollutants, *Environ. Pollut.* 112: 269-283

Semple K.T., Morriss A.W.J., Paton G.I. (2003) Bioavailability of hydrophobic organic contaminants in soils: fundamental concepts and techniques for analysis, *Eur. J. Soil Sci.* 564: 1-10

Semple K., Doick K., Jones K., Burauel P., Craven A, Harms H. (2004) Defining bioavailability and bioaccessibility of contaminated soil and sediments, *Environ. Sci. Technol.* 15: 229A-231A.

Sharma S., Radl V., Hai B., Kloos K., Fuka M.M., Engel M., Schauss K., Schloter M. (2007) Quantification of functional genes from procaryotes in soil by PCR, J. Microbiol. Methods 68:445-452.

Singer A.C., Jury W., Luepromchai E., Yahng C. S., Crowley D. E. (2001) Contribution of earthworms to PCB bioremediation, *Soil Biol. Biochem.* 33: 765-776.

Small J., Call D.R., Brockman F.J., Straub T.M., Chandler D.P. (2001) Direct detection of 16S rRNA in soil extracts by using oligonucleotide microarrays. *Appl. Environ. Microbiol.* 67: 4708-4716.

Smith J.J., Tow L.A., Stafford W., Cary C., Cowan D.A. (2006) Baterial diversity in there different Antarctic cold desert mineral soils. *Microbial Ecol.* 51: 413-421.

Southern E.M., Mir K., Shchepinov M. (1999) Molecular interactions on microarrays, *Nature Genetics Supplement* 21:5-9.

Stach J.E.M., Bathe S., Clapp J.P., Burns R.G. (2001) PCR-SSCP comparison of 16S rDNA sequence diversity in soil DNA obtained using different isolation and purification methods, *FEMS Microbiol. Ecol.* 36: 139-151.

Steffen K.T., Schubert S., Tuomela M., Hatakka A., Hofrichter M. (2007) Enhancement of bioconversion of high-molecular mass polycyclic aromatic hydrocarbons in contaminated non-sterile soil by litter-decomposing fungi, *Biodegradation* 18: 359-369.

Stoker C.R., M. Bullock A. (1997) Organic degradation under simulated Martian conditions, *J. Geophys. Res.*, *102*: 10,881-10,888.

Stralis-Pavese N., Sessitsch A., Weilharter A., Reichenauer T., Riesing J., Csontos J., Murrell J. C., Bodrossy L. (2004) Optimization of diagnostic microarray for application in analysing landfill methanotroph communities under different plant covers, *Environ. Microbiol.* 6: 347-363.

Thompson L., Troeh F. (1978) *Soils and soil fertility*, 4th Edition, McGraw-Hill, New York.

Tiquia S.M. (2005) Microbial community dynamics in manure composts based on 16S and 18S RDNA T-RFLP profiles, *Environ. Technol.* 26: 1101-1113.

Tringe S.G., von Mering C., Kobayashi A., Salamov A.A., Chen K., Chang H.W., Podar M., Short J.M., Mathur E.J., Detter J.C., Bork P., Hugenholtz P., Rubin E.M. (2005) Comparative metagenomics of microbial communities, *Science* 308: 554-557.

Uchiyama T., Abe T., Ikemura T., Watanabe K. (2005) Substrate-induced gene-expression screening of environmental metagenome libraries for isolation of catabolic genes, *Nat. Biotechnol.* 23:88-93.

USDA (1999) Soil taxonomy, a basic system of soil classification for making and interpreting soil surveys, U.S. Government Printing Office.

USEPA (1986) U.S. Environmental Protection Agency, *Test Method for Evaluating Solid Waste*, SW-846, 3th Edition, USEPA, Washington, DC.

USEPA (2002) U.S. Environmental Protection Agency, *Application, Performance, and costs of biotreatment technologies for contaminated soils* EPA/600/R-03/037, USEPA, Washington, DC.

USEPA (2003) U.S. Environmental Protection Agency, *Guide for Conducting Treatability Studies Under CERCLA: Biodegradation Remedy Selection* EPA/540/R-93/519b. USEPA, Washington, DC.

VanderGheynst J.S., Lei F. (2003) Microbial community structure dynamics during aerated and mixed composting, *Transactions American Society Agricultural Engineers* 46: 577-584.

Vianna M.E., Horz H.P., Gomes B.P.F.A., Conrads G. (2005) Microarrays complement culture methods for identification of bacteria in endodontic infections, *Oral Microbiol. Immunol.* 20: 253-258

von Wintzingerode F., Gobel U.B., Stackebrandt E. (1997) Determination of microbial diversity in environmental samples: Pitfalls of PCR-based rRNA analysis. *FEMS Microbiol. Rev.* 21: 213-229.

Wang, R.-F., Beggs M.L., Robertson L.H., Cerniglia C. E. (2002) Design and evaluation of oligonucleotide-microarray method for the detection of human intestinal bacteria in fecal samples, *FEMS Microbiol. Lett.* 213: 175-182.

Wessels J.G.H. (1997) Hydrophobins, proteins that change the nature of the fungal surface, *Adv. Microbiol. Physiol.* 38: 1-45.

Watanabe K., Hino S. (1996) Identification of a functionally important population in phenol-digesting activated sludge with antisera raised against isolated bacterial strains, *Appl. Environ. Microbiol.* 62:3901-3904

Wick L.Y., Remer R., Würz B., Reichenbach J., Braun S., Schäfer F., Harms H. (2007) Effect of fungal hyphae on the access of bacteria to phenanthrene in soil, *Environ. Sci. Technol.* 41: 500-505.

Woese C.R. (1987) Bacterial evolution, *Microbiol. Rev.* 51: 221-271.

Wong P.T.G., Griffin D.M. (1976) Bacterial movement at high matric potentials-II. In fungal colonies, *Soil Biol. Biochem.* 8: 219-223.

Wösten H.A.B., van Wetter M.A., Lugones I.G., van der Mei H.C., Busscher H.J., Wessels J.G.H. (1999) How a fungus escapes the water to grow into the air, *Current Biol.* 9: 85-88.

Zak J.C., Willig M.R., Moorhead D.L., Wildman H.G. (1994) Functional diversity of microbial communities: a quantitative approach, *Soil Biol. Biochem.* 26: 1101-1108.

Zhou J. (2003) Microarrays for bacterial detection and microbial community analysis, *Curr. Opin. Microbiol.* 6: 288-294.

In: Leading-Edge Environmental Biodegradation
Editor: L. E. Pawley, pp. 131-166
ISBN: 978-1-60021-903-0

Chapter 4

Recent Advancements in Microbial Biodegradation of Dioxins

K. Murugesan and Yoon-Seok Chang*
School of Environmental Science and Engineering,
POSTECH, San 31, Hyoja-dong, Nam-gu, Pohang, 790-784, Korea

Abstract

Polychlorinated dibenzo-*p*-dioxins (PCDDs) and polychlorinated dibenzofurans (PCDFs), commonly known as dioxins (PCDD/Fs), are toxic environmental pollutants, unintentionally formed from various sources. Due to their persistent and ubiquitous natures elimination of these pollutants from the environment is a difficult task. Removal of dioxins by biological degradation (biodegradation) is considered a reasonable method as an alternative to other expensive physico-chemical approaches. Several bacteria utilizing dioxins as sole carbon energy sources have been isolated from the environment. Biodegradation of dioxins has been extensively studied in several microorganisms, and details concerning biodiversity, biodegradation, biochemistry and molecular biology of this process have accumulated during the last three decades. There are several microbial mechanisms responsible for biodegradation of dioxins, including oxidative degradation by dioxygenase-containing aerobic bacteria, bacterial and fungal cytochrome P-450, fungal lignolytic enzymes, reductive dechlorination by anaerobic bacteria, and direct ether ring cleavage by fungi containing etherase-like enzymes. Many attempts have been made to bioremediate PCDD/Fs using this basic knowledge of microbial dioxin degradation. This review highlights the current knowledge and recent advancements in the microbial biotransformation, biodegradation, biosorption, and bioremediation of dioxins.

* E-mail: yschang@postech.ac.kr; Phone: +82-54-279-2281; Fax: +82-54-279-8299.

Abbreviations for Dioxin and Related Compounds

DD:	dibenzo-*p*-dioxin
DF:	dibenzofuran
CAR:	carbazole
FN:	fluorine
DBT:	dibenzothiophene
PCB:	polychlorinated biphenyl
MCDD/Fs:	monochloro-DDs and DFs
DCDD/Fs:	dichloro-DDs and DFs
TriCDD/Fs:	trichloro-DDs and DFs
TCDD/Fs:	tetrachloro-DDs and DFs
PeCDD/Fs:	pentachloro-DDs and DFs
HxCDD/Fs:	hexachloro-DDs and DFs
HeCDD/Fs:	heptachloro-DDs and DFs
OCDD/F:	octachloro-DD and DF
PCDD/Fs	Polychlorinated dibenzo-*p*-dioxins and Polychlorinated dibenzofurans
Lip	Lignin peroxidase
MnP	Manganese peroxidase
PAHs	Polycyclic aromatic hydrocarbons
BPDO	Biphenyl dioxygenase

Introduction

Polychlorinated dibenzo-*p*-dioxins (PCDDs) and dibenzofurans (PCDFs) belong to the class of two-ring aromatic compounds containing one to eight chlorine atoms (Figure 1). PCDDs contain 75 congeners, and PCDFs contain 135 congeners. The formation of large numbers of halogenated 1550-mixed PCDDs and 3050-mixed PCDFs are possible and are typically formed as undesired by-products of various processes [1,2]. PCDD/Fs are formed as a result of incineration processes and as by-products of chemical reactions such as chlorine bleaching of paper and pulp, and from the manufacture of some chlorinated pesticides, herbicides, fungicides, wood preservatives, and textile dyes [3-7]. From here on the above PCDDs and PCDFs are referred as dioxins, which can be formed as a result of some natural processes, such as volcanic activity and forest fires [8,9]. PCDD/Fs are among the most problematic environmental pollutants because of their chemical inertness and recalcitrance, lipid solubility, and toxicity [10,11]. The primary human health concerns associated with PCDD/Fs are immunotoxicity [12,13], developmental toxicity [14], neurotoxicity [15], carcinogenicity [16], and skin, liver, and gastrointestinal toxicity [17]. Large amounts of highly toxic 2,3,7,8-tetachlorodibenzo-*p*-dioxin (2,3,7,8-TCDD) and other PCDD/F congeners entered into the environment as a result of events such as the Vietnam War, and the Seveso industrial accident [18-20]. Research following these industrial disasters indicated that PCDD/Fs are not only extremely toxic but they can also be formed by incineration of organic compounds in the presence of chlorine. This research also indicated that the use of agrochemicals has likely caused widespread pollution of agricultural and terrestrial environments with PCDD/Fs [21]. In addition, natural formation of PCDD/Fs in soil sediments has been observed and is likely caused by polymerization of chlorophenol into dioxin in the presence of fungal peroxidase [4, 22].

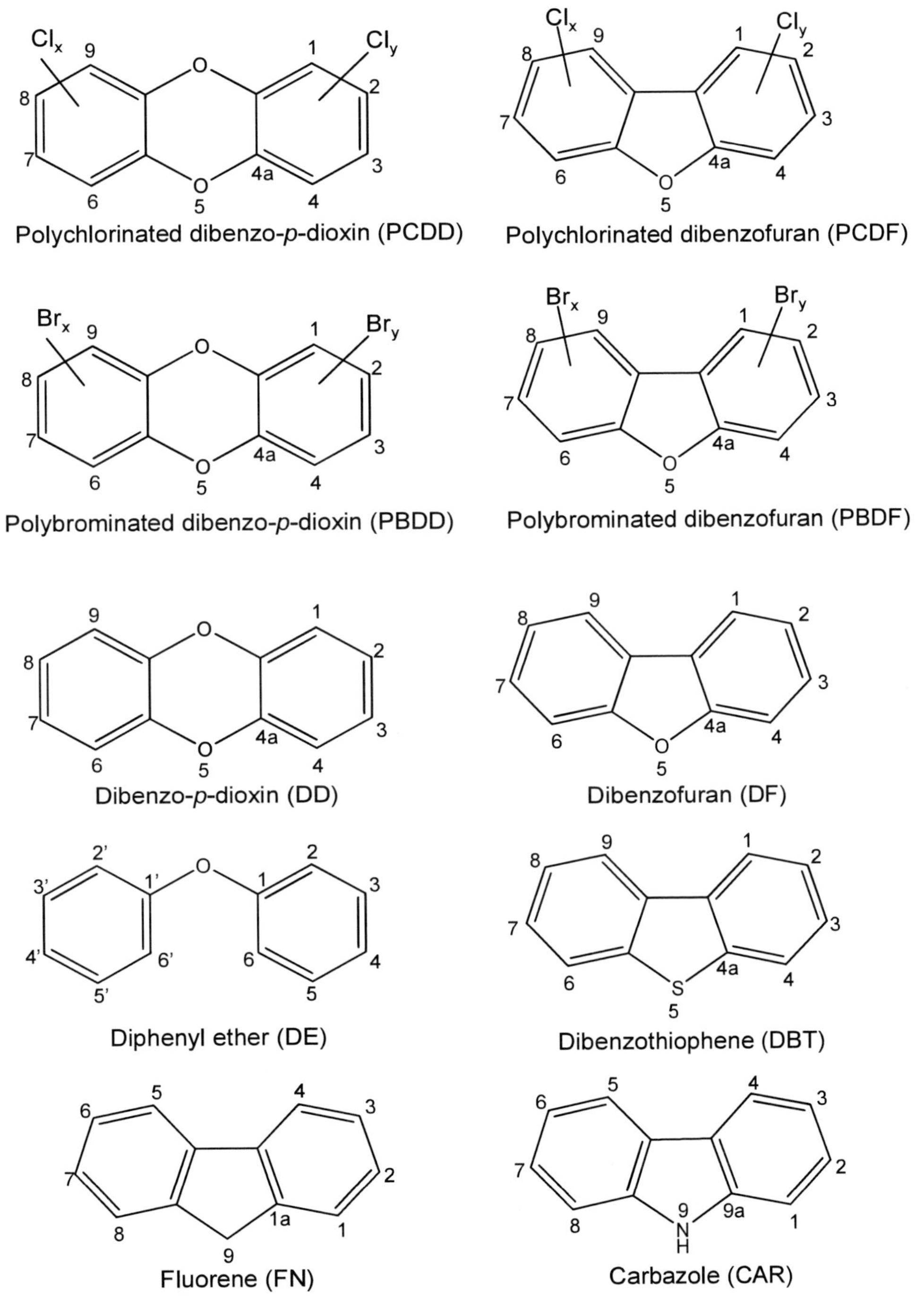

Figure 1. Chemical structure of dioxins and related compounds.

Clean-up of sites contaminated with these pollutants has been difficult due to their toxicity and long half-lives. However, many efforts have been made to develop efficient remediation technologies which remove, neutralize and/or decompose such compounds from the polluted sites. These remediation technologies can be classified into the following three types: biological (bioremediation), physical/chemical (physico-chemical) and thermal. Several effective physico-chemical methods have been developed to remove dioxins from the environment, but their successful application is currently not economically feasible. Efforts to use natural processes to eliminate this problem have been investigated, as microorganisms naturally destroy organic compounds and reintroduce them into the global carbon cycle. Using defined microorganisms or microbial consortia with the ability to degrade dioxins to eliminate pollution in the environment could have vast advantages over physico-chemical methods. Investigation into the microbial degradation of 2,3,7,8-TCDD was initiated by Kearny et al., Matsumura and Benezet, and Ward and Matsumura [23-25], and subsequent studies have defined the various aspects of dioxin degradation by several microorganisms. Biodegradation of dioxin has been reviewed periodically [26-31]. Dioxin degradation has been characterized in aerobic bacteria that contain aromatic ring hydroxylating dioxygenase enzymes. Dioxins are also degraded by anaerobic microorganisms through reductive dechlorination under anaerobic conditions. Dioxin degradation has also been characterized in fungi, where several oxidative mechanisms are responsible for dioxin degradation and employ cytochrome P-450, lignin peroxidase (Lip), manganese peroxidase (MnP), and laccase. Recently, mechanisms involved in the direct cleavage of ether bonds in dioxins have been reported in fungi. Several studies have described the bioremediation of dioxins in many different model systems. However, successful bioremediation techniques for field-scale application are still in their infancy. Considering the advantages and disadvantages of different methods of dioxin degradation, hybrid or sequential technologies such as the combination of nano-chemicals and biological methods could eliminate this bottleneck. This review focuses on recent developments in microbial degradation and biotransformation of dioxins.

Bacterial Degradation of Dioxins

Biodegradation of Non-chlorinated Dioxins

Generally, dioxin biodegradation studies use DD and DF as models for chlorinated DDs and DFs, which are of greater environmental concern (Figure 1). Under aerobic conditions, microorganisms degrade DD, DF, and their analogs through oxidative degradation by hydroxylating the aromatic rings and cleaving the aromatics structures of the compound. Table 1 shows the list of bacterial strains capable of degrading dioxins and related compounds. Initial dioxygenation of the aromatic ring occurs on either lateral positions (1,2 or 2,3 or 3,4 carbon atoms) or angular 4,4a positions. Lateral dioxygenation is similar to the well-characterized degradation pathways of naphthalene, biphenyl and other polycyclic aromatic hydrocarbons (PAHs) compounds which produce yellow-colored intermediate metabolites. On the other hand, angular dioxygenation takes place at 4,4a carbon atoms adjacent to an ether bridge, and is an attractive degradation mechanism because it ultimately results in the destruction of the planar structures from which the toxicity of dioxins is derived [27,30,32,33]

Dioxin Degradation by Lateral Dioxygenation

Pioneer works on dioxin degradation revealed that DD, DF and their chlorinated congeners were transformed to *cis*-dihydroxylated compounds in the initial step of degradation by naphthalene- and biphenyl-degrading *Pseudomonas* and *Beijerinckia* sp. strains (now reclassified as *Pseudomonas putida* and *Sphingobium yanoikuyae*, respectively) [34-36]. Bacterial degradation of these compounds was incomplete and produced dead-end products. These bacteria oxygenated the compounds at lateral carbons in positions 1,2 and 2,3 with high regioselectivity. The lateral dioxygenation reaction produced laterally-hydroxylated DF, which was further cleaved by *meta*-cleavage and produced polar yellow metabolites that had an absorption maximum around 460 nm at pH 7.0. Selifonov et al. [37] reported aromatic ring cleavage of DF using naphthalene-grown strains of *Pseudomonas fluorescence* and *Pseudomonas putida*. They identified a yellow metabolite as the enol of 2-oxo-4-[2'-(3'-hydroxy)benzofuranyl]-but-3-enoic acid. In another study, cometabolic degradation of DF by biphenyl-grown cells of *Ralstonia* (*Wautersia*) sp. strain SBUG 290 cleaved 1,2-dihydroxy-DF into 2-hydroxy-4-(3'-oxo-3'-H-benzofuran-2-yliden)-but-2-enoic acid (HOBB) [38]. The purified metabolites showed pH dependence (pH 3.0 to 7) with an absorbance maximum from 400 to 462 nm. This property was due to keto-enol tautomerism of *meta*-cleavage products observed in biphenyl [39]. Stope et al. [40] made a similar observation showing that 17 gram-positive biphenyl degrading isolates cometabolically transformed DF into 3'-HOBB during growth on biphenyl.

Recently, several bacteria have been identified showing both lateral and angular dioxygenation of dioxins (Table 1). The marine bacterium *Cycloclasticus pugetii* [41] grown on DF as substrate in the presence of yeast extract performed lateral dioxygenation similar to *Burkholderia cepacia* F297, and angular dioxygenation similar to *Sphingomonas* sp. HH69 [42]. The DF utilizing bacterium *Comamonas* sp. strain KD7, isolated from white clover roots, also showed similar activity [43]. The gram positive bacterium *Rhodococcus opacus* SAO101 also showed both lateral and angular dioxygenation; however, no additional ring cleavage products were detected in addition to that produced from the angular dioxygenation reaction [44]. Strain SAO101 has a high cooxidative potential for chlorinated DDs, and it utilizes many monocyclic and bicyclic aromatic compounds as its sole carbon and energy sources [45]. Kimura et al. [46] identified the genes (*narAaAb*) responsible for lateral dioxygenation of DF, and found that they share high homology with the naphthalene dioxygenase of *Rhodococcus* stain NCIMB 12038. Products of these genes (NarAaAb) are involved in the degradation pathway of many aromatic compounds.

Table 1. Bacteria capable of degrading nonchlorinated and chlorinated dioxins

Bacteria	Substrate degraded	Dioxygenation mode	References
Brevibacterium sp DPO1361	DF, FN	Angular	52
Burkholderia xenovorans LB400	DD, DF	L/A	47
Comamonas sp. strain KD7	DF	L/A	43
Comamonas testosteroni	DD, DF	L/A	47
Cycloclasticus pugetii	DF	L/A	41
Janibacter sp YA	DF,MCDDs	Angular	57
Janibacter sp YY1	FN, DE, DD, CAR	L/A	59,60
Janibacter sp. DAF3	DF	Angular	55
Microbacterium sp. YK18	DF	NA	91
Nocardioides aromaticivorans IC177	DF, DD, CAR	Angular	61,62
Paenibacillus sp. YK5	DF	Angular	63
Porphyrobacter sanguineus IAM 12620	DF	Lateral	49
Pseudomonas putida BS291	DF	Lateral	37
Pseudomonas putida G7(=NCIB 9816)	DF, FN, DBT	Lateral	34
Pseudomonas putida PH-01	DF	Angular	32
Pseudomonas resinovorans CA10	CAR, DD, DF	Angular	65-71
Pseudomonas veronii PH-03	DF, DD, MCDDs	Angular	54
Ralstonia (*Waustersia*) *eutropha* DSM 5536	MCDD/F, DCDD/F, TrCDD/F	NA	82
Ralstonia (*Waustersia*) sp. DSM 6708	MCDD/Fs, DCDD/Fs, TrCDD/Fs, TCDD/Fs	NA	82
Rhodococcus opacus SAO101	DF, DD, MCDD, DCDD	L/A	44-46
Sphingobium aromaticiconvertens	DF, MCDFs	Angular	81
Sphingobium yanoikuyae B1(=DSM 6900)	DD, DF, MCDD/Fs	Lateral	34-36, 49
Sphingomonas sp HH69 (=DSM 7135)	DD,DF, MCDD/Fs,DCDD/Fs, TrCDD/Fs,TCDD/Fs	Angular	42,79
Sphingomonas wittichii RW1	DD, DF, 9-FN, MCDD/Fs, DCDD/Fs, TrCDD/Fs, TCDD, PeCDD, HxCDD	Angular	26,27,33,72-77, 84-88, 149,150,153
Terrabacter (*Janibacter*) sp. DBF63	DF, DD, FN	Angular	68, 89, 90, 152,
Terrabacter (*Janibacter*) sp. DPO1361	DF	Angular	52
Terrabacter (*Janibacter*) sp. YK3	DF	Angular	91

L'Abbee et al. [47] comparatively studied the metabolism of DF and DD by the biphenyl dioxygenase (BPDO) of *Comamonas testosteroni* strain B-356 and *Burkholderia xenovorans* strain LB400. Both BPDOs oxygenated DF at a low rate, but the BPDO of clone LB400 transformed DF faster than the BPDO of strain B-356. Both BPDOs produced *cis*-dihydro-dihydroxy dibenzofurans as major metabolites during the course of the lateral dioxygenation reaction. These BPDOs also produced minor amounts of 2,2',3-trihydroxybiphenyl and 2,2',3-trihydroxydiphenyl ether from DF and DD, respectively, by angular dioxygenation. Kubota et al. [48] isolated seven novel gram-positive bacterial strains from dioxin-polluted sites. These strains grew by using DF as the sole carbon and energy source. They referred to these novel DF-degrading strains as the new species *Nocardioides aromaticivorans* (the type strain is strain H1; IAM 14992; JCM 11674; DSM15131). These strains produced soluble yellow metabolites which showed a unique pH-dependent change in absorption maximum from 400 to 507 nm, suggesting that the degradation pathway of DF in these isolates is different from typical lateral dioxygenation.

Several reports have indicated that lateral dioxygenation of DF and its analogs is a cometabolic process during growth with naphthalene or biphenyl as the primary carbon and energy source. No bacteria able to grow exclusively on DF by lateral dioxygenation have been reported except for those identified by Hiraishi et al. [49]. They found that the aerobic bacteriochlorophyll-containing bacterium *Porphyrobacter sanguineus* strain IAM 1260 and the *Sphingobium yanoikuyae* strain DSM 6900 (strain B1) were capable of growing with DF as the sole carbon and energy source, and that they produced yellow intermediates. This reaction was thought to be unproductive since lateral dioxygenation generally results in dead-end products [27]. For this reason lateral dioxygenation is not the most desirable method for eliminating DF, but the fact that several enzymes affect this reaction warrants further study of the regioselectivity and the degradation of any by-products of reactions mediated by these enzymes. Recently, Mohammadi and Sylvestre [50] developed an engineered biphenyl dioxygenase termed RR41, which evolved from strain *Burkholderia xenovorans* LB400. It attacked DF and chlorinated DF in the lateral position with a higher turnover rate than its parental strain. Initial dioxygenation and a second dioxygenation reaction produced ring cleavage products. All metabolites were unambiguously identified by ^{1}H-NMR. This study illustrates a nice example of how to detour a catabolic pathway in order to avoid dead-end products when chlorinated compounds are used [51].

Dioxin Degradation by Angular Dioxygenation

Elucidation of bacterial degradation pathways of dioxin-related compounds identified a new type of oxidative attack with high regioselectivity and specificity for the angular position. In this type of dioxygenation event, one carbon atom bound to the oxygen atom in DD and DF and the adjacent carbon atom in the aromatic ring are both oxidized. This reaction is termed angular dioxygenation, and is catalyzed by certain Riske non-heme iron oxygenases (ROSs), which are currently known as angular dioxygenases (Figure 2). Engesser et al. [52] and Fortnagel et al. [53] established that the initial angular dioxygenation reaction occurred on DF at the 4 and 4a carbon atoms. Wittich et al. [33] reported angular dioxygenation of DD. The angular reaction resulted in the concomitant cleavage of the C-O bond to generate 2,2'3-trihydroxybiphenyl or 2,2'3-trihydroxydiphenyl ether. The dihydroxylated rings of these products were then *meta*-cleaved by an extradiol dioxygenase. Hydrolysis of the *meta*-cleavage products in the DD and DF metabolic pathways yielded catechol and salicylic acid, respectively [32,33,42,54].

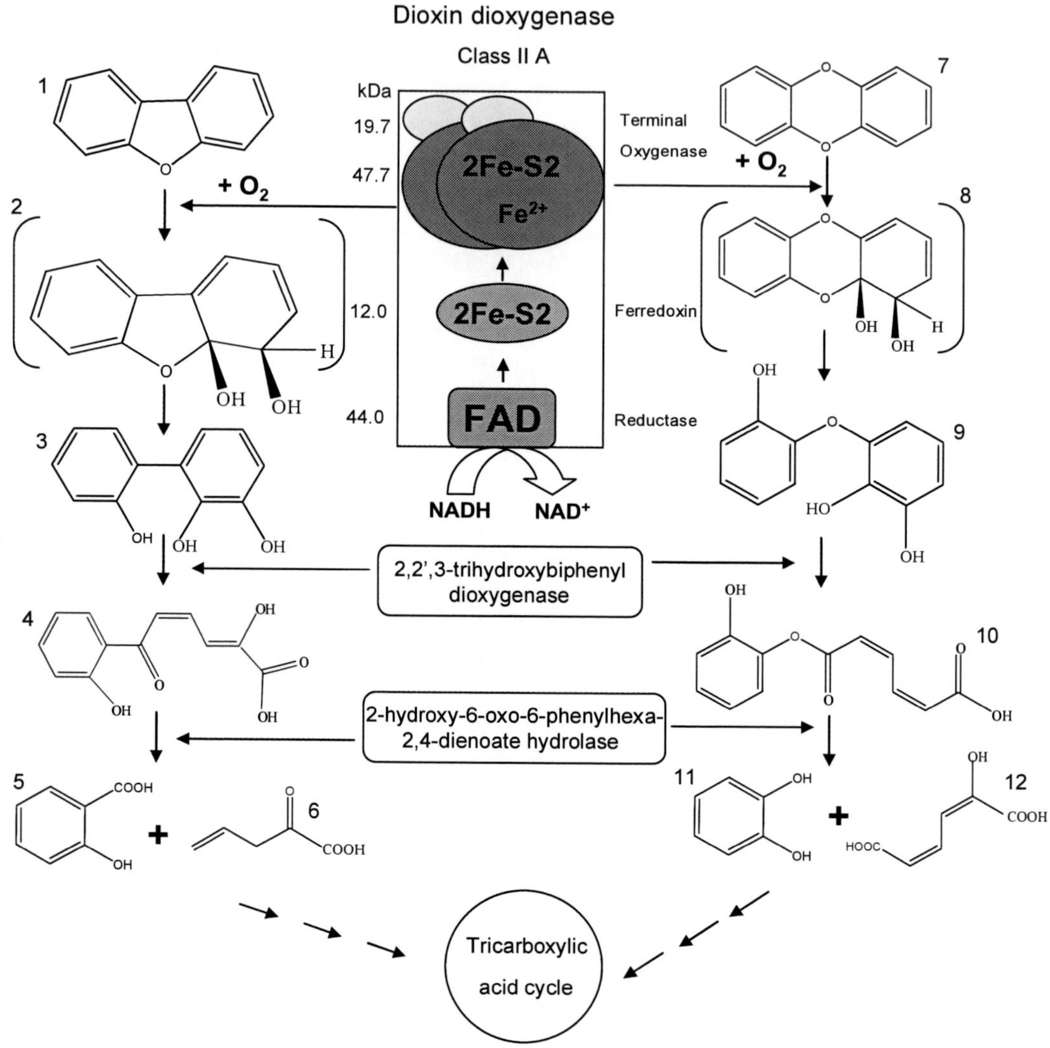

Figure 2. Degradative pathway for DF and DD by angular dioxygenation.
1. Dibenzofuran;
2. 4,4a-dihydro-dihydroxydibenzofuran;
3. 2,2',3-Trihydroxy-biphenyl;
4. 2-Hydroxy-6-oxo-6-(2-hydroxyphenyl)- hexa-2,4-dienoate
5. Salicylic acid; 6. 2-Oxopent-4-enoate 7. Dibenzo-*p*-dioxin; 8. 4,4a-dihydro-dihydroxydibenzo -*p*-dixoin; 9. 2,2',3-Trihydroxydiphenyl ether; 10. 2-hydroxy-6-oxo-6-(2-hydroxyphenoxy)hexa-2,4-dienoate; 11. Catechol; 12. 2-Hydroxy-muconate.

Numerous gram-negative and gram-positive bacteria capable of mineralizing DF, DD and/or their analogs have been isolated and characterized [*for review see Hiraishi [31]*]. *Sphingomonas wittichii* RW1 is one of the most intensively characterized bacteria involved in the angular dioxygenation pathway [33]. Kimura and Urushigawa [44] isolated the bacterium *Rhodococcus opacus* SAO101, which uses DD and DF as the sole carbon source. Other examples include *Pseudomonas veronii* PH-03 [54], *Rhodococcus* sp. strain DFA3 [55], *Janthinobacterium* [56], several strains of *Janibacter* sp. [57-60], *Nocardioides aromaticivorans* IC177 [61,62], and *Paenibacillus* sp strain YK5 [63]. Many of these bacteria effectively degraded some chlorinated congeners.

Angular dioxygenation of many analogs of DF has been reported. These analogs include carbazole (CAR), and 9-fluorenone and dibenzothiophene sulfone which are metabolic intermediates of fluorene (FN) and dibenzothiophene (DBT), respectively. Biodegradation of these compounds by angular dioxygenation and their biotechnological implications have been previously reviewed [27-30]. Yamazoe et al. [59,60] reported that *Janibacter* sp strain YY1, which utilizes DF and FN, exhibited angular dioxygenation against diphenyl ether, FN, DD and DF, but not with CAR. This strain also hydroxylated all of these compounds by lateral dioxygenation and produced corresponding dihydroxylated products. CAR is a nitrogen-containing analog of DF, and its biodegradation by angular dioxygenation has been reported in *Pseudomonas* sp CA06 and CA10 (currently known as *Pseudomonas resinovorans* CA10) by Ouchiyama et al. [64]. Subsequent studies concerning the molecular biology and the physiological and biochemical characteristics of carbazole degradation have been reported in *P. resinovorans* CA10 [65-71]. Recently, several other carbazole-degrading strains similar to strain CA10 have been isolated, and include *Janthinobacterium* sp. strain P, *Nocardioides* sp. strain IC177, and *Sphingomonas* sp. strain KA1 [56,61].

Biodegradation of Chlorinated Dioxins

Many DF- and DD-mineralizing bacterial stains have been extensively characterized. However, no single strain uses chlorinated dioxins as the carbon source for its growth except *Pseudomonas veronii* strain PH-03, which grows on 1 and 2-MCDD [54]. Many strains degrade low chlorinated DDs and DFs cometabolically. Strain *Sphingomonas wittichii* RW1 degraded several mono- and di-chlorinated DDs and DFs [72, 73]. However, our recent studies revealed that strain RW1 transformed selected congeners of the chlorinated dioxin HxCDD but was unable to degrade more highly chlorinated dioxins (Figure 3 and Figure 4) [74-76]. MCDD/Fs and DCDD/Fs were transformed to corresponding mono- and dichlorinated catechols and salicylates [72,74]. These findings indicate that *S. wittichii* RW1 performs dioxygenation reactions at both substituted and non-substituted aromatic rings. Despite the biotransformation ability for chlorinated dioxins, strain RW1 was unable to mineralize these chlorinated dioxins [72,74,76]. However, when strain RW1 and the 3,5-dichlorisalicylate-degrading bacterium *Burkholderia* sp. strain JWS were co-cultured, 4-MCDF was mineralized completely [77]. We found that *S. wittichii* RW1 transformed 2,7-DCDD, 1,2,3-TriCDD, 1,2,3,4-TCDD and 1,2,3,4,7,8-HxCDD into corresponding chlorocatechols such as 4-chloro-, 3,4,5-trichloro-, and tetrachlorocatechol [74,76], but was unable to biotransform 2,3,7-TriCDD and 1,2,3,7,8-PeCDD congeners. In the case of 1,2,3,4-TCDD and 1,2,3,4,7,8-HxCDD biotransformation, a second metabolite, 2-methoxy-3,4,5,6-tetrachlorocatechol, was produced from the original degradation product, 3,4,5,6-tetrachlorocatechol (Figure 3 and Figure 4). Biotransformation studies with mixtures of several chlorinated DDs showed that strain RW1 selectively transformed only the accessible congeners [76]. The above studies suggest that substitution patterns influence the degradability of chlorinated congeners.

Several other bacterial strains degrade chlorinated DDs and DFs. Harms et al. [78,79] reported that *Pseudomonas* sp. strain HH69, now reclassified as *Sphingomonas* sp., displayed cometabolic activity against DD and 3-MCDF. Wittich et al. [80] reported that 2-MCDF and 3-MCDF were completely mineralized in co-cultures containing the DF mineralizing strain *Sphingomonas* sp. RW16, very recently re-classified as *Sphingobium aromaticiconvertens* [81], the 4-chloro- and 5-chlorosalicylate mineralizing strain *Pseudomonas* sp. RW10, and two other strains of unknown function, which were necessary for the growth of the culture. Strain RW16 initiated the dioxygenation reaction on the halogenated and non-halogenated aromatic rings in this experiment in a similar manner as observed by Harms et al. [79]. Schreiner et al. [82] studied the influence of substitution patterns on the biodegradation of all 210 congeners of PCDD/Fs using different bacterial strains. In their study, *Sphingomonas* sp. strain HH69 depleted 31 % of 2,3,7,8-TCDF and 15 % of 2,3,7,8-TCDD within 84 days of incubation. *Ralstonia* sp. strain DSM 6708 (*Pseudomonas* sp. strain PSX) [83] degraded 64 % of 2,3,7,8-TCDF and 82 % of 2,3,7,8-TCDD during the same period of time. This result indicates that the rate of degradation of chlorinated dioxins by bacteria is inversely proportional to the number, and also influenced by the position of chlorine atoms in the congener. In particular, dioxygenation of chlorinated dioxins is hindered by chlorine substitution at positions 1, 4, 6, or 9, such that the degradation of highly chlorinated compounds by aerobic bacteria is difficult.

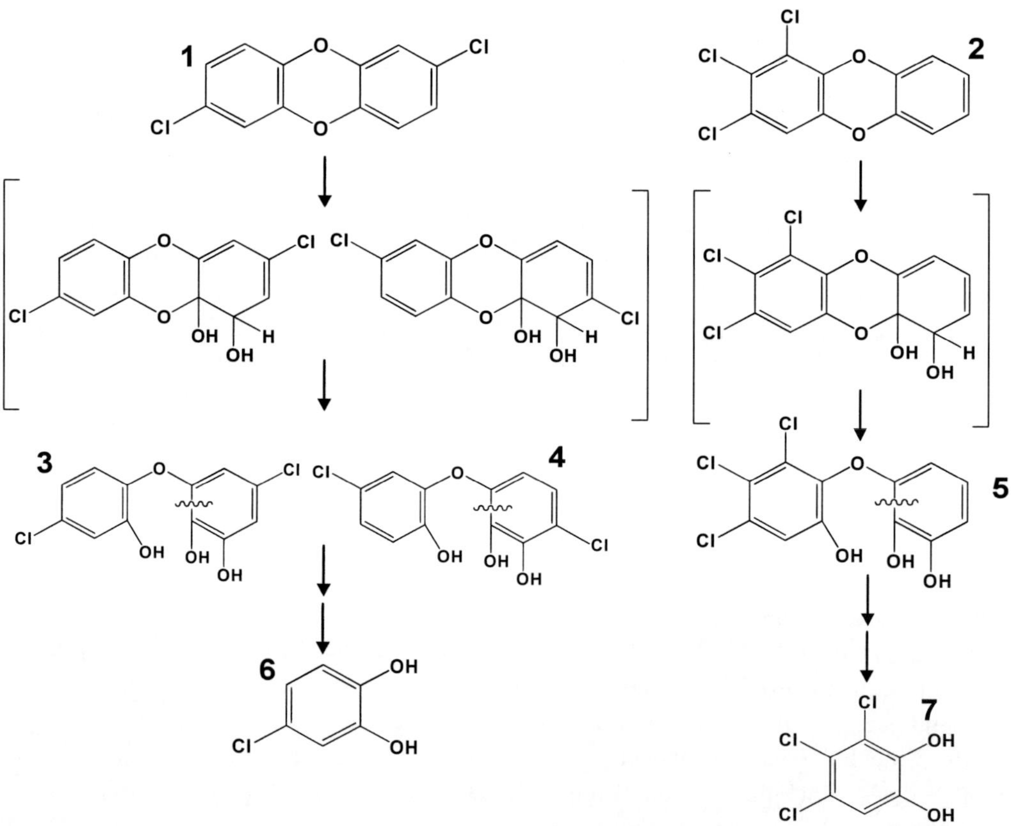

Figure. 3 Degradative pathway for 2,7-DCDD and 1,2,3- TriCDD by *Sphingomonas wittichii* RW1 [Hong et al., 2002 [74]; Nam et al., 2006 [75]].

1. 2,7-dichlorodibenzo-*p*-dixoin;

2. 1,2,3-trichlorodibnzo-*p*-dioxin;
3. 4',5-dichloro-2,2',3-trihydroxydiphenyl ether;
4. 4,5'-dichloro-2,2',3-trihydroxydiphenyl ether;
5. 4',5',6'-trichloro-2,2',3-trihydrxoydiphenyl ether;
6. 4-chlorocatechol;
7. 3,4,5-trichlorocatechol.

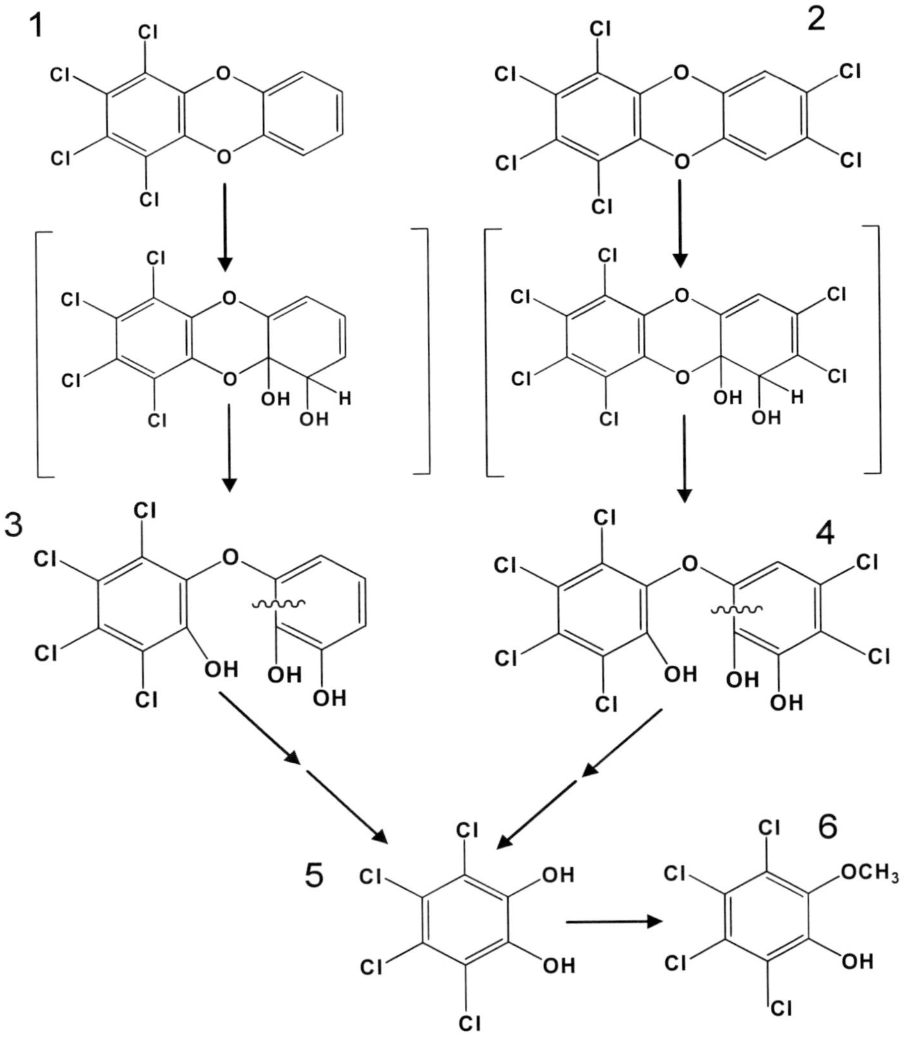

Figure 4. Degradative pathway for 1,2,3,4-TCDD and 1,2,3,4,7,8 HxCDD by *Sphingomonas wittichii* RW1 [Hong et al., 2002 [74]; Nam et al., 2006 [75]].

1. 1,2,3,4-Tetrachlorodibenzo-*p*-dixoin;
2. 1,2,3,4,7,8-hexachlorodibnzo-*p*-dioxin;
3. 2',3',4',5'-Tetrachloro-6,2',3'-trihydroxydiphenyl ether;
4. 3',4,4',5,5',6'-Hexachloro-2,2',3-trihydroxydiphenyl ether;
5. 3,4,5,6-Tetrachlorocatechol;
6. 2-methoxy-3,4,5,6-tetrachlorocatechol.

Enzymes Involved in Biodegradation of Dioxins

The biochemistry and molecular biology of dioxin degradation by various aerobic bacteria has been extensively studied. Bünz and Cook [84] purified and characterized the initial angular dioxygenase system for attacking DD and DF from *Sphingomonas wittichii* RW1 and termed it dioxin dioxygenase. This enzyme system is a multicomponent complex composed of electron transferring components and a terminal hydroxylating component (Fig 2). The electron transfer components consist of two isofunctional FAD-containing reductases of 44 kDa (RedA1 and Red A2), which transfer electrons from NADH to terminally hydroxylated components via a 12 kDa putidaredoxin-type (2Fe-2S) ferredoxin (Fdx1). The terminally hydroxylating component is composed of an oxygenase heterodimer containing a large 45 kDa (DxnA1) subunit and a small 23 kDa (DxnA2) subunit. The terminal oxygenase components introduce oxygen into the target substrate in angular positions. The genes coding for each component of this system have been cloned, sequenced, and functionally expressed [85-88]. Genes encoding enzymes related to DD and DF degradation, such as *dxnA1, dxnA2, fdx1,* and *redA2* were not clustered in one operon system but rather were spread over the entire chromosome.

Molecular and biochemical characterization of angular dioxygenases has also been reported in some gram-positive bacteria such as *Rhodococcus opacus* SAO101, *Terrabacter* (*Janibacter*) sp. strain YK3, and *Terrabacter* (*Janibacter*) sp. DBF63 [30,44, 89-91]. Carbazole 1,9a-dioxygenase (CARDO) from *Pseudomonas resinovorans* strain CA10 is another extensively characterized Rieske non-heme iron oxygenase capable of angular dioxygenation of DD and DF. Genes encoding CARDO were first cloned from strain CA10 and were found to comprise a three-component system [65,66]. CARDO has a broad substrate range and can catalyze diverse oxygenation reactions such as angular dioxygenation, *cis*-dihydroxylation, and monooxygenation, and can also *cis*-hydroxylate naphthalene and biphenyl. The diverse oxygenation reactions catalyzed by CARDO might reflect differences in substrate binding to the reaction center of the complex [67]. Nam et al. [70] crystallized the terminal oxygenase component (CarAa) of CARDO from strain CA10, and Nojiri et al. [92] determined the crystal structure of that from *Janthinobaterium* sp. strain J3. The overall α3 trimeric structure of these enzymes roughly corresponded to the α3 trimeric partial structure of other terminal oxygenase components of ROSs with α3 configuration.

The genes coding for different proteins of the three-component dioxygenase systems are similarly organized in different organisms. Relationships between the large subunit proteins of these systems have been reviewed previously [31], and phylogenetic tree analyses indicate that the riske non-heme iron dioxygenases (ROSs) can be grouped into three distinct clusters designated class I, II, and III. Although the angular dioxygenase system of *Sphingomonas wittichii* RW1 is similar to other three-component ring hydroxylating dioxygenases of monocyclic and bicyclic compounds, the reductase of this strain differs from those of class III and class I enzymes in that it lacks a Fe-S cluster [87]. Nam et al. [92] classified the ring hydroxylating oxygenase systems into classes I, II, III and IV. The angular dioxygenase CARDO system was classified into class I.

Degradation of Chlorinated Dioxins by Bacterial Cytochrome P-450 Monooxygenases

Cytochromes belong to a large family of enzymes widely distributed in eukaryotes and prokaryotes. Cytochrome P-450 monooxygenases [P-450] are nature's most versatile biocatalysts that introduce oxygen into a vast array of molecules. These enzymes catalyze diverse reactions in a regio- and stereoselective manner, and they have been used for drug development, bioremediation and the synthesis of fine chemicals and other useful compounds [94,95]. In addition to angular dioxygenation, P-450 monooxygenation of PCDDs was observed while engineering bacterial P-450 in *Bacillus magaterium* BM-3 [96]. Wild-type P-450 BM-3 is a fatty acid hydroxylase, and it and the corresponding AL(4)V mutant (Ala74Gly, Phe87Val, Leu188Gln triple mutant) metabolized 1-MCDD, 2,3-DCDD and 2,3,7-TriCDD, but failed to metabolize 2,3,7,8-TCDD. AL(4)V was two to twelve times more active than the wild-type P-450 BM-3 towards PCDDs. Products of these reactions were hydroxylated at unsubstituted positions, and the chloride moiety had migrated. They were also less toxic than other derivatives, demonstrating less than 10 % of the toxicity of the original compounds. These results indicate that mutant bacterial P-450 is a promising catalyst for detoxification of dioxins and suggests the possibility of constructing another mutant bacterial P-450 with high catalytic activity against PCDD/Fs to bioremediate dioxin-contaminated soil.

Prokaryotes with the ability to degrade dioxin are potential candidates for expressing the 2,3,7,8-TCDD-metabolizing enzyme derived from rat cytochrome P-450 [CYP1A1] to use in bioremediation applications. However, modifications of the N-terminal membrane-anchored region of P-450s are needed for successful expression in prokaryotic cells because mammalian P-450s, like rat CYP1A1, are associated with endoplasmic reticulum membranes [97]. Shinkyo et al. [98] successfully expressed N-terminally truncated CYP1A1 (Delta 1A1 and Delta F240A) in *Escherichia coli* cells. Kinetic analysis using membrane fractions prepared from recombinant *E. coli* cells revealed that Delta F240A had enzymatic properties similar to Delta F240A expressed in yeast, where 2,3,7-TriCDD was successfully converted into multiple metabolites. These results indicate the possible application of prokaryotic cells expressing Delta F240A in the bioremediation of PCDD-contaminated soil.

Reductive Dehalogenation of Dioxins

Anaerobic biodegradation of xenobiotic compounds has been subjected to extensive research during the past two decades. Many microorganisms use xenobiotic pollutants as electron acceptors during respiration, termed dehalorespiration or halorespiration. Several reviews have been published on anaerobic biodegradation of chlorinated xenobiotics [99-101], and although many studies have investigated reductive dechlorination mechanisms for chlorinated chemicals such as chlorinated ethylene and chlorobenzene, it has been studied to a lesser extent than oxidative biodegradation of dioxins. The first indication of reductive dechlorination of dioxins was observed by Townsent [102], where changes in the dioxin patterns and accumulation of low-chlorinated congeners in sediments were observed. Consequently, microbial reductive dechlorination increasingly attracted attention due to its activity against highly chlorinated congeners that were not attacked by aromatic hydrocarbon dioxygenases, and it is possibly the first step in transformation of polychlorinated pollutants in the environment. Many studies have demonstrated microbial reductive dechlorination of dioxins from contaminated sediments, sludge and soils [10, 103-114]. Unfortunately, microbial dechlorination produces highly toxic congeners such as 2,3,7,8-TCDD as an intermediates due to removal of chlorine from carbon atoms at 1, 4, 6 and/or 9 positions [10, 103, 105]. Barkovskii and Adriaens [105] reported that a microbial consortium isolated from dioxin-contaminated Passaic River sediments dechlorinated both freshly-spiked and aged PCDDs found in sediments. Freshly added OCDD was dechlorinated into HxCDD and MCDDs, and the maximum yield of cell protein coincided with the production of less-chlorinated DD isomers. Several other studies of sediment microcosms showed that dechlorination of 1,2,3,4-TCDD occurred at the lateral positions and resulted in 2-CDD as the final product [104,109].

Environmental conditions influence microbial dechlorination. Fu et al. [115] studied the influence of geochemical conditions such as freshwater, estuarine and marine environments for PCDD/F transformation in river sediments. They found that the amount of dechlorinated products increased as the salinity increased, where the presence of organic matter and the decrease in the levels of salinity increased the formation of 2,3,7,8-TCDD. Fu et al. [116] also studied the effect of respiratory conditions and priming compounds on dechlorination patterns of heptachlorodibenzo-*p*-dioxins (HpCDD) using estuarine sediment-eluted cultures in the presence and absence of 20 mM sulfate and 0.2 mM 2-bromodibenzo-*p*-dioxin (2-BrDD) as priming compound. It was observed that 2,3,7,8-TCDD formation decreased under sulfidogenic conditions, and the presence of 2-BrDD decreased the formation of 2,3,7,8-TCDD formation by an additional four- to five-fold. Ahn et al. [112] studied the effect of the addition of structurally similar compounds with dioxins to stimulate the activity of microbial dechlorination in sediments. They found that addition of 1,2,3,4-tetrachlorobenzene (TeCB) and 2,3,4,5-tetrachloroanisole (TeCA) enhanced the dechlorination of 1,2,3,4-TCDD and TCDF to tri- and dichlorinated daughter products. Several pure anaerobic strains have been isolated and characterized for degradation of different aromatic and aliphatic compounds, and are members of *Dehalococcoides, Dehalobacter, Desulfitobacterium, Desulfuromonas,* and *Methanosarcina* species [*for review see Zhang and Bennett [101]]*. Bunge et al. [117] reported the presence of *Dehalococcoides* species in dioxin-dechlorinating enrichment cultures from freshwater sediment highly contaminated with dioxins. They also observed dechlorination of selected PCDD congeners by the pure chlorobenzene-degrading anaerobic strain *Dehalococcoides* CBDB1 [118]. This strain dechlorinated the environmentally significant congeners 1,2,3,4-TCDD and 1,2,3,7,8-PeCDD to less chlorinated congeners (Figure 5). Although several dechlorinating *Dehalococcoides* species are known, strain CBDB1 is unique in that it is able to dechlorinate PCDDs. Recent molecular analysis revealed that this strain contained several reductive dechlorinating genes (RDH) absent in other strains [119].

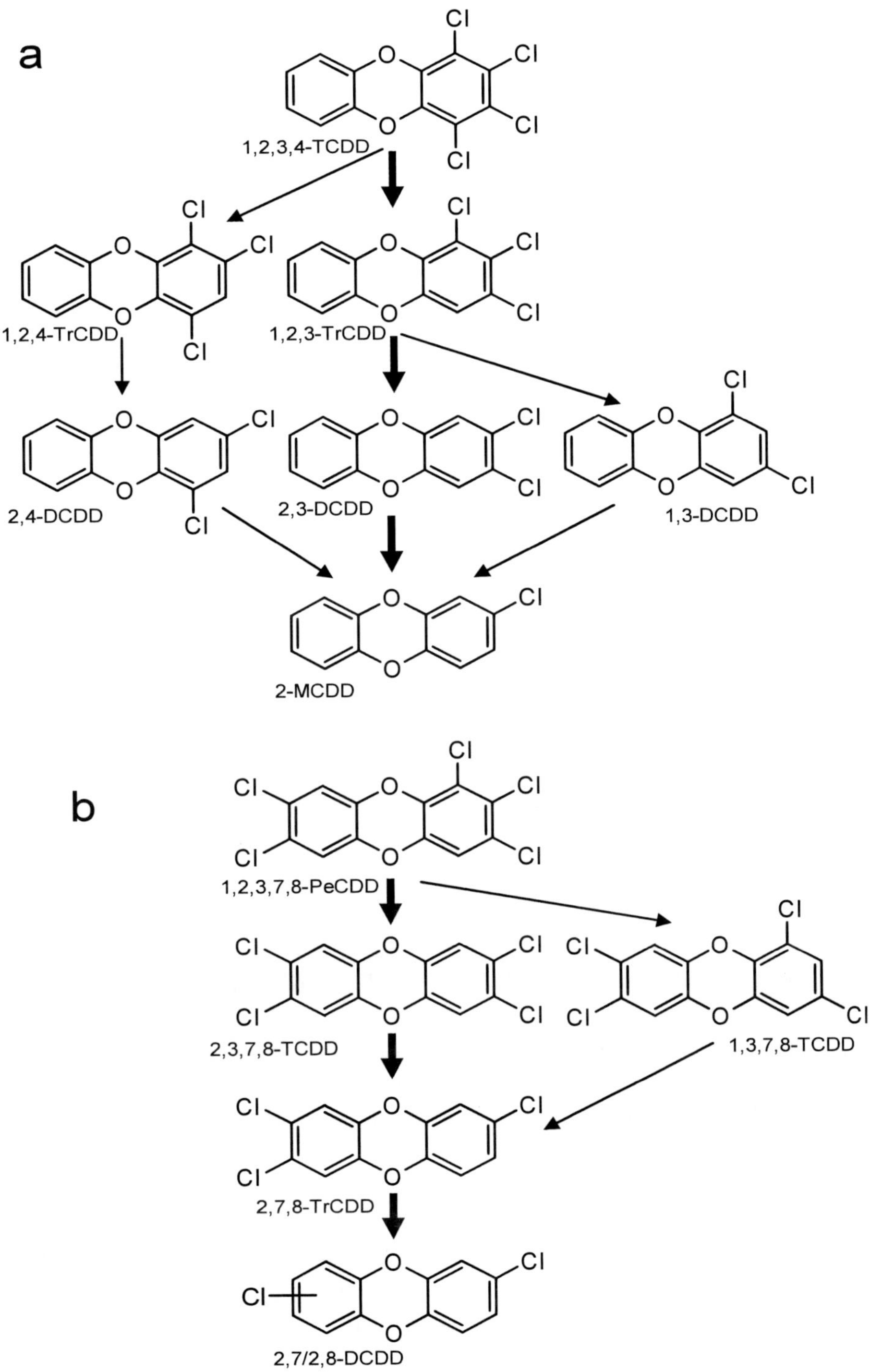

Figure 5. Proposed pathways of reductive dechlorination of spiked 1,2,3,4-TeCDD (a) and 1,2,3,7,8-PeCDD (b) by a pure culture of *Dehalococcoides* sp. strain CBDB1 [Bunge et al., 2003 [117]]. The major routes are marked with bold arrows.

Recent studies have focused on community analysis and characterization of dioxin dechlorination of enriched mixed cultures isolated from polluted environments. Ballerstedt et al. [111] identified putative dechlorinating bacteria from an anaerobic consortia enriched with TriCDDs in the presence of pyruvate and fumarate as co-substrates. These consortia contained members of *Acetobacterium, Desulfitobacterium, Desulfuromonas* and *Dehalococcoides*. *Dehalococcoides* was the most prevalent species positive for dioxin dechlorination. Yoshida et al. [113] also observed that *Dehalococcoides*-like bacteria were responsible for the complete dechlorination of dioxins in a microbial consortium. These *Dehalococcoides* strains were distinctly different from those of typical *Dehalococcoides* strains, such as *D. ethenogenes* strain 195 and *Dehalococcoides* sp. strain CBDB1. These studies also demonstrated the coexistence of reductive/oxidative degradation processes in an anaerobically incubated microcosm. Ubiquinone profiling indicated the coexistence and activity of ubiquinone-containing aerobic bacteria [113]. All of the dioxin congeners detected were equally reduced without the accumulation of significant amounts of less chlorinated congeners as intermediate or end products. They also observed large amounts of catechol and salicylic acid in the upper aqueous phase microcosm, suggesting that the dechlorination of dioxins and the oxidative degradation of the dechlorinated products likely occurs simultaneously. Microbial reductive dechlorination is currently the only potential method for the dechlorination of higher halogenated dioxins. Although it is expected to be successful in bioremediating dioxin-polluted environments, it is a very slow and time-consuming process and is therefore impractical for routine use. The use of zero valent iron (ZVI) for dechlorination of PCDD/Fs is an attractive alternative to microbial dechlorination [120]. In recent years there has been great interest in the exploration of hybrid technologies that combine anaerobic microbial processes with ZVI.

Fungal Degradation of Dioxins

Degradation of Dioxins by Cytochrome P-450 Monooxygenase

Cytochrome P-450 is a hemoprotein widely distributed in eukaryotic and prokaryotic organisms. Its ability to hydroxylate DF is widespread among yeast and filamentous fungi [27, 31]. Recently, much attention has being focused on protein engineering of P-450 in order to develop efficient dioxin bioremediation systems. Recombinant technology for P-450 expression has also provided a strategy for its application in dioxin degradation. Sakaki et al. [121] demonstrated the metabolism of DD, MCDD, 2,7-DCDD and 2,3,7-TriCDD by recombinant yeast cells expressing rat P-450 (CYPA1 and CYPA2). This metabolism employed multiple reactions such as hydroxylation at an unsubstituted position, hydroxylation with migration of a chloride substituent, hydroxylation with elimination of a chloride substituent, and opening of dioxin rings. A distinct difference was observed in the substrate and reaction specificities between CYP1A1 and CYP1A2. 2,7-DCDD and 2,3,7-TriCDD were good substrates for both CYP1A1 and CYP1A2. Most of the 2,3,7-TriCDD was first converted to 8-hydroxy-2,3,7-TriCDD, and was further metabolized to hydrophilic compounds with cleaved ethereal bridges. Shinkyo et al. [122] produced a novel P-450 enzyme capable of metabolizing 2,3,7,8-TCDD by modifying rat CYP1A1 through site-directed mutagenesis. Recently, Orihara et al. [123] produced recombinant white-rot fungus *Coriolus hirsutus* by transforming it with rat P-450 (CYPA1). Three transformants (ChTF5-2(CYP1A1), ChTF5-4(CYP1A1) and ChTF5-6(CYP1A1) produced similar amounts of rat CYPA1 enzyme and efficiently transformed 2,7- and 2,8-DCDDs. Rat CYP1A1 was also successfully expressed in a bacterial system for the degradation of chlorinated dioxins [98]. The studies described above indicate the possibility of using microorganisms expressing mammalian cytochrome P-450 to bioremediate soils contaminated with dioxins.

Degradation of Dioxins by Lignolytic Fungi

White-rot basidiomycetes are capable of degrading lignin, a naturally occurring recalcitrant phenylpropanoid polymer. Two families of lignolytic enzymes play a key role in the lignin degradation process, including phenol oxidases (laccase) and peroxidases (lignin peroxidase, manganese peroxidase). Unlike cytochrome P-450 monooxygenase, lignin peroxidases are highly nonspecific in their substrate range such that they are capable of degrading a wide variety of aliphatic and aromatic xenobiotics, including PCDD/Fs. *Phanerochaete chrysosporium* is one of the most characterized white-rot fungi. Bumpus et al. [124] first studied the dioxin degradation potential of *P. chrysosporium* using 2,4,7,8-TCDD. Hammel et al. [125] and Joshi and Gold [126] defined the detailed mechanism of DD degradation by lignin peroxidase (Lip) of *P. chrysosporium*. They also found that Lip of this fungus oxidized 2-MCDD. A subsequent study by Valli et al. [127] revealed that *P. chrysosporium* Lip catalyzed the oxidative cleavage of the aromatic ring of 2,7-DCDD, and Takada et al. [128] demonstrated the biodegradation ability of the white-rot fungus *P. sordida* YK-624 for the degradation of PCDD/Fs, including 2,3,7,8-TCDD. When culturing this fungus in low-nitrogen medium modified with a mixture of PCDD/Fs, 40 – 76 % of PCDDs (TCDD-HxCDDs) and 45 – 70 % of PCDFs (TCDF-HxCDFs) were degraded. In this study 4,5-dichlorocatechol and 3,4,5,6-tetrachlorocatechol were the major metabolites from 2,3,7,8-TCDD and OCDD, respectively. This is the first study demonstrating the potential use of white-rot fungi for the degradation of PCDD/Fs.

Extensive screening studies have been performed to select for potential fungi having dioxin degradation capabilities from various environments such as contaminated soils and temperate and tropical forests. The screening for and selection of potential fungi was based on various methods such as dye decolorizing ability and DD, 2,7-DCDD and TCDD degradation. These studies revealed that several wood rotting fungi degraded DD and low-chlorinated DDs, while other fungi exclusively degraded OCDD. These fungi belong to the species *Aleurodiscus*, *Bjerkandera, Ceriporia*, *Panellus, Phanerochaete*, *Phlebia* and *Pleurotus*, and other unidentified strains [129-139]. Likewise, phylogenetic studies of dioxin-degrading fungi indicated that the species *Phlebia* are very effective at degrading chlorinated compounds [135]. *Phlebia brevispora* simultaneously degraded the commercial herbicide 2,4,6-trichlorophenyl-*p*-nitrophenol and its contaminant 1,3,6,8-TCDD, and dechlorinated some coplanar polychlorinated biphenyls [138,140]. Laccases are one of the components of the fungal lignolytic enzyme system and have recently attracted attention due to their ability to oxidize a wide variety of recalcitrant environmental pollutants [141]. Unlike lignin peroxidase, details on laccase-mediated dioxin degradation are limited. Early studies showed that *Polyporus versicolor* laccase did not significantly oxidize 2,3,7,8-TCDD [142], whereas that of *Trametes versicolor* and *Pycnoporus cinnabarinus* transformed 2-hydroxy-DF [143]. In addition, laccase of *P. cinnabarinus* transformed 2-hydroxydiphenylether and 2-hydroxybiphenyl, and dechlorinated chlorinated hydroxybiphenyls [144,145].

Degradation of Dioxins by Soil Fungi

Several non-wood rotting fungi were isolated from soil and sludge and have the potential for PCDD/F degradation. Nakamiya et al. [146] isolated the fungus *Acrymonium* sp strain 622 (later identified as *Pseudallescheria boydii*) from denitrifying activated sludge. This strain grew under aerobic conditions and effectively degraded highly chlorinated dioxins [TCDD-OCDD] by oxidation of aromatic rings and by dechlorination. The enzymes responsible for the degradation of chlorinated dioxins in strain 622 were membrane-associated, and their mechanisms differ from that of typical fungi [147]. Recently, a new dioxin degradation pathway [Figure 6] has been reported in the cyclic ether degrading fungus *Cordyceps sinensis* strain A [148]. This fungus degraded DD, 2,3,7-TriCDD, and OCDD. Degradation of DD yielded catechol, which was further metabolized to *cis-cis*-muconate; 2,3,7-TriCDD degradation resulted in mono- and dichlorinated catechol and non-chlorinated catechol. OCDD degradation yielded mono- to tri-chlorinated catechols and non-chlorinated catechol. For all the dioxins tested, dechlorination resulted in replacement of chlorine atoms with hydrogen atoms, and no hydroxylation occurred. These data suggest that it is possible to detoxify PCDD/Fs using soil fungi and indicates another promising way for bioremediation. However, the enzymatic mechanisms responsible for these pathways remain still unclear.

Bioremediation of Dioxins-Polluted Matrices

Bioremediation has distinct advantages over physico-chemical methods as it can be cost-effective and eco-friendly. For example, many studies have investigated the bioremediation of dioxins using bacteria and fungi in the form of axenic cultures or consortia. Many aerobic strains such as *Sphingomonas wittichii* RW1, *Sphingomonas* sp. strain KA1, *Pseudomonas resinovorans* CA10, and *Terrabacter* sp. strain DBF63 degraded DD, DF, and their low-chlorinated congeners in soil slurry systems [68,149-152]. Although these studies indicate the possibility of using such strains to remove PCDD/Fs from the environment, these experiments were conducted in laboratory soil systems by augmenting pure cultures with externally added dioxins. It remains necessary to study their bioremediation potential in real environmental samples contaminated with PCDD/Fs.

Recent research in our laboratory has investigated the use of aerobic bacteria to reduce the toxicity of PCDD/Fs in municipal solid waste incinerator fly ash by bioaugmentation [75]. In this study, *Sphingomonas wittichii* RW1 survived in fly ash and mineralized DD, DF and their metabolic products. In actual fly ash systems, PCDDs were reduced by 75 % and 20 % in live and dead cell-supplied samples, respectively, during incubation for 15 days. These results indicate that adsorption onto biomass is an important factor in dioxin elimination. The removal of PCDD/Fs can be enhanced by various strategies. Nam et al. [75] studied the effect of repeated inoculation, pre-adaptation of cultures, and the addition of surfactants to the contaminants. Repeated inoculation of strain RW1 enhanced the removal of PCDD/Fs by up to 10.3 %. Halden et al. [150] found that pre-adapted RW1 cells showed superior survival and efficient rates of degradation in soil slurries as compared to non-adapted cells. Although the removal of PCDD/Fs was primarily due to adsorption, biotransformation and the analysis of metabolites requires further study in order to ready the process for real-site bioremediation [75]. The identification of bacterial catalytic activity in complex matrices by conventional techniques is difficult. As an alternative, Halden et al. [153] recently developed an identification method based on peptide mass fingerprinting, which holds promise for monitoring catalytic activity in bioremediation studies.

Biostimulation is another way to achieve bioremediation by providing favorable conditions for indigenous microbial populations in polluted sites. Hiraishi et al. [107] reported that the addition of autoclaved compost to static bottles containing soil contaminated with PCDD/Fs resulted in a 22 % reduction in PCDD/Fs during incubation for three months. The indigenous microbial population was suggested to have the potential for reductive dechlorination of PCDD/Fs and for oxidative degradation of lower chlorinated products once produced. Recently, Yoshida et al. [113] demonstrated the bioremediation of river water sediment contaminated with PCDD/Fs in an anaerobic microcosm incubated with an organic medium. They found that all of the PCDD/F congeners detected were equally reduced without the accumulation of significant amounts of less-chlorinated congeners as the intermediate or end products.

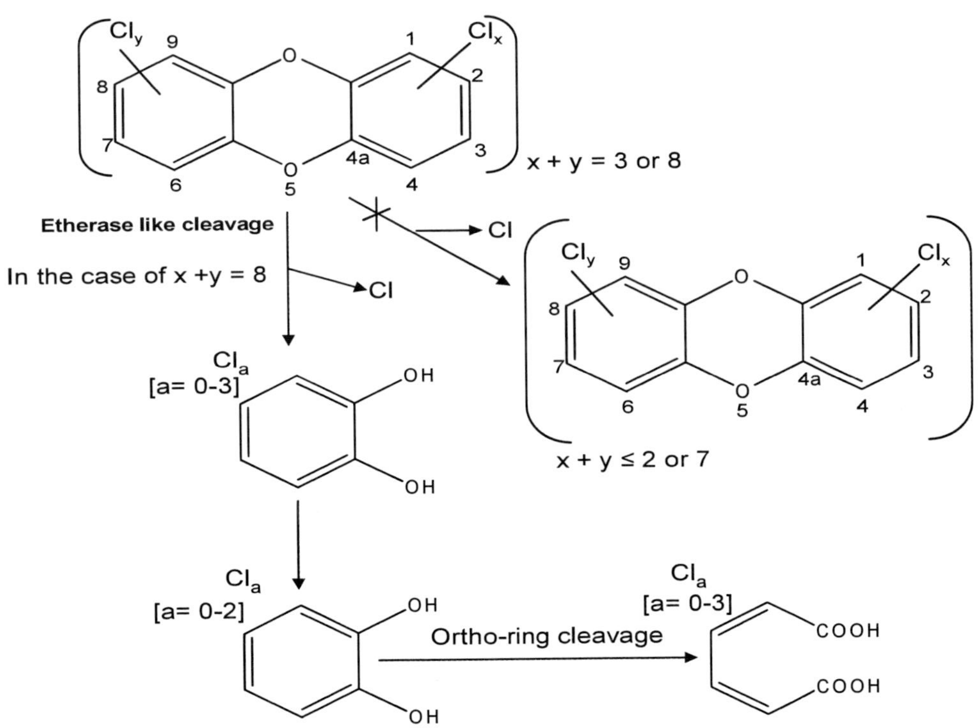

Figure 6. Proposed degradative pathway for chlorinated dioxins by cyclic ether degrading fungus *Cordyceps sinensis*. The crossed arrow indicates that dechlorination of intact dioxins does not occur [Nakamiya et al., 2005 [148]].

Since many fungi show promising potential to degrade chlorinated dioxins, fungal-based remediation technology has received great attention in recent years. Extensive research has been conducted concerning the remediation of PCDD/Fs and several patents have been filed regarding this topic. Nakamiya et al. [146] monitored the degradation of a mixture of PCDD/Fs congeners from fly ash under denitrifying conditions in 700 liter bioreactors. They found that approximately 90 % of PCDD/Fs were removed within seven days, and subsequently a dioxin-degrading fungus was isolated from the activated sludge in this bioreactor. This strain also degraded OCDD under aerobic conditions. Suhara et al. [154] developed a system for fly ash treatment with the white-rot fungus *Ceriporia* sp. strain M2. This fungus reduced the PCDD/F content up to 46.5 % during incubation for 12 weeks. Similarly, Tachibana et al. [2006] reported the effective degradation of the toxic congener 2,3,7,8-TCDD in soil by fungal isolates obtained from nature. Degradation of 2,3,7,8-TCDD was confirmed by detecting the metabolite 4,5-dichlorocatechol.

Compost-assisted bioremediation of contaminated soil is an attractive method for the elimination of environmental toxins. Hacketta et al. [155] composted dioxin-containing paper mill fly ash with wastewater treatment sludge; this resulted in a 45 % reduction of dioxin levels over the eight week incubation. The dioxin concentration in the final soil/compost mixture was 3 pg/g TEQ, and allowed the soil/compost mixture to be classified as agricultural soil. Sites contaminated with many different chemicals present special problems for removal of target pollutants. In such cases it may be necessary to use multiple microbial cultures or consortia for treatment. The use of co-cultures containing defined bacteria and fungi for bioremediation of PAH has been reported [156]. Our recent study also demonstrated the effectiveness of mixed cultures containing dioxin-degrading strains of five fungi and four bacteria for the efficient removal of PCDD/Fs from incinerator fly ash [Nam et al., 2005b]. Mixed cultures incubated with fly ash compost resulted in a 68.7 % reduction in PCDD/Fs during 21 days of incubation. The results of these composting studies reveal that its remediation efficiency depends on the microbial content of the compost. Although this method could potentially be used for bioremediation of dioxin-contaminated sites, information on the fate of dioxins and their degradation mechanisms is still lacking.

Phytoremediation is an efficient technique to remove or degrade various pollutants in soils, water and sediments, and provides some new possibilities for reducing risks associated with dioxins and PCBs [158]. Recently, Sonoki et al. [159] developed a transgenic plant capable of degrading dioxins. They cloned genes of laccase, lignin peroxidase and manganese peroxidase from white-rot fungi *Phanerochaete chrysosporium* and *Trametes versicolor* and recombinantly expressed the gene products in *Arabidopsis thaliana*.

Bioremediation of persistent organic pollutants (POPs) using genetically modified microorganisms (GMOs) has become a rapidly growing area of research [160]. GMOs have the potential to bioremediate soil, groundwater, and activated sludge, exhibiting enhanced degradative capabilities against a wide range of chemical contaminants [161-163]. Recently, Mohammadi and Sylvestre [50] engineered the biphenyl dioxygenase of *Burkholderia xenovorans* LB400 to degrade chlorinated dioxins. Aso et al. [164] incorporated superchannels, or specialized membrane structures that enable the direct incorporation of macromolecules into cells, from *Sphingomonas* sp A1 into the membrane of *Sphingomonas wittichii* strain RW1. The engineered strain showed substantially enhanced bioremediation capacity. In addition, the GEM *Cupriavidus necator* RW112 grew in two commercial PCB formulations, including Aroclor 1221 and Aroclor 1232. This is the first report concerning the aerobic growth of a genetically improved bacterial strain using technical Aroclor mixtures as the substrate [163]. The studies described above indicate the possibility for the genetic development of aerobic bacterial strains that use PCDD/Fs as a nutrient source.

PCDD/Fs Removal by Biosorption

Biosorption is the uptake or accumulation of chemicals by microbial biomass, and has the potential to remove a variety of hazardous chemicals from the environment [165]. The use of dead biomass has several advantages. Dead biomass is more superior than live biomass for the elimination of chemicals that are not readily degradable [165, 166]. Our laboratory recently demonstrated the removal of PCDD/Fs and other halogenated compounds by biosorption [75,76,167-170]. It was observed that the dead biomass of the gram-positive bacterium *Bacillus pumilus* PH-01 adsorbed PCDD/Fs better than corresponding live biomass [167,169]. When *B. pumilus* was exposed to high temperature, it released a protein and carbohydrates, exclusively. We purified a novel protein identified to be responsible for the adsorption of PCDD/Fs, PCBs, chlorobenzenes and chlorinated naphthalenes [169]. The adsorption profile of standard chlorinated chemicals is shown in Figure 7. It was observed that the highly chlorinated PCDD/Fs were adsorbed to a greater extent than the lowly chlorinated ones. The adsorption mechanism of this protein was similar to other passive physico-chemical phenomena. To extent its adsorption capability to real pollutants a study was conducted with PCDD/Fs extracted from fly ash. As shown in Figure 8. The general trend observed is that highly chlorinated congeners are adsorbed to a greater extent than lowly chlorinated congeners (Figure 8a). However, the amount adsorbed is proportional to the initial concentration below the maximum equilibrium concentration. Thus, the adsorption ratio (adsorbed amount/initial amount) instead of the adsorption amount should be considered for a more precise interpretation of the adsorption patterns. The adsorption ratios of highly chlorinated congeners are greater than those of lowly chlorinated ones, but the difference is slight. In addition, the adsorption ratios of PCDDs and PCDFs in the fly ash extract are similar (Figure 8b). Among the 17 toxic congeners, isomers in the same homologue are three Hexa-CDDs (HxCDDs), two Penta-CDFs (PeCDFs), four Hexa-CDFs (HxCDFs), and two Hepta-CDFs (HpCDFs). The adsorption ratios of isomers in the same homologue are different. In the case of HxCDDs, the adsorption ratio is high if the chlorines are located at ortho positions of 1,9 or 4,6. For PeCDFs, HxCDFs, and HpCDFs, the adsorption ratios are high if the chlorines are located at ortho positions of 4,6. This suggests that even in the same homologue the adsorption ratios can vary with the isomers, depending on the positions of the substituted chlorines.

Reducing PCDD/Fs emission from municipal solid waste incineration is growing in importance because of more stringent emission standards and because of public concerns. In recent years, air biofiltration and gas phase remediation have received much attention due to their removal of volatile organic compounds and halogenated hydrocarbons from polluted air streams [171-173]. The dioxin adsorption capacity of the novel protein *Bacillus pumilus* PH-01 suggests that it is possible to use it to remove halogenated hydrophobic compounds from polluted air, and as an adsorbing agent in air samplers collecting PCDD/Fs. Further studies are needed to explore these possibilities.

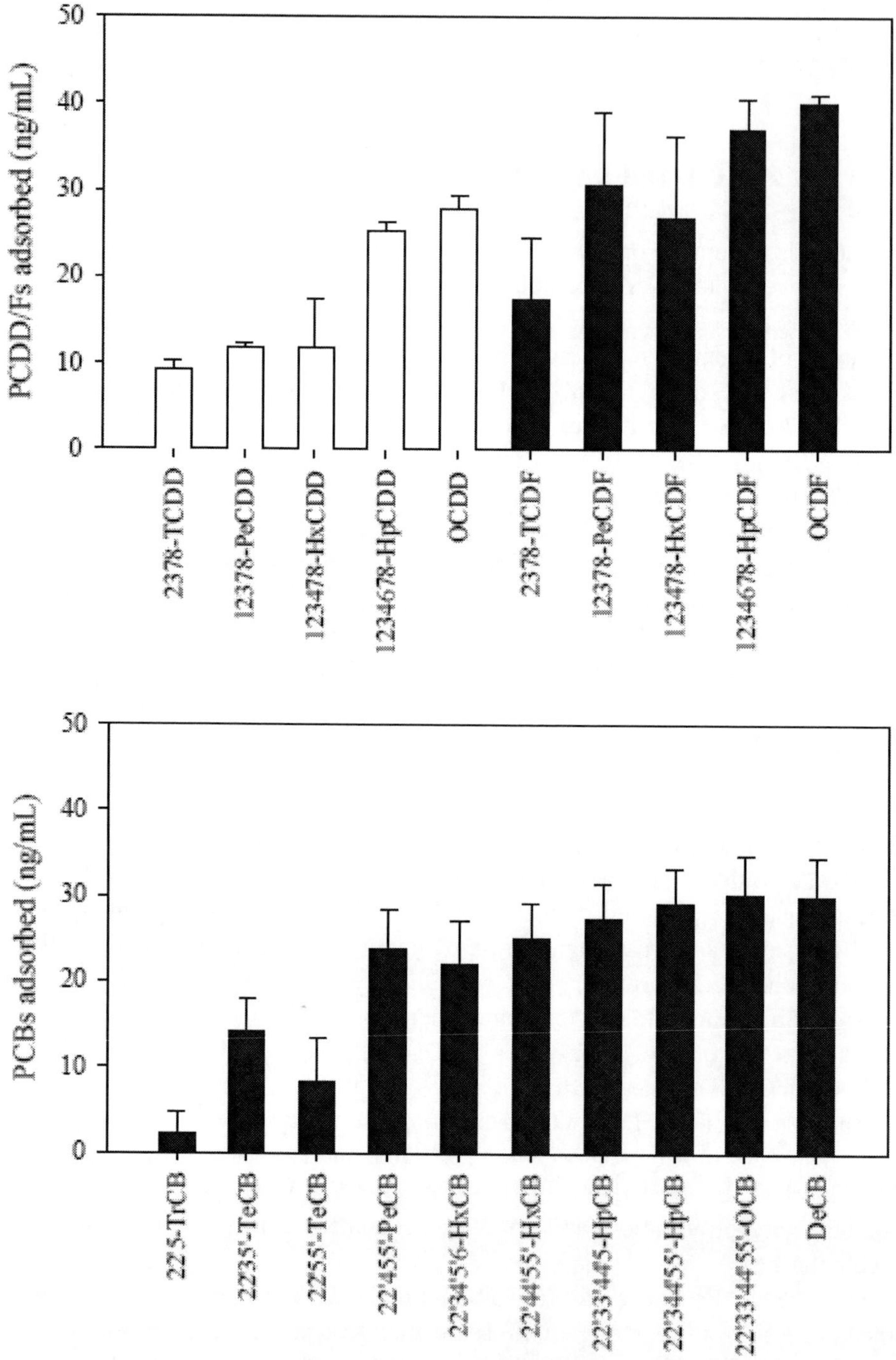

Figure 7. Adsorbed amounts of PCDDs and PCDFs (a); and PCBs standard mixture. Each standard mixture has five congeners. The initial concentration of each congener is 50 ng/mL [Choi et al. [170]].

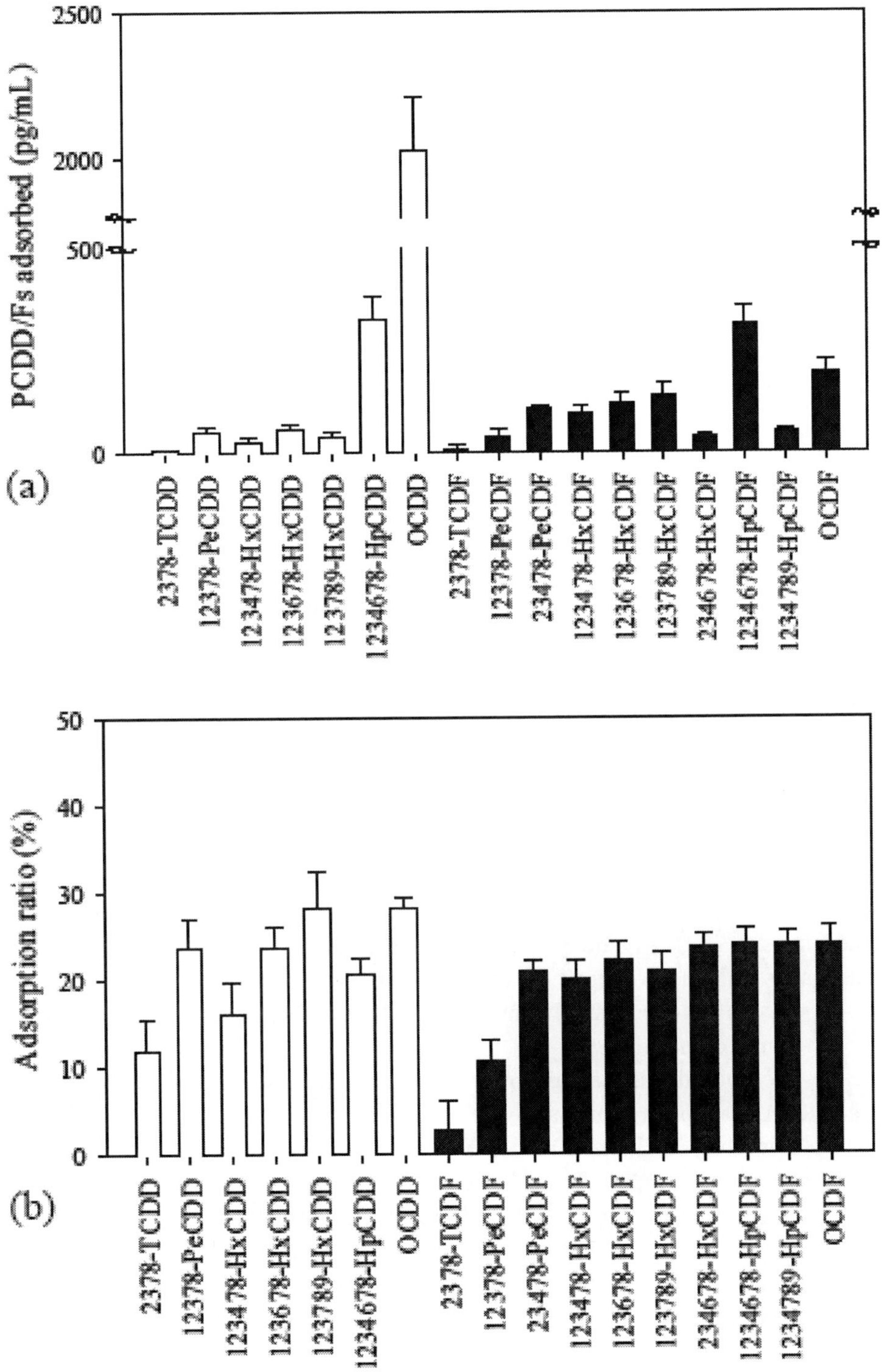

Figure 8. Adsorbed amounts of toxic PCDD/Fs congeners extracted from fly ash (a) and their adsorption ratios (adsorbed amount/ initial amount) (b) [Choi et al. [170]].

CONCLUSION

The ultimate goal of dioxin biodegradation research is to develop bioremediation systems to restore contaminated sites. Since PCDD/Fs are present in a wide range of environmental media, it is necessary to develop multiple bioremediation approaches appropriate for individual contaminated sites. Aerobic biodegradation of dioxins by bacterial dioxygenase systems is well understood and is only useful for the elimination of low-chlorinated congeners from MCDD/Fs to TCDD/Fs and some extend to HxCDD. Unfortunately, toxic chlorocatechols are unwanted by-products of the degradation reaction, since it remains difficult to mineralize toxic PCDD/Fs only by native bacterial dioxygenase enzymes. Therefore, it is necessary to engineer the bacterial dioxygenase system in conjunction with the chlorocatechol pathway to mineralize PCDD/Fs. Degradation of dioxins by lignolytic enzyme systems of white-rot fungi is another potential method for bioremediation. Several studies have proved the efficacy of white-rot fungi for the degradation of PCDD/Fs and to bioremediate contaminated matrices under laboratory conditions. However, their effectiveness in authentic contaminated sites remains to be elucidated. A new degradation mechanism has been proposed for the direct cleavage of ether bonds of dioxins by some fungi. Although such mechanisms seem to be effective in destroying the dioxin structure, their applicability in bioremediation remains unclear. The engineering of microorganisms and their enzymes to improve their efficacy for bioremediation has been of recent interest. The 2,3,7,8-TCDD detoxifying mammalian cytochrome P-450 monooxygenase has been successfully expressed in bacteria, yeast and white-rot fungi. Several transgenic plants have been developed to express dioxin degrading enzymes of white-rot fungi. Although these recombinant strains show detoxifying activity against chlorinated dioxins, their effectiveness in authentic environmental samples contaminated with real PCDD/Fs has yet to be elucidated. Furthermore, the use of engineered strains for the treatment of polluted sites remains controversial and extremely limited. Compost-assisted bioremediation is also an attractive possibility for the effective removal of PCDD/Fs. Although it is likely sufficient for less-contaminated sites, its effectiveness is highly dependent on the chemicals present in the contaminated sites and the composition of microorganisms in those sites. In addition, this method is not suitable for use in sites contaminated with chlorophenols, as this may increase PCDD/F formation in compost.

Reductive microbial dechlorination by anaerobic microorganisms is a promising way to dechlorinate PCDD/Fs. Since many anaerobic microorganisms are widespread in soils and sediments, the microbial dechlorination of PCDD/Fs is an economically and ecologically feasible way for removing PCDD/Fs from contaminated soils and sediments. However, the rate of anaerobic chlorination is too slow to accommodate widespread PCDD/F contamination. However, the transformation rate can be enhanced by stimulating favorable conditions. One such strategy is to combine chemical methods such as ZVI. Recently, the combined ZVI/anaerobic microorganism reductive dechlorination method proved effective for the dechlorination of chlorinated solvents such as TCE. Such technologies may be useful to enhance the reductive dehalogenation of PCDD/Fs. The use of microbial consortiums that show both reductive dechlorination and oxidative degradation may be useful. However, research into this area is limited, and further information concerning the interactions between these microbes is lacking. As an alternative to slow microbial dechlorination, hybrid chemical and biological methods, such as sequential ZVI-mediated reductive dechlorination and aerobic oxidative degradation by potential microorganisms, could be a promising technique to remediate PCDD/F-polluted sites.

REFERENCES

[1] Buser, H.R. (1987a). Brominated and brominated chlorinated dibenzodioxins and dibenzofurans - potential environmental contaminants. *Chemosphere. 16*, 713-732.

[2] Buser, H.R. (1987b). Brominated and brominated chlorinated dibenzodioxins and dibenzofurans - potential environmental contaminants. *Chemosphere. 16*, 1873-1876.

[3] Olie, K. (1980). Chlorodibenzo-*p*-dioxins and chlorodibenzofurans are trace components of fly ash and flue gas of some municipal incinerators in the Netherlands. *Chemosphere*, *9*, 501-522.

[4] Laine, M.M., Ahtiainen, J., Wagman, N., Oberg, L.G., & Jorgensen, K.S. (1997). Fate and toxicity of chlorophenols, polychlorinated dibenzo-*p*-dioxins, and dibenzofurans during composting of contaminated sawmill soil. *Environ Sci Technol. 31*, 3244-3250.

[5] Gilpin, R.K., Wagel, D.J., & Solch, J.G. (2003). Production, distribution, and fate of polychlorinated dibenzo-*p*-dioxins, dibenzofurans, and related organohalogens in the environment. In: Schecter, A., Gasiewicz, T.A. (Eds.), Dioxins and Health. Wiley, Hoboken, NJ, pp. 89–136.

[6] Evans, C.S., & Dellinger, B. (2005). Mechanisms of dioxin formation from the high-temperature oxidation of 2-bromophenol. *Environ Sci Technol*, *39*, 2128-2134.

[7] Krizanec, B., Le Marechal. A.M., Voncina, E., & Brodnjak-Voncina, D. (2005). Presence of dioxins in textile dyes and their fate during the dyeing processes. *Acta Chimica Slovenicai*, *52*, 111-118.

[8] Gribble, G.W. (2003). The diversity of naturally produced organohalogens. *Chemosphere*, *52*, 289-297.

[9] Kim, E.J., Oh, J.E., & Chang,Y.S. (2003). Effects of forest fire on the level and distribution of PCDD/Fs and PAHs in soil. *Sci Total Environ*, *311*, 177-189.

[10] Adriaens, P., & Grbic-Galic, D. (1994). Reductive dechlorination of PCDD/F by anaerobic cultures and sediments. *Chemosphere*, *29*, 2253-2259.

[11] Mandal, P.K. (2005). Dioxin: a review of its environmental effects and its aryl hydrocarbon receptor biology. *J Compar Physiol B-Biochem Syst Environ Physiol*, *175*, 221-230.

[12] Lorenzen, A., & Okey, A.B. (1991). Detection and Characterization of Ah Receptor in Tissue and Cells from Human Tonsils. *Toxicol Appl Pharmacol*, *107*: 203-214.

[13] Vos, J.G., De Heer, C., & Van Loveren, H. (1997). Immunotoxic effects of TCDD and toxic equivalency factors. *Teratogen Carcinog Mutat*, *17*, 275-284.

[14] Sweeney, M.H., Calvert, G.M., Egeland, G.A., Fingerhut, M.A., Halperin, W.E., & Piacitelli, L.A. (1997). Review and update of the results of the NIOSH medical study of workers exposed to chemicals contaminated with 2,3,7,8-tetrachlorodibenzodioxin. *Teratogenesis Carcinog Mutat 17*, 241-247.

[15] Peper, M., Klett, M., Frentzelbeyme, R., & Heller, W.D. (1993) Neuropsychological effects of chronic exposure to environmental dioxins and furans. *Environ Res*, *60*, 124-135.

[16] Kogevinas, M., (1999). Carcinogenicity of dioxins. *Lancet*, *354*, 429-430.

[17] Geusau, A., Abraham, K., Geissler, K., Sator, M.O., Stingl, G., & Tschachler, E. (2001). Severe 2,3,7,8-tetrachlorodibenzo-*p*-dioxin (TCDD) intoxication: Clinical and laboratory effects. *Environ Health Perspec*, *109*, 865-869.

[18] Hay, A. (1977). Seveso - Dioxin Damage. *Nature*, *266*, 7-8.

[19] Hay, A. (1978). Vietnams Dioxin Problem. *Nature*, *271*, 597-598.

[20] Bertazzi, P.A., Bernucci, I., Brambilla, G., Consonni, D., Pesatori, A.C. (1998). The Seveso studies on early and long-term effects of dioxin exposure: A review. *Environ Health Perspect*, *106*, 625-633.

[21] Masunaga, S., Takasuga, T., & Nakanishi, J. (2001). Dioxin and dioxin-like PCB impurities in some Japanese agrochemical formulations. *Chemosphere*, *44*, 873-885.

[22] Oberg, L.G., Glas, B., Swanson, S.E., Rappe, C., & Paul, K.G. (1990). Peroxidase-catalyzed oxidation of chlorophenols to polychlorinated dibenzo-*p*-dioxins and dibenzofurans. *Arch Environ Contam Toxicol*, *19*, 930-938.

[23] Kearny, P.C., Wollson, E.A., & Ellington, Jr.C.P. (1972). Persistence and metabolism of chlorodioxins in soils. *Environ Sci Technol*, *6*, 1017-1019.

[24] Matsumura, F., & Benezet, H.J. (1973). Studies on the bioaccumulation and microbial degradation of 2,3,7,8-tetra- chlorodibenzo-*p*-dioxin. *Environ Health Perspect*, *5*, 253-258.

[25] Ward, C.T., & Matsumura, F. (1978). Fate of 2,3,7,8-tetrachlorodibenzo-*p*-dioxin (TCDD) in a model aquatic environment. *Arch Environ Contam Toxicol*, *7*, 349-357.

[26] Armengaud, J., Timmis, K.N. (1997). Biodegradation of dibenzo-*p*-dioxin and dibenzofuran by bacteria. *Microbiology*, *35*, 241-252.

[27] Wittich, R.M. (1998). Degradation of dioxin-like compounds by microorganisms. *Appl Microbiol Biotechnol*, *49*, 489-499.

[28] Bressler, D.C., & Fedorak, P.M. (2000) Bacterial metabolism of fluorene, dibenzofuran, dibenzothiophene, and carbazole. *Can J Microbiol*, *46*, 397-409.

[29] Nojiri, H., Habe, H., & Omori, T. (2001). Bacterial degradation of aromatic compounds via angular dioxygenation. *J Gen Appl Microbiol*. *47*, 279-305.

[30] Nojiri, H., & Omori, T. (2002). Molecular bases of aerobic bacterial degradation of dioxins: Involvement of angular dioxygenation. *Biosci Biotechnol Biochem*, *66*, 2001-2016.

[31] Hiraishi, A. (2003). Biodivesity of dioxin-degrading microorganisms and potential utilization in Bioremediation. *Microb Environ*, *18*, 105-125.

[32] Hong, H.B., Chang, Y.S., Choi, S.D., & Park, Y.H. (2000b). Degradation of dibenzofuran by *Pseudomonas putida* PH-01. *Water Res*, *34*, 2404-2407.

[33] Wittich, R.M., Wilkes, H., Sinnwell, V., Francke, W., & Fortnagel, P. (1992). Metabolism of dibenzo-*p*-dioxin by *Sphingomonas* sp. strain RW1. *Appl Environ Microbiol*, *58*, 1005-1010.

[34] Cerniglia, C.E., Morgan, J.C., & Gibson, D.T. (1979). Bacterial and fungal oxidation of dibenzofuran. *Biochem J*, 180, 175-185

[35] Klecka, G.M., & Gibson, D.T. (1979). Metabolism of dibenzo[1,4] dioxan by a *Pseudomonas* species. *Biochem J*, *180*, 639-645.

[36] Klecka, G.M., & Gibson, D.T., (1980). Metabolism of dibenzo-*p*-dioxin and chlorinated dibenzo-*p*-dioxins by a *Beijerinckia* species. *Appl Environ Microbiol*, *39*, 288-296.

[37] Selifonov, S.A., Slepen'kin, A.V., Adanin, V.M., Nefedova, M.Y.U., & Starovoitov, I.I. (1991). Oxidation of dibenzofuran by *Pseudomonas* strains harbouring plasmids of naphthalene degradation. *Mikrobiologiya*, *60*, 67-71.

[38] Becher, D., Specht, M., Hammer, E., Francke, W., & Schauer, F. (2000). Cometabolic degradation of dibenzofuran by biphenyl-cultivated *Ralstonia* sp. strain SBUG 290. *Appl Environ Microbiol*, *66*, 4528-4531.

[39] Catelani, D., & Colombi, A. (1974). Metabolism of biphenyl - structure and physicochemical properties of 2-hydroxy-6-oxo-6-phenylhexa-2,4-dienoic acid, *meta*-cleavage product from 2,3-dihydroxybiphenyl by *Pseudomonas putida*. *Biochem J*, *143*, 431-434.

[40] Stope, M.B., Bacher, D., Hammer, E., & Schauer, F. (2002). Cometabolic ring fission of dibenzofuran by gram-negative and gram-positive biphenyl-utilizing bacteria. *Appl Microbiol Biotechnol*. *59*, 62-67.

[41] Fuse, H., Takimura, O., Murakami, K., Inoue, H., & Yamaoka, Y. Degradation of chlorinated biphenyl, dibenzofuran, and dibenzo-*p*-dioxin by marine bacteria that degrade biphenyl, carbazole, or dibenzofuran. *Biosci Biotechnol Biochem*, *67*, 1121-1125.

[42] Fortnagel, P., Harms, H., Wittich, R.M., Krohn, S., Meyer, H., Sinnwell, V., Wilkes, H., Francke, W. (1990). Metabolism of dibenzofuran by *Pseudomonas* sp. strain HH69 and the mixed culture HH27. *Appl Environ Microbiol*, *56*, 148-1156.

[43] Wang, Y., Yamazoe, A., Suzuki, S., Liu, C.T., Aono, T., & Oyaizu, H. (2004). Isolation and characterization of dibenzofuran-degrading *Comamonas sp*. strains isolated from white clover roots. *Curr Microbiol*, *49*, 288-294.

[44] Kimura, N., & Urushigawa, Y. (2001). Metabolism of dibenzo-*p*-dioxin and chlorinated dibenzo-*p*-dioxin by a gram-positive bacterium, *Rhodococcus opacus* SAO101. *J Bioscience Bioeng*, 92, 138.

[45] Kitagawa, W., Kimura, N., & Kamagata, Y. (2004). A novel *p*-nitrophenol degradation gene cluster from a gram-positive bacterium, *Rhodococcus opacus* SAO101. *J Bacteriol*, *186*, 4894-4902.

[46] Kimura, N., Kitagawa, W., Mori, T., Nakashima, N., Tamura, T., & Kamagata, Y. (2006). Genetic and biochemical characterization of the dioxygenase involved in lateral dioxygenation of dibenzofuran from *Rhodococcus opacus* strain SAO101. *Appl Microbiol Biotechnol*, *73*, 474-484.

[47] L'Abbee, J.B., Barriault, D., & Sylvestre, M. (2005). Metabolism of dibenzofuran and dibenzo-*p*-dioxin by the biphenyl dioxygenase of *Burkholderia xenovorans* LB400 and *Comamonas testosteroni* B-356. *Appl Microbiol Biotechnol*, *67*, 506-514.

[48] Kubota, M., Kawahara, K., Sekiya, K., Uchida, T., Hattori, Y., Futamata, H., & Hiraishi, A. (2005). *Nocardioides aromaticivorans* sp. nov., a dibenzofuran-degrading bacterium isolated from dioxin-polluted environments. *Syst Appl Microbiol*, *28*, 165-174.

[49] Hiraishi, A., Yonemitsu, Y., Matsushita, M., Shin, Y.K., Kuraishi, H., (2002). Kawahara, K. Characterization of *Porphyrobacter sanguineus* sp. nov., an aerobic bacteriochlorophyll-containing bacterium capable of degrading biphenyl and dibenzofuran. *Arch Microbiol*, *178*, 45-52.

[50] Mohammadi, M., & Sylvestre, M. (2005). Resolving the profile of metabolites generated during oxidation of dibenzofuran and chlorodibenzofurans by the biphenyl catabolic pathway enzymes. *Chem Biol*, *12*, 835-846.

[51] Buckel, W. (2005). On the road to bioremediation of "Dioxin". *Chem Biol*, *12*, 723-724.

[52] Engesser, K.H., Strubel, V., Christoglou, K., Fischer, P., & Rast, H.G. (1989). Dioxygenolytic cleavage of aryl ether bonds: 1,10-dihydro-1,10-dihydroxyfluoren-9-one, a novel arene dihydrodiol as evidence for angular dioxygenation of dibenzofuran. *FEMS Microbiol Lett*, *65*, 205-210.

[53] Fortnagel, P., Wittich, R.M., Harms, H., Schmidt, S., Franke, S., Sinnwell, V., Wilkes, H., Francke. W. (1989). New bacterial degradation of the biaryl ether structure. Regioselective dioxygenation prompts cleavage of the ether bonds. *Naturwissenschaften*, *76*, 523-524.

[54] Hong, H.B., Nam, I.H., Murugesan, K, Kim, Y.M., & Chang, Y.S. (2004). Biodegradation of dibenzo-*p*-dioxin, dibenzofuran, and chlorodibenzo-*p*-dioxins by *Pseudomonas veronii* PH-03. *Biodegradation*, *15*, 303-313.

[55] Noumura, T., Habe, H., Widada, J., Chung, J.S., Yoshida, T., Nojiri, H., & Omori, T. (2004). Genetic characterization of the dibenzofuran-degrading *Actinobacteria* carrying the *dbfA1A2* gene homologues isolated from activated sludge. *FEMS Microbiol Lett*, *239*, 147-155.

[56] Inoue, K., Widada, J., Nakai, S., Endoh, T., Urata, M., Ashikawa, Y., Shintani, M., Saiki, Y., Yoshida, T., Habe, H., Omori, T., & Nojiri, H. (2004). Divergent structures of carbazole degradative car operons isolated from gram-negative bacteria. *Biosci Biotechnol. Biochem*, *68*, 1467-1480.

[57] Iwai, S., Yamazoe, A., Takahashi, R., Kurisu, F., & Yagi, O. (2005). Degradation of monochlorinated dibenzo-*p*-dioxins by *Janibacter* sp. strain YA isolated from river sediment. *Current Microbiol*, *51*, 353-358.

[58] Jin, S.W., Zhu, T., Xu, X.D., Xu, Y. (2006). Biodegradation of dibenzofuran by *Janibacter terrae* strain XJ-1. *Current Microbiol*, *53*, 30-36.

[59] Yamazoe, A., Yagi, O., & Oyaizu, H. (2004a). Biotransformation of fluorene, diphenyl ether, dibenzo-*p*-dioxin and carbazole by *Janibacter* sp. *Biotechnol Lett*, *26*, 479-486.

[60] Yamazoe, A., Yagi, O., & Oyaizu, H. (204b). Degradation of polycyclic aromatic hydrocarbons by a newly isolated dibenzofuran-utilizing *Janibacter* sp. strain YY-1. *Appl Microbiol Biotechnol*, *65*, 211-218.

[61] Inoue, K., Habe, H., Yamane, H., Omori, T., & Nojiri, H. (2005). Diversity of carbazole-degrading bacteria having the *car* gene cluster: isolation of a novel gram-positive carbazole-degrading bacterium. *FEMS Microbiol Lett*, *245*,145-153.

[62] Inoue, K., Habe, H., Yamane, H., & Nojiri, M. (2006). Characterization of novel carbazole catabolism genes from gram-positive carbazole degrader *Nocardioides aromaticivorans* IC177. *Appl Environ Microbiol*, *72*, 3321-3329.

[63] Iida, T., Nakamura, K., Izumi, A., Mukouzaka, Y., & Kudo, T. (2006). Isolation and characterization of a gene cluster for dibenzofuran degradation in a new dibenzofuran-utilizing bacterium, *Paenibacillus* sp. strain YK5. *Arch Microbiol*, *184*, 305-315.

[64] Ouchiyama, N., Zhang, Y., Omori, T., Kodama, T. (1993). Biodegradation of carbazole by *Pseudomonas* spp CA06 and CA10. *Biosci Biotechnol Biochem*, *57*, 455-460.

[65] Sato, S.I., Nam, J.W., Kasuga, K., Nojiri, H., Yamane, H., & Omori, T. (1997a). Identification and characterization of genes encoding carbazole 1,9a-dioxygenase in *Pseudomonas* sp. strain CA10. *J Bacteriol*, *179*, 4850-4858.

[66] Sato, S., Ouchiyama, N., Kimura, T., Nojiri, H., Yamane, H., & Omori, T. (1997b). Cloning of genes involved in carbazole degradation of *Pseudomonas* sp. strain CA10: Nucleotide sequences of genes and characterization of meta-cleavage enzymes and hydrolase. *J Bacteriol*, *179*, 4841-4849.

[67] Nojiri, H., Nam, J.W., Kosaka, M., Morii, K.I., Takemura, T., Furihata, K., Yamane, H., & Omori, T. (1999). Diverse oxygenations catalyzed by carbazole 1,9a-dioxygenase from *Pseudomonas* sp. strain CA10. *J Bacteriol*, *181*, 3105-3113.

[68] Habe, H., Chung, J.S., Lee, J.H., Kasuga, K., Yoshida, T., Nojiri, H., & Omori, T. (2001a). Degradation of chlorinated dibenzofurans and dibenzo-*p*-dioxins by two types of bacteria having angular dioxygenases with different features. *Appl Environ Microbiol*, *67*, 3610-3617.

[69] Habe, H., Ide, K., Yotsumoto, M., Tsuji, H., Hirano, H., Widada, J., Yoshida, T., Nojiri, H., & Omori, T. (2001b). Preliminary examinations for applying a carbazole-degrader, *Pseudomonas* sp. strain CA10, to dioxin-contaminated soil remediation. *Appl Microbiol Biotechnol*, *56*, 788-795.

[70] Nam, J.W., Fujimoto, Z., Mizuno, H., Yamane, H., Yoshida, T., Habe, H., Nojiri, H., Omori, T. (2002). Crystallization and preliminary crystallographic analysis of the terminal oxygenase component of carbazole 1,9a-dioxygenase of *Pseudomonas resinovorans* strain CA10. *Acta Crystallographica Section D-Biological Crystallography, 58*, 1350-1352.

[71] Maeda, K., Nojiri, H., Shintani, M., Yoshida, T., Habe, H., & Omori, T. (2003). Complete nucleotide sequence of carbazole/dioxin-degrading plasmid pCAR1 in *Pseudomonas resinovorans* strain CA10 indicates its mosaicity and the presence of large catabolic transposon Tn4676. *J Mol Biol*, *326*, 21-33.

[72] Wilkes, H., Wittich, R.M., Timmis, K.N., Fortnagel. P., & Francke, W. (1996). Degradation of chlorinated dibenzofurans and dibenzo-*p*-dioxins by *Sphingomonas* sp. strain RW1. *Appl Environ Microbiol*, *62*, 367-371.

[73] Keim, T., Francke, W., Schmidt, S., & Fortnagel, P. (1999). Catabolism of 2,7-dichloro- and 2,4,8-trichlorodibenzofuran by *Sphingomonas* sp. strain RW1. *J Ind Microbiol Biotechnol*, *23*, 359-363.

[74] Hong, H.B., Chang, Y.S., Nam, I.H., Fortnagel, P., & Schmidt, S. (2002). Biotransformation of 2,7-dichloro- and 1,2,3,4-tetrachlorodibenzo-*p*-dioxin by *Sphingomonas wittichii* RW1. *Appl Environ Microbiol*, *68*, 2584-2588.

[75] Nam, I.H., Hong, H.B., Kim, Y.M., Kim, B.H., Murugesan, K., & Chang, Y.S. (2005a) Biological removal of polychlorinated dibenzo-*p*-dioxins from incinerator fly ash by *Sphingomonas wittichii* RW1. *Water* Res, *39*, 4651-4660.

[76] Nam, I.H., Kim, Y.M., Schmidt, S., & Chang, Y.S. (2006). Biotransformation of 1,2,3-Tri- and 1,2,3,4,7,8-hexachlorodibenzo-*p*-dioxin by *Sphingomonas wittichii* strain RW1. *Appl Environ Microbiol*, *72*, 112-116.

[77] Arfmann, H.A., Timmis, K.N., & Wittich, R.M. (1997). Mineralization of 4-chlorodibenzofuran by a consortium consisting of *Sphingomonas* sp. strain RW1 and *Burkholderia* sp. Strain JWS. *Appl Environ Microbiol*, *63*, 3458-3462.

[78] Harms, H., Wittich, R.M., Sinnwell, V., Meyer, H., Fortnagel, P., & Francke, W. (1990). Transformation of dibenzo-*p*-dioxin by *Pseudomonas* sp. strain HH69. *Appl Environ Microbiol, 56*, 1157-1159.

[79] Harms, H., Wilkes, H., Sinnwell, V., Wittich, R.M., Figge, K., Francke, W., & Fortnagel, P. (1991). Transformation of 3-chlorodibenzofuran by *Pseudomonas* sp. HH69. *FEMS Microbiol Lett, 81*, 25-30.

[80] Wittich, R.M., Strömpl, C., Moore, E.R.B., Blasco, R., Timmis, K.N. (1999). Interaction of *Sphingomonas* and *Pseudomonas* strains in the degradation of chlorinated dibenzofurans. *J Ind Microbiol Biotechnol, 23*, 353-358.

[81] Wittich, R.M., Busse, H.J., Kämpfer, P., Tiirole, M., Wieser, M., Macedo, A., & Abraham, W.R. (2007). *Sphingobium aromaticiconvertens* sp. nov., a xenobiotic-compound-degrading bacterium from polluted river sediment. *Int J Syst Evol Microbiol, 57*, 306-310.

[82] Schreiner, G., Wiedmann, T., Schimmel, H., & Ballschmitter, K. (1997). Influence of the substitution pattern on the microbial degradation of mono- to tetrachlorinated dibenzo-*p*-dioxins and dibenzofurans. *Chemosphere, 34*, 1315-1331.

[83] Sander, P., Wittich, R.M., Fortnagel, P., Wilkes, H., Francke, W. (1991). Degradation of 1,2,4-trichloro-and 1,2,4,5-tetrachlorobenzene by *Pseudomonas* strains. *Appl Environ Microbiol, 57*, 1430-1440.

[84] Bunz, P.V., & Cook, A.M. (1993). Dibenzofuran 4,4a-dioxygenase from *Sphingomonas* sp. strain RW1 - angular dioxygenation by a 3-component enzyme system. *J Bacteriol, 175*, 6467-6475.

[85] Armengaud, J., Timmis, K.N. (1997). Biodegradation of dibenzo-*p*-dioxin and dibenzofuran by bacteria. *Microbiology, 35*, 241-252.

[86] Armengaud, J., & Timmis, K.N. (1998). The reductase RedA2 of the multi-component dioxin dioxygenase system of *Sphingomonas* sp. RW1 is related to class-I cytochrome P-450-type reductases. *Eur J Biochem, 253*, 437-444.

[87] Armengaud, J., Happe, B., & Timmis, K.N. (1998). Genetic analysis of dioxin dioxygenase of *Sphingomonas* sp. strain RW1: Catabolic genes dispersed on the genome. *J Bacteriol. 180*, 3954-3966.

[88] Armengaud, J., Gaillard, J., & Timmis, K.N. (2000). A second ferredoxin from *Sphingomonas* sp. strain RW1 can function as an electron donor for the dioxin dioxygenase. *J Bacteriol, 182*, 2238-2244.

[89] Kasuga, K., Nojiri, H., Yamane, H., Kodama, T., & Omori, T. (1997). Cloning and characterization of the genes involved in the degradation of dibenzofuran by *Terrabacter* sp. strain DBF63. *J Fermentation Bioeng. 84*, 387-399.

[90] Kasuga, K., Habe, H., Chung, J.S., Yoshida, T., Nojiri, H., Yamane, H., & Omori, T. (2001). Isolation and characterization of the genes encoding a novel oxygenase component of angular dioxygenase from the gram-positive dibenzofuran-degrader *Terrabacter* sp. strain DBF63. *Biochem Biophy Res Commun, 283*, 195-204.

[91] Iida, T., Mukouzaka, Y., Nakamura, K., & Kudo, T. (2002). Plasmid-borne genes code for an angular dioxygenase involved in dibenzofuran degradation by *Terrabacter* sp. strain YK3. *Appl Environ Microbiol. 68*, 3716-3723.

[92] Nojiri, H., Ashikawa, Y., Noguchi, H., Nam, J.W., Urata, M., Fujimoto, Z., Uchimura, H., Terada, T., Nakamura, S., Shimizu, K., Yoshida, T., Habe, H., & Omori, T. (2005).

Structure of the terminal oxygenase component of angular dioxygenase, carbazole 1,9 a-dioxygenase. *J Mol Biol*, *351*, 355-370.

[93] Nam, J.W., Nojiri, H., Yoshida, T., Habe, H., Yamane, H., & Omori, T. (2001). New classification system for oxygenase components involved in ring-hydroxylating oxygenations. *Biosci Biotechnol Biochem*, *65*, 254-263.

[94] Coon, M.J. (2005). Cytochrome P450: Nature's most versatile biological catalyst. *Annu Rev Pharmacol Toxicol*, *45*, 1-25.

[95] Urlacher, V.B., & Eiben, S. (2006). Cytochrome P-450 monooxygenases: perspectives for synthetic application. *Trends Biotechnol*, *24*, 324-330.

[96] Sulistyaningdyah, W.T., Ogawa, J., Li, Q.S., Shinkyo, R., Sakaki, T., Inouye, K., Schmid, R.D., & Shimizu, S. (2004). Metabolism of polychlorinated dibenzo-*p*-dioxins by cytochrome P450BM-3 and its mutant. *Biotechnol Lett*, *26*, 1857-1860.

[97] Mast, N., Andersson, U., Nakayama, K., Bijorkhem, I., (2004). Pikuleva, A. Expression of human cytochrome *P*450 46A1 in *Escherichia coli*: effects of N- and C-terminal modifications. *Arch Biochem Biophys*, *428*, 99-108.

[98] Shinkyo, R., Kamakura, M., Ikushiro, S., Inouye, K., & Sakaki, T. (2006). Biodegradation of dioxins by recombinant *Escherichia coli* expressing rat CYP1A1 or its mutant. *Appl Microbiol Biotechnol*, *72*, 584-590.

[99] Lee, M.D., Odom, J.M., & Buchanan, R.J. (1998). New perspectives on microbial dehalogenation of chlorinated solvents: Insights from the field. *Annu Rev Microbiol*, *52*, 423-452.

[100] Chen, G. (2004). Reductive dehalogenation of tetrachloroethylene by microorganisms: Current knowledge and application strategies. *Appl Microbiol Biotechnol*, *63*, 373-377.

[101] Zhang, C.L., & Bennett, G.N. (2005). Biodegradation of xenobiotics by anaerobic bacteria. *Appl Microbiol Biotechnol*, 67, 600-618.

[102] Townsend, D.I. (1983). Change of isomer ratio and fate of polychlorinated-*p*-dioxins in the environment. *Chemosphere*, *12*, 637-643.

[103] Adriaens, P., Fu, Q.Z., & Grbic-Galic, D. (1995). Bioavailability and transformation of highly chlorinated dibenzo-*p*-dioxins and dibenzofurans in anaerobic soils and sediments. *Environ Sci Technol*, *29*, 2252-2260.

[104] Beurskens, J.E.M., Toussaint, M., Dewolf, J., Vandersteen, J.M.D., & Slot, P.C. (1995). Commandeur LCM, Parsons JR: Dehalogenation of chlorinated dioxins by an anaerobic microbial consortium from sediment. *Environ Toxicol Chem*, *14*, 939-943.

[105] Barkovskii, A.L., & Adriaens, P. (1996). Microbial dechlorination of historically present and freshly spiked chlorinated dioxins and diversity of dioxin-dechlorinating populations. *Appl Environ Microbiol*, *62*, 4556-4562.

[106] Ballerstedt, H., Kraus, A., & Lechner, U. (1997). Reductive dechlorination of 1,2,3,4-tetrachlorodibenzo-*p*-dioxin and its products by anaerobic mixed cultures from Saale river sediment. *Environ Sci Technol*, *31*, 1749-1753.

[107] Hiraishi, A., Miyakoda, H., Lim, B.R., Hu, H.Y., Fujie, K., & Suzuki, J. (2001). Toward the bioremediation of dioxin-polluted soil: structural and functional analyses of in situ microbial populations by quinone profiling and culture-dependent methods. *Appl Microbiol Biotechnol*, *57*, 248-256.

[108] Kao, C.M., Chen, S.C., Liu, J.K., & Wu, M.J. (2001). Evaluation of TCDD biodegradability under different redox conditions. *Chemosphere*, *44*, 1447-1454.

[109] Vargas, C., Fennell, D.E., & Haggblom, M.M. (2001). Anaerobic reductive dechlorination of chlorinated dioxins in estuarine sediments. *Appl Microbiol Biotechnol*, *57*, 786-790.

[110] Bunge, M., Ballerstedt, H., & Lechner, U. (2001). Regiospecific dechlorination of spiked tetra- and trichlorodibenzo-*p*-dioxins by anaerobic bacteria from PCDD/F-contaminated Spittelwasser sediments. *Chemosphere*, *43*, 675-681.

[111] Ballerstedt, H., Hantke, J., Bunge, M., Werner, B., Gerritse, J., Andreesen, J.R., & Lechner, U. (2004) Properties of a trichlorodibenzo-*p*-dioxin-dechlorinating mixed culture with a *Dehalococcoides* as putative dechlorinating species. *FEMS Microbiol Ecol*, *47*, 223-234.

[112] Ahn, Y.B., Haggblom, M.M., & Fennell, D.E. (2005). Co-amendment with halogenated compounds enhances anaerobic microbial dechlorination of 1,2,3,4-tetrachlorodibenzo-*p*-dioxin and 1,2,3,4-tetrachlorodibenzofuran in estuarine sediments. *Environ Toxicol Chem*. *24,* 2775-2784.

[113] Yoshida, N., Takahashi, N., & Hiraishi, A. (2005). Phylogenetic characterization of a polychlorinated-dioxin-dechlorinating microbial community by use of microcosm studies. *Appl Environ Microbiol*, *71*, 4325-4334.

[114] Adriaens, P., Li, M.Y., & Michalak, A.M. (2006). Scaling methods of sediment bioremediation processes and applications. *Eng Life Sci*, *6*, 217-227.

[115] Fu, Q.S., Barkovskii, A.L., & Adriaens, P. (2001). Dioxin cycling in aquatic sediments: the Passaic River Estuary. Chemosphere, *43*, 643-648.

[116] Fu, Q.S., Barkovskii, A.L., & Adriaens, P. (2005). Microbial dechlorination of dioxins in estuarine enrichment cultures: effects of respiratory conditions and priming compound on community structure and dechlorination patterns. *Marine Environ Res*, *59*, 177-195.

[117] Bunge, M., Adrian, L., Kraus, A., Opel, M., Lorenz, W.G., Andreesen, J.R., Gorisch, H., & Lechner, U. (2003). Reductive dehalogenation of chlorinated dioxins by an anaerobic bacterium. *Nature*, *421*, 357-360.

[118] Adrian, L., Szewzyk, U., Wecke, J., & Görisch, H. (2000). Bacterial dehalorespiration with chlorinated benzenes. *Nature*, *408*, 580-583.

[119] Holscher, T., Krajmalnik-Brown, R., Ritalahti, K.M., von Wintzingerode, F., Görisch, H., Löffler, F.E., & Adrian, L. (2004). Multiple nonidentical reductive-dehalogenase-homologous genes are common in *Dehalococcoides*. *Appl Environ Microbiol*, *70*, 5290-5297.

[120] Kluyev, N., Cheleptchikov, A., Brodsky, E., Soyfer, V., & Zhilnikov, V. (2002). Reductive dechlorination of polychlorinated dibenzo-*p*-dioxins by zerovalent iron in subcritical water. *Chemosphere*, *46*, 1293-1296.

[121] Sakaki, T., Shinkyo, R., Takita, T., Ohta, M., & Inouye, K. (2002). Biodegradation of polychlorinated dibenzo-*p*-dioxins by recombinant yeast expressing rat CYP1A subfamily. *Arch Biochem Biophy*, *401*, 91-98.

[122] Shinkyo, R., Sakaki, T., Takita, T., Ohta, M., Inouye, K. (2003). Generation of 2,3,7,8-TCDD-metabolizing enzyme by modifying rat CYP1A1 through site-directed mutagenesis. *Biochem Biophy Res Commun*, *308*, 511-517.

[123] Orihara, K., Yamazaki, T., Shinkyo, R., Sakaki, T., Inouye, K., Tsukamoto, A., Sugiura, J., & Shishido, K. (2005). Rat cytochrome *P*450-mediated transformation of

dichlorodibenzo-*p*-dioxins by recombinant white-rot basidiomycete *Coriolus hirsutus*. *Appl Microbiol Biotechnol*, *69*, 22-28.

[124] Bumpus, J.A., & Tien, M., Wright, D., & Aust, S.D. (1985). Oxidation of persistent environmental pollutants by a white rot fungus. *Science*, *228*, 1434-1436.

[125] Hammel, K.E., Kalyanaraman, B., & Kirk, T.K. (1986). Oxidation of polycyclic aromatic hydrocarbons and dibenzo-*p*-dioxins by *Phanerochaete chrysosporium* ligninase. *J Biol Chem*, *261*, 6948-6952.

[126] Joshi, D.K., & Gold, M.H. (1994). Oxidation of dibenzo-*p*-dioxin by lignin peroxidase from the basidiomycete *Phanerochaete chrysosporium*. *Biochem*, *33*, 10969-10976.

[127] Valli, K., Wariishi, H., & Gold, M.H. (1992). Degradation of 2,7-dichlorodibenzo-*p*-dioxin by the lignin degrading basidiomycete *Phanerochaete chrysosporium*. *J Bacteriol,* 174:2131-2137.

[128] Takada, S., Nakamura, M., Matsueda, T., Kondo, R., & Sakai, K. (1996). Degradation of polychlorinated dibenzo-*p*-dioxins and polychlorinated dibenzofurans by the white rot fungus *Phanerochaete sordida* YK-624. *Appl Environ Microbiol*, *62*, 4323-4328.

[129] Mori, T., & Kondo, R. (2002a). Degradation of 2,7-dichlorodibenzo-*p*-dioxin by wood-rotting fungi, screened by dioxin degrading ability. *FEMS Microbiol Lett*, *213*, 127-131.

[130] Mori, T., & Kondo, R. (2002b). Oxidation of dibenzo-*p*-dioxin, dibenzofuran, biphenyl, and diphenyl ether by the white-rot fungus *Phlebia lindtneri*. *Appl Microbiol Biotechnol*, *60,* 200-205.

[131] Mori, T., & Kondo, R. (2002c). Oxidation of chlorinated dibenzo-*p*-dioxin and dibenzofuran by white-rot fungus, *Phlebia lindtneri*. *FEMS Microbiol Lett*, *216*, 223-227.

[132] Manji, S., & Ishihara, A. (2004). Screening of tetrachlorodibenzo-*p*-dioxin-degrading fungi capable of producing extracellular peroxidases under various conditions. *Appl Microbiol Biotechnol*, *63*, 438-444.

[133] Sato, A., Watanabe, T., Watanabe, Y., Harazono, K., & Fukatsu, T. (2002) Screening for basidiomycetous fungi capable of degrading 2,7-dichlorodibenzo-*p*-dioxin. *FEMS Microbiol Lett*; *213*, 213-217.

[134] Kamei, I., & Kondo, R. (2005). Biotransformation of dichloro-, trichloro-, and tetrachlorodi-benzo-*p*-dioxin by the white-rot fungus *Phlebia lindtneri*. *Appl Microbiol Biotechnol*, *68*, 560-566.

[135] Kamei, I., Suhara, H., & Kondo, R. (2005). Phylogenetical approach to isolation of white-rot fungi capable of degrading polychlorinated dibenzo-*p*-dioxin. *Appl Microbiol Biotechnol*, *69*, 358-366.

[136] Miyoshi, S., Kimura, K., Matsumoto, R., Itoh, K., & Tachibana, S. (2005). Biodegradation of 2,8-dichlorodibenzo-*p*-dioxin by fungi screened from nature. *Pakistan J Biol. Sci*, *8*, 1265-1271.

[137] Tachibana, S., Kiyota, Y., & Koga, M. (2005). Biodegradation of 2,7-dibenzo-*p*-dixoin and 2,4,8-trichlorodibenzofuran in soil by fungi screened from nature. *Pakistan J Biol Sci*, *8*, 1751-1757.

[138] Kamei, I., Sonoki, S., Haraguchi, K., & Kondo, R. (2006). Fungal bioconversion of toxic polychlorinated biphenyls by white-rot fungus, *Phlebia brevispora*. *Appl Microbiol Biotechnol*, *73*, 932-940.

[139] Tachibana, S., Kiyota, Y., & Koga, M. (2006). Bioremediation of 2,3,7,8-tetrachlorodibenzo-*p*-dioxin in soil by fungi screened from nature. *Pakistan J Biol Sci, 9*, 217-222.

[140] Kamei, I., & Kondo, R. (2006). Simultaneous degradation of commercially produced CNP herbicide and of contaminated dioxin by treatment using the white-rot fungus *Phlebia brevispora. Chemosphere, 65*, 1221-1227.

[141] Couto, S.R., & Herrera, J.L.T. (2006). Industrial and biotechnological applications of laccases: A review. *Biotechnol Advan, 24*, 500-513.

[142] Camoni, I., Dimuccio, A., Pontecorvo, D., Rubbiani, M., Silano, V., Vergori, L., Vonhunolstein, C., Antonini, G., Orsi, N., & Valenti, P. (1983). Lack of *in vitro* oxidation of 2,3,7,8-tetrachlorodibenzo-*p*-dioxin (TCDD) in the presence of laccase from *Polyporus versicolor* fungus. *Chemosphere, 12*, 945-949.

[143] Jonas, U., Hammer, E., Schauer, F., & Bollag, J.M. (1997). Transformation of 2-hydroxydibenzofuran by laccases of the white rot fungi *Trametes versicolor* and *Pycnoporus cinnabarinus* and characterization of oligomerization products. *Biodegradation, 8*, 321-328.

[144] Jonas, U., Hammer, E., Haupt, E.T.K., & Schauer, F. (2000). Characterisation of coupling products formed by biotransformation of biphenyl and diphenyl ether by the white rot fungus *Pycnoporus cinnabarinus*. *Arch Microbiol, 174*, 393-398.

[145] Schultz, A., Jonas, U., Hammer, E., & Schauer, F. (2001). Dehalogenation of chlorinated hydroxybiphenyls by fungal laccase. *Appl Environ Microbiol. 67*, 4377-4381.

[146] Nakamiya, K., Furuichi, T., & Ishii, K. (2002). Isolation of a fungus from denitrifying activated sludge that degrades highly chlorinated dioxins. *J Mater Cycles Waste Manag, 4*, 127-134.

[147] Ishii, K., Furuichi, T., Koike, K., & Kuboshima, M. (2004) Purification of highly chlorinated dioxins degrading enzyme. *Organohal Compo, 66*, 1262-1266.

[148] Nakamiya, K., Hashimoto, S., Ito, H., Edmonds, J.S., Yasuhara, A., & Morita, M. (2005) Degradation of dioxins by cyclic ether degrading fungus, *Cordyceps sinensis. FEMS Microbiol Lett*; *248*, 17-22.

[149] Megharaj, M., Wittich, R.M., Blasco, R., Pieper, D.H., & Timmis, K.N. (1997). Superior survival and degradation of dibenzo-*p*-dioxin and dibenzofuran in soil by soil-adapted *Sphingomonas* sp. strain RW1. *Appl Microbiol Biotechnol. 48*, 109-114.

[150] Halden, R.U., Halden, B.G., Dwyer, D.F. (1999). Removal of dibenzofuran, dibenzo-*p*-dioxin, and 2-chlorodibenzo-*p*-dioxin from soils inoculated with *Sphingomonas* sp. strain RW1. *Appl Environ Microbiol, 65*, 2246-2249.

[151] Habe, H., Ashikawa, Y., Saiki, Y., Yoshida, T., Nojiri, H., & Omori, T. (2002a). *Sphingomonas* sp. strain KA1, carrying a carbazole dioxygenase gene homologue, degrades chlorinated dibenzo-*p*-dioxins in soil. *FEMS Microbiol Lett, 211*, 43-49.

[152] Habe, H., Ide, K., Yotsumoto, M., Tsuji, H., Yoshida, T., Nojiri, H., Omori, T. (2002b) Degradation characteristics of a dibenzofuran-degrader *Terrabacter* sp. strain DBF63 toward chlorinated dioxins in soil. *Chemosphere, 48*, 201-207.

[153] Halden, R.U. Colquhoun, D.R., & Wisniewski, E.S. (2005). Identification and phenotypic characterization of *Sphingomonas wittichii* strain RW1 by peptide mass fingerprinting using matrix-assisted laser desorption ionization-time of flight mass spectrometry. *Appl Environ Microbiol, 71*, 2442-2451.

[154] Suhara, H., Daikoku, C., Takata, H., Suzuki, S., Matsufuji, Y., Sakai, K., & Kondo, R. (2003) Monitoring of white-rot fungus during bioremediation of polychlorinated dioxin-contaminated fly ash. *Appl Microbiol Biotechnol*, *62*, 601-607.

[155] Hacketta, G.A.R., Eastonb, C.A., & Du, S.J.B. (1999). Composting of pulp and paper mill fly ash with wastewater treatment sludge. *Bioresource Technol*, *70*, 217-224.

[156] Boonchan, S., Britz, M.L., & Stanley, G.A. Degradation and mineralization of high-molecular-weight polycyclic aromatic hydrocarbons by defined fungal-bacterial cocultures. *Appl Environ Microbiol*, *66*, 1007-1019.

[157] Nam, I.H., Kim, Y.M., Murugesan, K., Chang, Y.S. (2005b) New microbial agent for removal of PCDD/Fs from municipal solid waste incinerator fly ash. *Organohal Compo*, *67*, 1044-1047.

[158] Campanella, B.F., Bock, C., & Schroder, P. (2002). Phytoremediation to increase the degradation of PCBs and PCDD/Fs - Potential and limitations. *Environ Sci Pollut Res*, *9*, 73-85.

[159] Sonoki, S., Hisamatsu, S., & Takagi, S. (2006). Transgenic plants capable of dioxin degradation, transformed with dioxin degrading enzyme coding genes of white rot fungi. *Jpn. Kokai Tokkyo Koho* CODEN: JKXXAF JP 2006333758 Patent.

[160] Sayler, G.S., & Ripp, S. (2000). Field applications of genetically engineered microorganisms for bioremediation processes. *Curr Opin Biotechnol*, *11*, 286-289.

[161] Liu, Z., Hong, Q., Xu, H.H., Jun, W., Li, S.P. (2006). Construction of a genetically engineered microorganism for degrading organophosphate and carbamate pesticides. *Int Biodeterio Biodegradation*, *58*, 65-69.

[162] Urgun-Demirtas, M., Stark, B., & Pagilla, K. (2006). Use of genetically engineered microorganisms (GEMs) for the bioremediation of contaminants. *Crit Rev Biotechnol*, *26*, 145-164.

[163] Wittich, R.M., & Wolff, P. (2007). Growth of the genetically engineered strain *Cupriavidus necator* RW112 with chlorobenzoates and technical chlorobiphenyls. *Microbiol*, *153*, 186-195.

[164] Aso, Y., Miyamoto, Y., Harada, K.M., Momma, K., Kawai, S., Hashimoto, W., Mikami, B., & Murata, K.(2006). Engineered membrane superchannel improves bioremediation potential of dioxin-degrading bacteria. *Nature Biotechnol*, *24*, 188-189.

[165] Tsezos, M., Bell, J.P. (1989). Comparison of the biosorption and desorption of hazardous organic pollutants by live and dead biomass. *Water Res*, *23*, 561-568.

[166] Schorer, M., & Eisele, M. (1997). Accumulation of inorganic and organic pollutants by biofilms in the aquatic environment. *Water Air Soil Pollut*, *99*, 651-659.

[167] Hong, H.B., Hwang, S.H., & Chang, Y.S. (2000a). Biosorption of 1,2,3,4-tetrachlorodibenzo-*p*-dioxin and polychlorinated dibenzofurans by *Bacillus pumilus*. *Water Res*. *34*, 349-353.

[168] Chang, Y.S., Hong, H.B., Kim, B.H., Hall, S.C., & Settineri, T.A. (2001). Sequencing of novel protein from *Bacillus pumilus* PH-01 using a high-resolution hybrid quadrupole-time-of-flight mass spectrometer. *Int J Mass Spectrom*, *209*, 47-55.

[169] Hong, H.B., Chang, Y.S., Choi, S.D., Nam, I.H., & Lee, Y.E. (2001). Isolation and characterization of a cell-associated protein of *Bacillus pumilus* PH-01. *Appl Microbiol Biotechnol*, *56*, 402-405.

[170] Choi, S.D., Hong, H.B., & Chang, Y.S. (2003). Adsorption of halogenated aromatic pollutants by a protein released from *Bacillus pumilus*. *Water Res*, *37*, 4004-4010.

[171] Erable, B., Goubet, I., Lamare, S., Legoy, M.D., & Maugard, T. (2004). Haloalkane hydrolysis by *Rhodococcus erythropolis* cells: Comparison of conventional aqueous phase dehalogenation and nonconventional gas phase dehalogenation. *Biotech Bioeng*, *86*, 47-54.

[172] Erable, B., Goubet, I., Lamare, S., Seltana, A., Legoy, M.D., & Maugard, T. (2005a). Nonconventional hydrolytic dehalogenation of 1-chlorobutane by dehydrated bacteria in a continuous solid-gas biofilter. *Biotech Bioeng*, *91*, 304-313.

[173] Erable, B., Maugard, T., Goubet, I., Lamare, S., & Legoy, M.D. (2005b). Biotransformation of halogenated compounds by lyophilized cells of *Rhodococcus erythropolis* in a continuous solid-gas biofilter. *Process Biochem*. *40*, 45-51.

In: Leading-Edge Environmental Biodegradation
Editor: L. E. Pawley, pp. 167-198 ISBN: 978-1-60021-903-0

Chapter 5

ALGAL-BASED PHOTOBIOREACTORS: A NEW TECHNOLOGY PLATFORM FOR RESOURCE-EFFICIENT POLLUTION CONTROL?

Raúl Muñoz[1*] and Benoit Guieysse[2†]

[1]Department of Chemical Engineering and Environmental Technology, Valladolid University, Paseo del Prado de la Magdalena, s/n, Valladolid, Spain

[2]Division of Environmental and Water Resources Engineering, Nanyang Technological University, Block N1, Nanyang Avenue, Singapore, Singapore

ABSTRACT

Microalgae-based environemntal technologies offer attractive and reliable alternatives to conventional energy-demanding and green house gases emitting wastewater treatment processes. For instance, photosynthetic oxygenation is more economical (no need for external aeration) and safer (no CO_2 emission and low risks of pollutant and microorganism spreading). In addition, nutrient removal can be coupled with the mineralization of organic comounds in a single stage process, with a subsequent reduction in plant construction and operation costs. Microalgae also support cost-efficient technologies for the removal of toxic heavy metals or in the fixation of CO_2 and the biomass produced from algae-based process is a potential source of valuable chemicals, fertilizers or biofuels (methane, hydrocarbon or biodiesel). Unfortunately, despite the merits of this approach, the limitations imposed by the light distribution to the photosynthetic cells, together with the high cost and scale-up problems of photobioreactors have constrained the use of algae-based technologies in environmental applications. However, the massive demand for algal-based products in the food, biotechnology and agricultural sectors has contributed to generate a considerable knowledge on these processes. More recently, increasing energy prices and CO_2-mitigation policies even boosted research and development on algal-based processes for

* Phone: +34983184934; Fax: +34983423013
† Phone: +6567911744; Fax: +65-67910676.

bioenergy production, an application still economically unviable a decade ago. With mature technologies now becoming available and a highly favorable political and energetic conjuncture, time has come to reevaluate algal-based photobioreactors as energy-efficient and waste-to-resource pollution control processes.

1. Introduction

Microalgae constitute the basis of the food chain and contribute to half of the total photosynthetic O_2 production in the world (Day et al., 1999). Since microalgal cultivation yields higher areal productivities than traditional extensive agriculture, microalgae constitute an interesting source of proteins for human and animal (Oswald, 1995; Nurdogan and Oswald, 1995). As examples, microalgae such as *Tetraselmis* sp., *Spirulina* sp. or *Chaetoceros* sp. are currently employed as food source in aquaculture for shrimps or salmonids production (Borowitzka, 1997; Day et al., 1999). In regards to human nutrition, products such as microalgal tablets, microalgal drinks or even microalgae noodles have been recently marketed as health products (Liang et al., 2004). Microalgae also support a significant biotechnological market due to the large variety of their metabolites (i.e. astaxanthin, ascorbic acid, β-carotene, glycerol, H_2 or poly-β-hydroxibutyrate) (Ghirard et al., 2000; Tsygankov, 2001).

Green microalgae and cyanobacteria have also a great potential in wastewater bioremediation due to their ability to supply the O_2 necessary to support aerobic biodegradation processes (Laliberte et al., 1994; Tison and Lingg, 1979), degrade toxic organic pollutants (Hirooka et al., 2003; Lima et al., 2003), absorb heavy metals (Oswald, 2003; Travieso et al., 1999), absorb and degrade dyes (Azig and Ng, 1993), and remove nutrients and pathogens from contaminated streams (Aiba, 1982; Mallick, 2002; Robinson, 1998; Schumacher et al., 2003). They are also an important tool in environmental research, being commonly used in toxicity tests or in studies of microbial ecology, as they are sensitive indicators of ecological changes (Day et al., 1999). In addition, the residual biomass from the biological degradation of organic pollutants in wastewater treatment processes can be used for energy generation or as biofertilizer (Eisenberg et al., 1981; Oswald, 1976). Thus, several processes producing liquid fuel by thermochemical liquefaction or pyrolysis of microalgae have been developed (Sawayama et al., 1998).

This chapter aims to provide an extensive evaluation of the potentials and limitations of microalgae in wastewater treatment. Recommendations for microalgae selection, process start-up, photobioreactor design, and process operation are provided.

2. The Potential of Microalgae in Wastewater Treatment

2.1. Microalgae in the Removal of Organic Pollutants

Microalgae can play a very important role during the aerobic biodegradation of organic compounds by supplying O_2 via photosynthesis (Borde et al., 2003; Grobbelaar et al., 1988; Guieysse et al., 2002). More precisely, microalgae produce the O_2 required by the aerobic bacteria to mineralize organic pollutants using in turn the CO_2 released during organic matter oxidation (Figure 1) (McGriff and McKinney, 1972; Oswald, 1988). This symbiotic relationship indeed accounts for a large part of the O_2 supply and BOD removal in ponds during municipal wastewater treatment (Abeliovich, 1986; Mara and Pearson, 1986).

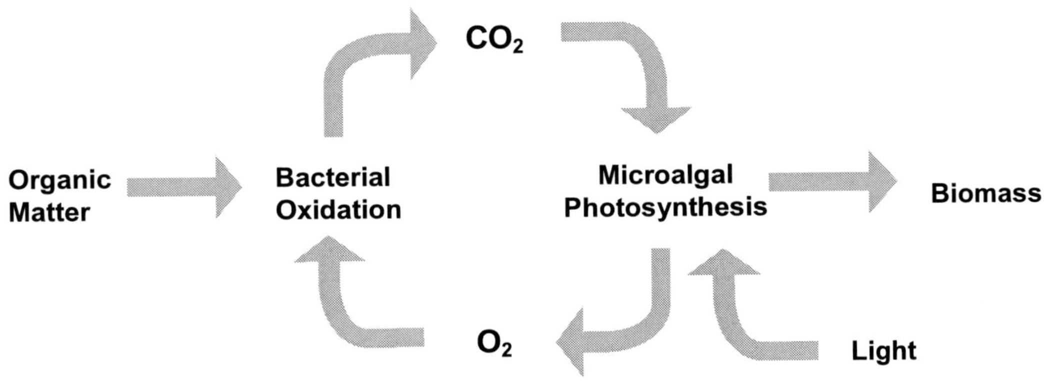

Figure 1.Principle of photosynthetic oxygenation in BOD removal processes.

Because the O_2 produced during the photosynthetic assimilation of CO_2 originates from the biochemical H_2O oxidation, the theoretical O_2 supply of microalgal-based processes is almost unlimited (Madigan et al., 1997). Several authors have studied the oxygenation potential of microalgae, concluding that during photoautotrophic growth microalgae release between 1.5 to 1.92 Kg of O_2 per Kg of microalgae produced, depending on the microalgal composition and age of the cells (Grobbelaar et al., 1988; Martinez Sancho et al., 1993; Oswald, 1988). Oxygenation rates ranging from 0.48 to 1.85 kg $O_2 \cdot m^{-3} \cdot day^{-1}$ have thus been reported in photosynthetically oxygenated pilot-scale ponds or laboratory scale photobioreactors degrading municipal or artificially contaminated wastewater (McGriff and McKinney, 1972; Muñoz et al., 2004). However, these rates are far from meeting the maximum capacities of 4.3 kg $O_2 \cdot m^{-3} \cdot day^{-1}$ achieved in tubular reactors culturing *Spirulina platensis* (Torcillo et al., 2003). Likewise, Tredici and coworkers (1991) estimated O_2 production rates ranging from 8.4 to 12 kg $O_2 \cdot m^{-3} \cdot day^{-1}$ under artificial illumination (1500-2600 $\mu E \cdot m^{-2} \cdot s^{-1}$) using *Spirulina platensis* in a vertical alveolar panel reactor, which is higher than the maximum oxygenation capacity of 3 kg $O_2 \cdot m^{-3} \cdot day^{-1}$ of mechanical surface aerators (Boon, 1983). Thus, taking into account that an efficiency of 1.2-2.4 Kg $O_2 \cdot kWh^{-1}$ is normally achieved in mechanically aerated ponds, a power consumption ranging from 0.15 to 0.29 $kWh \cdot m^{-3}$ could be saved using photosynthetic oxygenation at a photosynthetic O_2 production of 8.4 kg $O_2 \cdot m^{-3} \cdot day^{-1}$.

Although microalgae-based processes have been traditionally used for the treatment of domestic wastewaters (Oswald, 1988), they are also suitable for the treatment of toxic industrial wastes. Thus, Guieysse and coworkers have recently demonstrated the ability of the green microalgae *Chlorella sorokiniana* to support the aerobic degradation of phenanthrene, acetonitrile, phenol, and salicylate without any external oxygen supply (Borde et al., 2003, Guieysse et al., 2002; Muñoz et al., 2005a,b). This symbiotic combination of microalgae and bacteria was also successfully used to degrade 1% black oil and detoxify a real industrial wastewater (Safonova et al., 1999, 2004). Photosynthetic oxygenation is especially advantageous in the case of toxic organic volatile contaminants because it also improves process safety by reducing the risks of hazardous release of pollutants and aerosols, which might occur during mechanical aeration.

In addition to supplying O_2 for pollutant mineralization, various photosynthetic microorganisms are capable of simultaneous or alternative heterotrophic metabolism. This accounts for a large part of the BOD assimilation in stabilisation ponds and enhances the bioremediation of recalcitrant organic pollutants (Hirooka et al., 2003; Ogbonna, 2000a). Indeed, Abeliovich and Weisman (1978) estimated that up to 50 % of the carbon assimilated in microalgal-based ponds is due to heterotrophic nutrition. Some green microalgae and cyanobacteria are able to use toxic recalcitrant compounds as carbon, nitrogen, sulphur or phosphorous source (Semple et al., 1999; Subaramaniana and Uma, 1997). Thus, microalgae from the genus *Chlorella, Ankistrodesmus* or *Scenedesmus* are able to assimilate phenolic compounds and have been successfully used for the treatment of olive oil mill wastewater and paper industry wastewater (Pinto et al., 2002; Pinto et al., 2003; Tarlan et al., 2002). Microalgae can also bioaccumulate or biotransform pollutants like hormones or low molecular weight phenols (Greca et al., 1997; Pinto et al., 2003). Higher aromatic structures like naphthalene or phenanthrene can also be oxidized by cyanobacteria under photoautotrophic conditions as reported by Narro (1987). In some cases monooxygenases or dioxygenases were present in microalgal catabolic pathways, although the degradation mechanisms remain still unclear (Semple et al., 1999).

2.2. Microalgae in Nutrient Removal

Microalgae offer the possibility to remove simultaneously nitrogen and phosphorous in a single stage process (De la Noüe et al., 1992) by metabolic assimilation and/or photosynthetically induced physical-chemical processes (Nurdogan and Oswald, 1995; McGriff and McKinney, 1972). Microalgae use large amounts of nitrogen for protein synthesis (which constitute about 45-60 % of the microalgae dry weight) and phosphorous for nucleic acid and phospholipids synthesis (Laliberte et al., 1994). Thus, although microalgae composition greatly varies depending on the culture conditions, a typical microalgal cells may contain approx. 9.2 % and 1.3 % (in g per g dry weight) of nitrogen and phosphorous, respectively (Oswald, 1988; Vollenweider, 1985). Based on the microalgal volumetric productivities reported for inclined flat-plate enclosed photobioreactors, it is therefore possible to achieve nitrogen and phosphorous metabolic assimilation up to 396 mg $N\text{-}NH_4^+\cdot l^{-1}\cdot day^{-1}$ and 56 mg $P\cdot l^{-1}\cdot day^{-1}$, respectively (Lee, 2001). The actual removal rates can be much higher due to the contribution of NH_3 stripping or phosphate precipitation caused by the raise in the pH associated with photosynthesis (Nurdogan and Oswald, 1995; McGriff and McKinney, 1972, Craggs et al., 1996). García et al., (2000a) reported nitrogen removal by stripping of up to 47 % and concluded that the relative importance of the NH_3 stripping depended on the total nitrogen input and the environmental variables controlling algal growth.

Reported values of nitrogen and phosphate removal using microalgae in laboratory and field studies are still lower than the rates achieved using novel biological (ANAMMOX) or physical/chemical processes (Table 1). Nevertheless, microalgal based nutrient removal is simpler and more cost-effective since there is no need for external BOD or electron acceptor. Indeed, 4.6 $kgO_2\cdot kgN^{-1}$, 3.5 $kgCH_3OH\cdot kgN^{-1}$, and two reaction tanks (aerobic and anoxic) are needed in the conventional nitrification/denitrification process, which increases the total process costs (Van Loosdrecht and Jetten, 1998). In addition, noxious greenhouse gases such as nitrous oxide (N_2O), whose greenhouse effects are 300 times greater than that of CO_2, are produced and released into the atmosphere (Osada et al., 1995; Su, 2003). Likewise, 1.1 KgO_2, 20 kg COD and two tanks (anaerobic and aerobic/anoxic) are required for the biological removal of 1 kg of Phosphorous (Kuba et al., 1993; Van Loosdrecht et al., 1997). Algal-bacterial systems are therefore especially advantageous for the treatment of N-containing contaminants such as acetonitrile or acrylonitrile because in addition to the reduction of hazardous volatilizations, they allow combining C and N removal in a single stage. Muñoz et al., (2005a,b) thus showed that NH_4^+ removal during the batch biodegradation of acetonitrile increased from 26 % under mechanical aeration to 53 % under photosynthetic aeration. This percentage further increased up to 83 % when the process was operated in continuous mode at a HRT of 3.5 days.

2.3. Microalgae for Heavy Metal Removal

Microalgae offer an interesting alternative to physical/chemical processes for the removal and recovery of heavy metals due to their ability to adsorb, uptake or enhance their precipitation. Traditionally, inexpensive and readily available bacterial or fungal biomass from industrial fermentations or biological wastewater treatment processes is used in biosorption processes. However, algae often support higher metal uptake capacities than bacterial or fungal biomass (Cañizares-Villanueva, 2000; Volesky, 2001; Muñoz et al., 2006). Algae biomass from algal blooms or algal-bacterial biomass from microalgae-based wastewater treatment processes therefore constitutes an inexpensive and efficient biosorbent for heavy metal removal (Muñoz et al., 2006). Several authors have suggested that biosorption can compete current applied technologies only if they support removal efficiencies and specific metal uptakes higher than 99 % and 15 $mg \cdot g_{Biomass}^{-1}$, respectively (Cañizares-Villanueva, 2000). The performance of microalgal biomass is above this threshold and sometimes even higher than conventional resins (Table 2). For instance, Peric et al., (2004) measured a zinc uptake of 13 $mg \cdot g^{-1}$ using a natural zeolite, which is much lower than the reported uptake by *Scenedesmus subspicatus* (72.1 $mgZn \cdot g^{-1}$) (Schmitt et al., 2001). Likewise, Villaescusa and coworkers (2004) measured copper uptakes of 10 $mg \cdot g^{-1}$ using grape stalks wastes as biosorbent, which is in the range of the algal uptake capacities for this metal (Table 2).

The accumulation of metals by photosynthetic microorganisms is a complex process involving passive and active mechanisms (Figure 2). Passive mechanisms are fast, reversible cell surface associated processes that are mainly based on physical adsorption, ion exchange and chemisorption. It is now well accepted that the principal mechanism of passive metal biosorption involves their exchange with counterions present in the biomass such as protons from the weak acidic carboxyl (R-COOH), hydroxyl (R-OH) or amino groups (R-NH/R-NH_2) (Chojnacka et al., 2005). Commercially available bioremoval processes such as AlgaSORB™ and AMT-Bioclaim™, that use the algal biomass as ion-exchange resins in flow-through packed bed sorption columns, are based on passive biosorptive mechanisms (Kratochvil and Volesky, 1998; Volesky, 2001). This passive biosorption is often followed by much slower irreversible metal binding processes that include covalent bonding, surface precipitation, redox reactions or crystallization on the cell surface (Wilde and Benemann, 1993). Active metal uptake often involves the transport of the metals into the cell interior, which requires the expense of metabolic energy. It is either a cell defensive tool to avoid poisoning or a mean to accumulate essential elements present at trace level concentrations (Schmitt et al., 2001; Wilde and Benemann, 1993). Once a metal ion has diffused into the cell interior, it binds to intracellular proteins, sites or polyphosphate bodies. Certain microalgae for instance synthesize intracellular metal-binding peptides (Class III metallothioneins) as a defensive response to metal toxicity (Kaplan et al., 1995). Likewise, Yu and Wang (2004) have recently reported the increase in the Cd^{2+} and Zn^{2+} intracellular uptake due to the binding capacity of polyphosphate bodies accumulated inside a *Sc. obliquus* strain in phosphate rich environments. Microalgae are also able to remove heavy metals by releasing extracellular metabolites, such as polysaccharides, which are capable of chelating metal ions (Kaplan et al., 1987; Van Hille et al., 1999). In addition, the generation of alkalinity associated with microalgal growth can enhance heavy metal precipitation and contribute significantly to the detoxification of metal-bearing effluents (Rose et al., 1998; Van Hille et al., 1995). Rose et al., (1998) also demonstrated that *Spirulina sp.* biomass was a suitable carbon source for sulphate reducing bacteria for further precipitation of heavy metals with sulphide.

Microalgae-based processes can advantageously combine BOD and heavy metal removal. When heavy metals are present at subinhibitory concentrations, they can be simultaneously adsorbed or taken-up by the actively growing algal-bacterial biomass generated during the mineralization of organic pollutants. When present at inhibitory concentrations, biosorption can take place prior to the biomass-producing biological BOD removal stage. Muñoz et al., (2006) thus reported the successful utilization of an adsorption column packed with residual biomass placed before a photobioreactor used for the treatment of an effluent containing 500 mg salicylate l^{-1} and 2 mg $Cu^{2+} \cdot l^{-1}$. The use of actively growing photosynthetic microorganisms is however generally preferable due to its self-replenishment capacity, the continuous metabolic uptake and bioprecipitation effects it offers, and the diminution of the risk associated to metal release back to the environment. However, practical limitations arise from the fact that living microalgae are very sensitive to metals and usually difficult to harvest. In addition, nutrients are needed for cell growth. These limitations can be overcome by the use of metal-resistant species created by genetical engineering or isolated from metal contaminated sites, by the use of flocculent strains that enhance microalgal harvesting and by the exploitation of locally available inexpensive carbon sources (Malik, 2004; Wilde and Benemann, 1993).

Table 1. Nitrogen and Phosphorous removal rates

Experimental system	Immobilization	Microorganisms	P (mg P·l^{-1}day^{-1})	N (mg N·l^{-1}day^{-1})	References
Continuous combined light-dark reactor (15 l)	No	*Chlorella, Nitzchia*	22	144	Mc Griff & McKinney, 1972
Continuous HRAP[1] (570 l)	No	*Dictyosphaerium, Chlorella*	-	8.1	Garcia et al., 2000
Batch conical flasks (0.5 l)	No	*Chlorella*	0.36	3.5	Lau et al., 1995
Batch chemostat	Alginate	*Chlorella* coimmobilized with *Azospirillum brasilense*	4.5	1.4	De-Bashan et al., 2002
Batch conical flasks (0.5 l)	Alginate	*Chlorella*	4.7	-	Yan et al., 1996
Batch conical flasks (0.5 l)	No	Mixed algal culture	0.6	0.5	Aziz & Ng, 1993
Continuous hollow fibres tubular reactor	Cellusose hollow fibres	*Phormidium laminosum*	3.3	3.4	Sawayama et al., 1998
-	Chitosan	*Phormidium laminosum*	1.5	8.9	De la Noüe & Proux, 1988
Column reactor (1 l)	Polyurethane bed	*Chlorella*	11.8	86	Travieso et al., 1996
Column reactor (3 l)	Alginate	*Scenedesmus*	1.9	4.0	Cañizares et al., 2000
Batch conical flasks	Alginate	*Chlorella*	74.4	-	Robinson et al., 1988
Batch tank reactor (3.5 l)	Alginate	*Chlorella*	0.4	5.1	Tam et al., 1994
Pyrex flasks (6 l)	Recirculation	*Scenedesmus*	13.4	224.0	De la Noüe & Eidhin, 1988
Round flasks (1 l)	k-carrageenan	*Scenedesmus*	8.18	67.2	Chevalier & De la Noüe, 1985
Column reactor (6 l)	Alginate	*Chlorella*	33.4	45.0	Tam & Wong, 2000
Fluidized bed reactor (ANAMMOX)	No	*Planctomycetales*	-	4800	Van Loosdrecht & Jetten, 1998
Anaerobic/Anoxic batch reactor	No	P- accumulating mixed culture	80-120	-	Kuba et al., 1993
Biological-Ion exchange column	Zeolite	Nitrifiers Culture	6000	-	Green et al., 1996
Wetland with recirculating subsurface flow	Gravel	Plants	-	1800	Sikora et al., 1995

1-HRAPs- High Rate Algal Ponds.

Table 2. Reported studies on heavy metal accumulation by microalgae

Metal	Biomass	Accumulation capacity ($mgMe \cdot g_{Biomas}^{-1}$)	Adsorption removal rate ($mgMe \cdot l^{-1} \cdot day^{-1}$)	Experimental set-up	Reference
Zn	*Chlorella. vulgaris*	-	114.2	Column reactor (1-L), microalgae immobilized in κ-carrageenan	Travieso et al., 1999
Cr	*Scenedesmus acutus*	-	3.5		
Cd	*Chlorella vulgaris*	-	2.5		
Co	*Scenedesmus obliquus*	-	0.82	Rotary biofilm reactor (BIOALGA)	Travieso et al., 2002
Zn	*Euglena gracilis*	7.5	-	500-ml E-Flasks, free microorganisms	Fukami et al., 1988
Cd	*Chlorella Homosphaera*	8.4	1.44	500-ml E-Flasks, free microorganisms	Costa et al., 1990
Zn	*Chlorella Homosphaera*	15.6	2.67		
Cd	*Chlorella vulgaris*	2.6	-	1-L E-Flasks, free microorganisms	Khoshmanesh et al., 1996
	Chlorella pyrenoidosa	2.8	-		
	Chlamydomonas reinhardtii	2.3	-		
Al	*Scenedesmus subspicatus*	6.8	-	50-ml polyethylene-flasks, free microorganisms	Schmitt et al., 2001
Cd		7.3	-		
Cu		13.2	-		
Hg		9.2	-		
Pb		38.7	-		
Zn		72.1	-		
Cu	*Scenedesmus obliquus*	10.5	-	500-ml E-flasks, free microorganisms	Mattuschka et al., 1993
Cd	*Chlorella sorokiniana*	192	-	Column reactor, microalgae immobilized on a vegetable sponge	Akhtar et al., 2003
Zn	*Scenedesmus obliquus*	6.7	-	-	Omar, 2002
	Scenedesmus quadricauda	5.0	-		

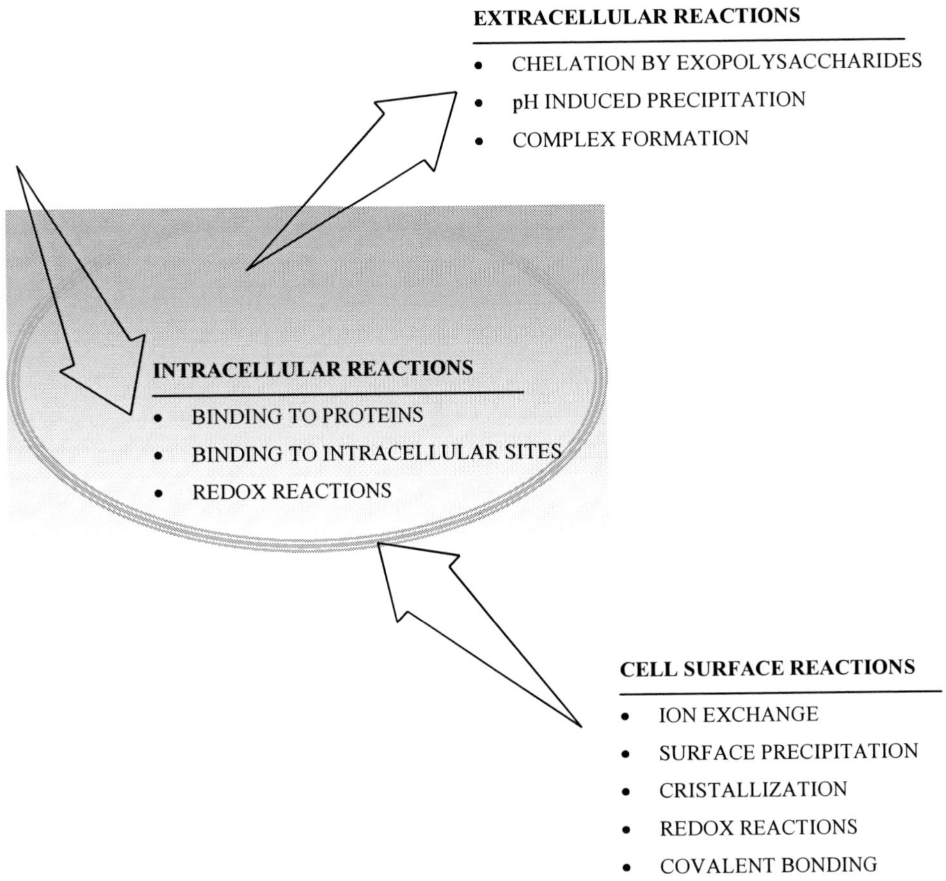

Figure 2. Underlying mechanisms of heavy metal removal in microalgal cultures.

3. Microalgae Selection

3.1. Interaction within the Algal-bacterial Microcosm

Microalgae are found together with bacteria in photobioreactors treating BOD, nutrient and heavy metals bearing effluents. Therefore, the interactions between microalgae and bacteria are crucial to the stability and efficiency of the process. Microalgal and bacterial communities do not limit their interactions to a simple CO_2/O_2 exchange, but can also exert positive or negative mutual influences (Figure 3). Thus, microalgae can have a detrimental effect on bacterial activity by increasing the pH, the dissolved oxygen concentration (DOC) or the temperature of the cultivation broth, or by excreting inhibitory metabolites (Oswald, 2003, Schumacher et al., 2003). In fact, these antibacterial mechanisms are responsible for the removal of pathogens such as *Escherichia coli* and *Vibrio cholerae* in waste stabilization ponds (Mara and Pearson, 1986; Mezrioui et al., 1994; Oswald, 2003). On the other hand, microalgae can have a positive effect on bacterial activity by releasing extracellular matter that improves the degradation of pollutants within the microcosms (Wolfaardt et al., 1994).

Likewise, bacterial growth can either enhance microalgal metabolism by releasing growth-promoting factors (Fukami et al., 1997; De-Bashan et al., 2002) or by reducing O_2 concentration in the medium (Mouget et al., 1995; Paerl and Kellar, 1978). For instance, De-Bashan et al., 2002 reported that the presence of the bacterium *Azospirillum brasilense* enhanced the removal of ammonium, and phosphorous by *C. vulgaris* when compared with a control conducted only with the green microalga. Bacteria can also inhibit microalgae by producing algicidal extracellular metabolites (Fukami et al., 1997).

3.2. Microalgae Growth Rate

In photosynthetically oxygenated processes, carbon removal is often limited by O_2 supply and is directly linked to the microalgae activity because O_2 production and C-fixation are biochemically linked (Muñoz et al., 2004; Muñoz et al., 2005). Consequently, rapidly growing microalgae also exhibit high O_2 production rates and should be preferred as O_2 suppliers. High specific growth rates are also beneficial in terms of nutrient assimilation and heavy metal removal. In fact, the rapidly growing *Chlorella* and *Scenedesmus* sp. dominate most of the continuous microalgal-based treatment systems (Garcia et al., 2000a; Martinez-Sancho et al., 1993). Likewise, in two studies, *Chlorella. sorokiniana* exhibited higher growth rates and supported faster pollutant than *Chlorella vulgaris*, *Scenedesmus obliquus*, *Selenastrum capricornutum* and *Scotiellopsis oocystiformis* during the photosynthetically oxygenated degradation of salicylate and acetonitrile by specific degrading bacteria (Muñoz et al., 2003; Muñoz et al., 2005a).

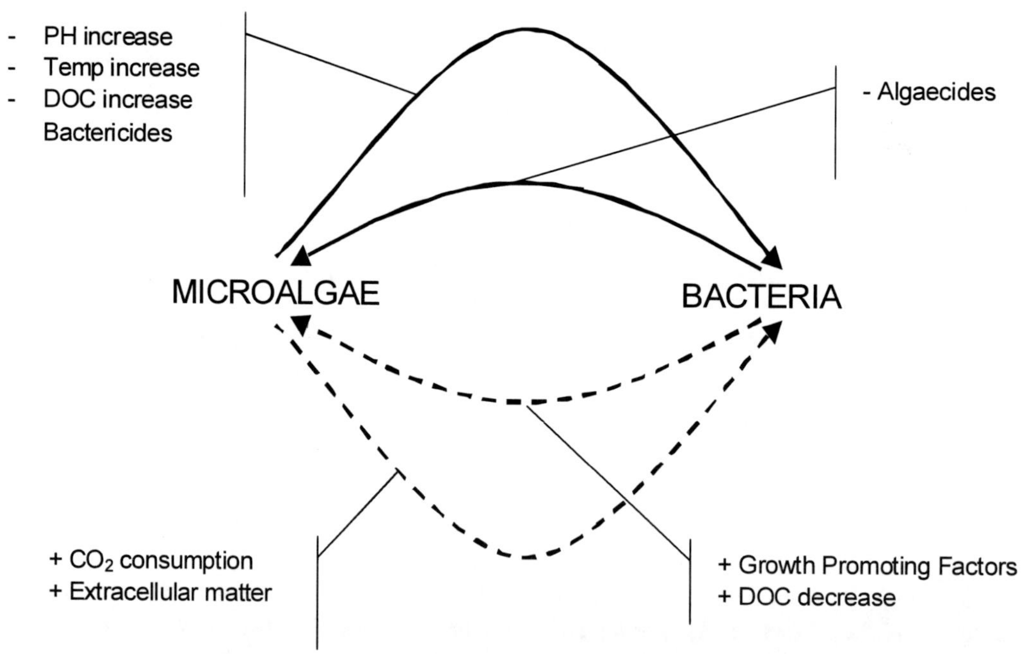

Figure 3. Illustration of the positive (dashed lines) and negative (plain lines) interactions between microalgae and bacteria.

3.3. Microalgae Inhibition

Inhibition of photosynthetic microorganisms during industrial wastewater treatment is a major problem during algae-based water treatment (Aksmann and Tukaj, 2004; Borde et al., 2003).

The response of photosynthetic microorganisms to inhibitory compounds is highly species and condition dependant (Muñoz et al., 2003; Muñoz et al., 2005a). In our laboratory, *C. sorokiniana* exhibited a higher tolerance towards acetonitrile and salicylic acid compared to *C. vulgaris*, *Sc. obliquus*, *S. capricornutum* and *Sco. oocystiformis*. Likewise, based on a compilation of 269 reports, Palmer (1969) also classified microalgae from the genus *Chlorella* as highly pollutant-tolerant microalgae.

The combination of high NH_3 concentrations and high pH values can also inhibit microalgal photosynthesis (Azov and Goldman, 1982). For instance, Abeliovich and Azov (1976) observed a decline in the efficiency of a High Rate Algal Pond (HRAP) due to the inhibition of the microalgal photosynthetic apparatus when NH_3 concentration and pH were simultaneously above 2 mM and 8, respectively. NH_3 can uncouple electron transport or compete with H_2O in the oxidation reactions leading to O_2 generation. pH determines the degree of NH_4^+ dissociation and therefore NH_3 availability ($NH_4^+ + OH^- \leftrightarrow NH_3 + H_2O$; $pK_a = 9.25$ at 25 °C). The use of NH_3 tolerant microalgae can increase the stability of the process as for instance, Ogbonna et al., (2000a) reported no significant effect on the growth of *C. sorokiniana* at NH_4^+ concentrations of up to 22 mM, while the growth of *Sp. platensis* nearly stopped when NH_4^+ was present at 11 mM. Likewise Muñoz et al., 2005b reported complete inhibition of the photosynthetic process at a total NH_3/NH_4^+ concentration of 15 mM and pH values of 8.7 during the degradation of 2 $g{\cdot}l^{-1}$ of acetonitrile in a 50-L column photobioreactor oxygenated by *C. sorokiniana.* However, after the pH was restored to its initial value, the biodegradation process resumed until complete acetonitrile depletion.

Finally, heavy metals are the third major group of algae inhibitors due to their capacity to disrupt photosynthesis at various physiological levels (Clijsters and Vanassche, 1985). Heavy metals are also able to generate morphological modifications as gigantism or changes in the shape of the microalgae cell wall (Peña-Castro et al., 2004; Travieso et al., 1999). In our studies, the performance of a salicylate removal process supported by *C. sorokiniana* was severely diminished when Cu^{2+} was present at concentrations of 2 $mg{\cdot}l^{-1}$ (Muñoz et al., 2006).

Hence, a high tolerance towards the target pollutant is an important selection criterion for microalgae, since fluctuations in the influent inhibitor concentration can dramatically affect the stability of the treatment process (Muñoz et al., 2003). More resistant strains can be obtained by genetic manipulation, by cell adaptation to progressively higher pollutant concentrations, or by isolation from contaminated sites where the whole ecosystem has been exposed to a high selection pressure (Malik, 2004). For instance, Essam et al., (2005) reported the continuous photosynthetically oxygenated degradation of phenol using a phenol adapted *C. vulgaris* strain able to tolerate concentrations of up to 300 $mg{\cdot}l^{-1}$. This microalgae was isolated from a stabilisation lagoon of a cooking factory near Cairo.

3.4. Microalgae Removal

Although technically possible, none of the common industrial approaches for biomass removal (filtration, centrifugation, microstraining, etc.) are economically suitable for large-scale microalgal removal (Hoffmann, 1998). High harvesting costs can even affect the economic viability of mass algal cultivation systems and represent a major limitation to the use of microalgae-based processes (Olguin, 2003).

Microalgal flocculation followed by gravity sedimentation remains therefore the most common harvesting technique during wastewater treatment because of the large volumes treated and the low value of the biomass generated (Nurdogan and Oswald, 1996; Molina et al., 2003). Unfortunately, this approach is not always efficient, especially in the case of the small, rapidly growing *Chlorella* or *Scenedesmus* sp. (García et al., 2000b). Thus, the selection of microalgae with a natural tendency to self-aggregate and sediment is strongly recommended (Olguin, 2003; Hoffmann, 1998). For instance, multicellular cyanobacteria of the genus *Spirulina* or the self-aggregating *Phormidium bohneri* have been successfully applied in wastewater treatment of farm effluents (Olguin, 2003).

Muñoz *et al.,* 2004 reported that the sedimentation of the residual biomass generated during the continuous degradation of salicylate (supported by a *C. sorokiniana* strain) was enhanced at high HRT, when the system was operated at high pH and high DOC concentrations (due to the autoflocculation of the biomass). Microalgal flocculation can occur naturally via autoflocculation and bioflocculation. Autoflocculation is due to the electrostatic interactions among the cell walls as a result of Ca/Mg carbonate or ortophosphate precipitation, while bioflocculation occurs due to the microalgal release of long-chain polymers (Garcia et al., 1998). At high pH values (8-11), Ca, Mg, and other polyvalent cations tend to precipitate and become incorporated in the algal biomass, which explains the enhancement in biomass sedimentation observed at high HRT (Oswald, 1988).

The addition of chemical flocculants such as lime, alum or polyferric sulfate is efficient and reliable. However the use of chemical is poorly suitable because of their costs and the fact the sludge mass increases with a non-biodegradable fraction. Chitosan is an edible, non-expensive and non-toxic flocculant that is efficient for the removal of freshwater microalgae (Divakaran and Sivasankara, 2002). In our lab biomass removal efficiencies of up to 90 % were obtained using chitosan at 15 $mg \cdot l^{-1}$ during the batch degradation of acetonitrile (data not published). The use of bioflocculants from bacteria present within the microcosms has also been tested (Oh et al., 2001).

3.5. Microalgae Predominance

Long-term predominance of the selected microalgae is another important aspect that influences the stability of algal-bacterial processes. In open systems, the predominance of slow growing microalgae is especially difficult to maintain due to contamination by the small, rapidly growing microalgae (Hoffmann, 1998). In this regard, closed photobioreactors offer a better species control (Tredici, 1999).

Variables such as the composition of the wastewater, the seasonal environmental conditions, or the potential interactions within the microcosms influence the composition of the microalgal population (Fukami et al., 1997; Mara and Pearson, 1986). For instance, *Euglena* and *Chlamydomonas* are dominant at high organic loads in sewage treatment while *Chlorella* and *Scenedesmus* are the most abundant species at medium loads (Martinez Sancho et al., 1993). Likewise, *Euglena* and *Scenedesmus* species predominate over *Chlorella* below 15 °C due to their higher tolerance to low temperatures (Mara and Pearson, 1986). Control temperature of the process by external heat exchanger or water spray has been proposed as a solution to ensure a stable microalgal population. However their implementation is often too expensive, even for high quality algal mass cultivation (Tredici, 1999). An alternative to temperature control is the combination of microalgal strains with similar characteristics (in terms of O_2 supply, inhibition and harvesting) but with different optimum growth temperatures (Morita et al., 2001).

Efforts to sustain a certain microalgal population in open cultivation systems by controlling the operational variables have not always been successful. For instance, Benemann et al., (1980) failed to establish the predominance of *Oscillatoria* sp. and *Micractinium* sp. by microscreening of the biomass and subsequent recirculation into the photobioreactor. However, Wood (1987) successfully maintained the predominance of a *Stigeoclonium* strain by combining a short hydraulic retention time (HRT), to wash the freely-suspended microalgae, with crossflow microscreening of the target strain.

4. Process Design

Biological reactions involving microalgae are carried out in photobioreactors, which can be classified according to the mode of operation (complete-mix *vs* plug flow; suspended *vs* immobilized) and the reactor configuration (open *vs* closed bioreactors). These basic design features determine the final quality of the treated effluent. Therefore, a detailed understanding of these variables and their influence on process performance is required for successful wastewater treatment.

4.1. Hydrodynamic Regime and Biomass Retention

Plug flow operation is not generally recommended in microalgae-based processes due to the inherent need for intense turbulence to achieve high frequency light/dark cycles (that enhance photosynthetic activity) and to maintain microalgae in suspension. In addition, it might be difficult to photosynthetically meet the biological oxygen demand near the influent entrance point, and algae inhibition is more likely to occur in plug flow system. Instead, homogenisation in complete mix reactors generates a turbulent flow, which enhances the rates of photosynthetic oxygenation.

Suspended cultures are generally used for microalgae production and microalgal-based bioremediation. They allow for better process control, medium homogenization and therefore higher light utilization efficiency (defined as the energy stored as new biomass per unit of light absorbed, Janssen et al., 2003). However, suspended microalgae-based processes are often limited by the problems associated with microalgae harvesting. Microalgae immobilization can overcome this limitation by producing a cell-free effluent (Nicolella et al.,

2000), which improves the final quality of the effluent (Hoffmann, 1998). Immobilization in polymeric matrices such as carrageenan, chitosan, or alginate has been the most commonly used approach (Chevalier and De la Noüe, 1985; Lau et al., 1995; Robinson et al., 1988). However, the weakness of these polymeric matrices during long term operation and their high cost hinders their full scale application (Hoffmann, 1998). Another approach consists on using enclosed photobioreactor with the algal-bacterial microcosm attached onto the reactor walls. Apart from the advantages mentioned above, microalgae are more protected towards pollutant toxicity in biofilms than in suspended cultivations. Indeed, in enclosed biofilm reactors the pollutant diffuses from the bulk liquid stream and O_2 from the surface of the photobioreactor through the biofilm, where biodegradation takes place (Figure 4). Diffusion transport limitations and the fate of the pollutant through the biofilm decrease the concentration of the contaminant through the biofilm. Microalgae are therefore exposed to lower pollutant concentrations than in the bulk phase. However, biofilm-based photobioreactors have 2 major limitations: 1) photoinhibition in outdoors cultures can occur due to the fact that immobilized microalgae at the reactor surface are constantly exposed to very high photon flux density, and 2) the risk of photobioreactor clogging at high biomass densities.

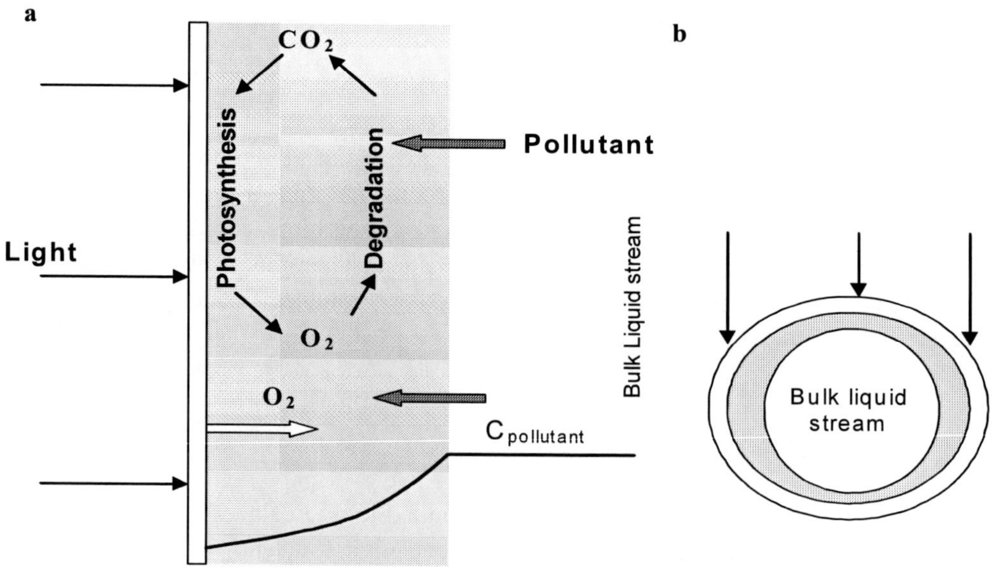

Figure 4. Proposed mechanism for pollutant removal in a biofilm attached onto the tubular photobioreactor wall (a), cross sectional view of the tubular photobioreactor illustrating the growth of the algal-bacterial biofilm onto the inner part of the tube (b).

4.2. Process Configuration

Photobioreactors for microalgal mass cultivation and for bioremediation of pollutant-laden effluents share the same basic design criteria: an efficient photobioreactor should exhibit a high light utilization efficiency, good scalability, efficient mixing and control over the reaction conditions, and should avoid hydrodynamic stress on the photosynthetic cells (Borowitzka, 1999).

Open Photobioreactors

Open photobioreactors have been extensively used for both microalgae cultivation and for domestic wastewater treatment. Four different designs for microalgae cell cultivations are currently used: shallow ponds, tanks, circular ponds and raceway ponds (Pulz, 2001; Tredici, 1999; Lee and Lee, 2003). In the late 1950s, Oswald and co-workers modified the design of a conventional stabilisation pond by matching algal growth and O_2 production with the O_2 demand of the receiving domestic wastewater, originating the first high rate algal pond (HRAP) (Oswald, 1988). HRAPs are traditionally designed as shallow open tanks with a raceway configuration. Raceway ponds are shallow channels 2-3 meters wide, and 10-30 cm depth, lined with PVC, clay or asphalt to avoid infiltration, that occupy a surface ranging from 1000 to 5000 m^2 (Abeliovich, 1986; Molina Grima, 1999). These systems are mixed with paddle wheels or pumps, (although paddle wheels have been proven to be the most cost-effective mode of mixing). Under optimal conditions, HRAPs can be loaded with up to 350 $kg{\cdot}BOD{\cdot}ha^{-1}{\cdot}d^{-1}$ and operated at residence times of 2-6 days (Mara and Pearson, 1986).

Given the merits of HRAPs, there are nowadays only a few full-scale systems in operation (De la Noüe et al., 1992; Mara and Pearson, 1986). The high water losses and space-requirements for construction, together with the poor process control, low productivity and high risk of contamination by predators remain important limitations (Pulz, 2001). Another approach to HRAPs consists on the use of microbial attachment to surfaces. For instance, Craggs et al. (1996) engineered a shallow open photobioreactor, namely algal-turf scrubber, where the algal-bacterial biomass attached onto a polymeric liner. This system was successfully operated for the treatment of agricultural run-off and domestic wastewater. Unfortunately, there is limited experience available on the design, maintenance and long term stability of these biofilm-based processes (Craggs et al., 1996; Hoffmann, 1998).

Closed Photobioreactors

The increasing research on enclosed photobioreactors carried out in the two decades (Figure 5) has resulted in a wide variety of designs and in a significant increase in microalgal productivity (Table 3). Tubular or flat plate photobioreactors arranged in a horizontal, inclined, vertical or spiral manner and connected via manifolds or in a serpentine configuration are considered as the most efficient enclosed photobioreactors (Tredici, 1999) (Figure 6). These bioreactors are constructed in transparent materials such as Plexiglas, glass, PVC, etc which substantially increases the manufacture costs.

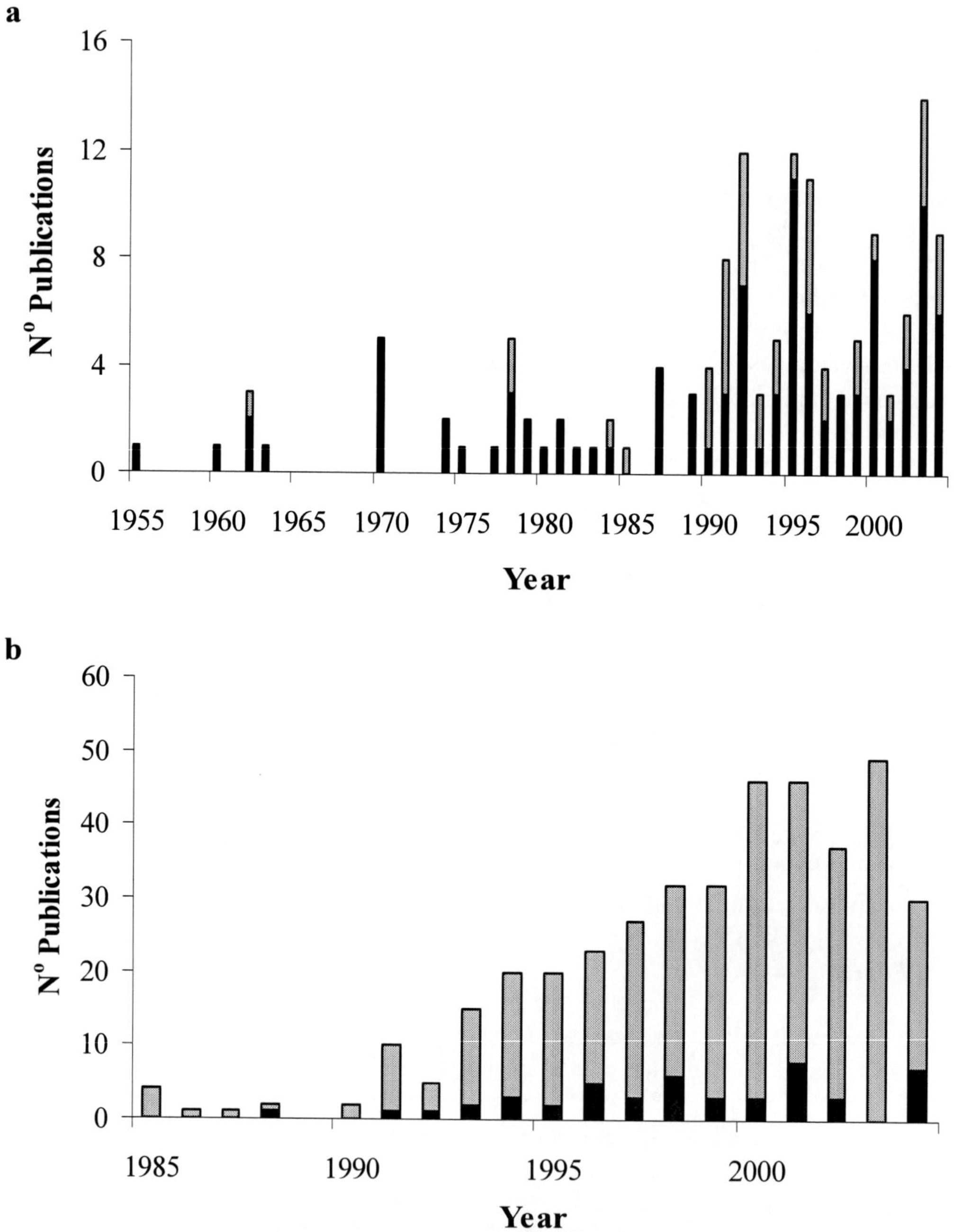

Figure 5. Number of papers yearly reported in Web of Science on open (a) enclosed (b) photobioreactors for environmental applications (solid) or biomass and chemical production (empty). Data obtained from Web of Science using the query words "photobioreactor" and "algal AND pond".

Table 3. Comparison of large-scale photobioreactors for algal-mass production

Reactor type	Maximum Biomass productivity ($g \cdot l^{-1} \cdot day^{-1}$)	Light utilisation efficiency	Scala-bility	Reactor volume (m^3)	Design criteria[1]	Reference
HRAP	0.2	Low	Easy	1000	Channel width 2-3 m, light path 10-30 cm, surface 1000-5000 m^2	Molina Grima, 1999
Tubular	3.6	Very high	Easy	700	Tube length 10-100 m, tube diameter 3-6 cm	Lee and Low, 1991
Flat plate	4.3	Very high	Difficult	1	Light path 1-5 cm	Hu et al., 1996
Tubular (coil)	1.2	Very high	Easy	10	Tube diameter 2-3 cm, cylindrical structure 8 m height, 2 m diameter	Borowitzka, 1999
Vertical column	1.6	High	Difficult	0.2	Diameter 30-50 cm, height 2-4 m	Miron et al., 1999

[1] According to Tredici (1999), Borowitzka (1999) and Janssen et al., 2003.

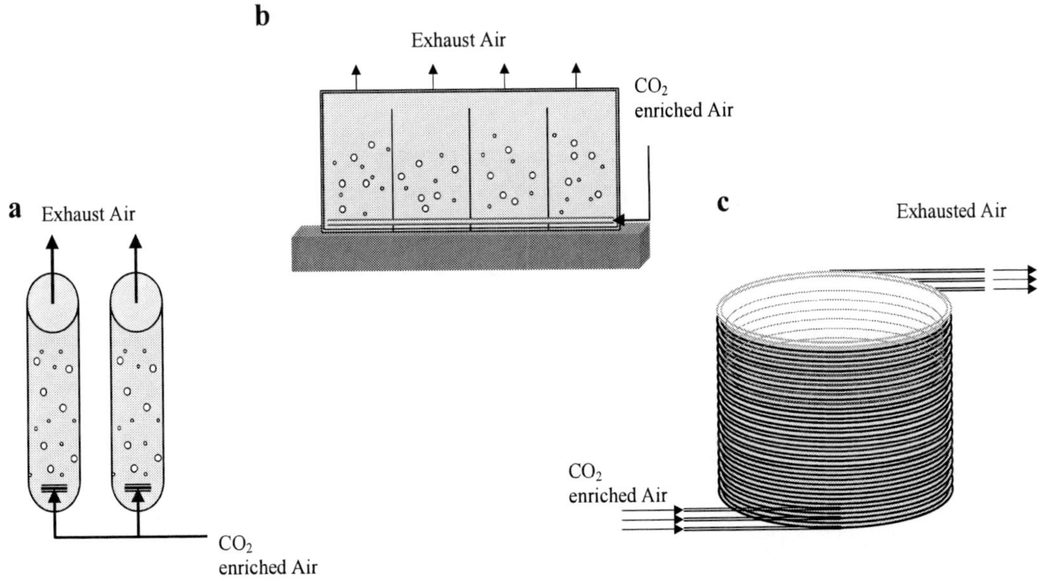

Figure 6. Schematic representation of a vertical column reactor (a), a flat-plate reactor (b), and a tubular photobioreactor arranged in spiral (Biocoil) (c) devised for mass algal cultivation and fed with air enriched with CO_2 to supply C and to create mixing.

The most important design parameter in enclosed photobioreactors is the illuminated surface to volume ratio. This variable determines the volumetric microalgae growth rate and therefore, the volumetric O_2 production and volumetric nutrient and heavy metal removals. Therefore, efficient photobioreactors are based on large surface to volume ratios, which comes at the expense of the land requirement. The use of compact enclosed photobioreactor in vertical configuration could eventually overcome this limitation. Although enclosed photobioreactors are generally difficult to scale-up, tubular photobioreactors offer the simplest scale up procedure by increasing the length and number of tubes and by the connection of several units via manifolds (Borowitzka, 1999). The risk of a dangerous O_2 build-up (that usually limits the scale up of tubular reactors when designed for algal mass production) is significantly reduced due to the continuous O_2 take-up by heterotrophic bacteria. In addition, tubular bioreactors have been proven to exhibit higher light utilization efficiencies than flat plate photobioreactors because of the higher dilution of the light as a consequence of the larger reactor surface area per unit of occupied land (Tredici and Zittelli, 1998).

The use of enclosed photobioreactors for pollution control is still at an embryonic stage as research on microalgae for wastewater treatment has traditionally focused on HRAP for sewage treatment (Oswald, 1988). Data from full-scale closed photobioreactors for wastewater detoxification is scarce with only two commercial systems reported for wastewater treatment: the Bio-Fence manifold tubular reactor (Applied Photosynthetic Limited, Manchester, United Kingdom) with a total volume ranging from 50 to 1000 L, and a helical tubular reactor called Biocoil (Biotechna-Graesser A.P. Ltd, Australia) with working volumes of up to 10 m^3 (Tredici, 1999).

5. Operation of Microalgae-based Systems

The major limitation of microalgae-based processes, inherent to any biological technology, is the high sensitivity of the process to environmental, biotic and operational factors.

5.1. Environmental Factors

pH

The effect of pH is complex since it is difficult to dissociate the direct effects on microbial growth from the collateral effects such as the modification in the $CO_2/HCO_3^-/CO_3^{-2}$ and NH_3/NH_4^+ equilibria or on phosphorus or heavy metal availability (Laliberte et al., 1994). In HRAPs, the pH of the medium increases up to 10-11 due to microalgal CO_2 uptake. Likewise, pH values up to 9 were measured during salicylate biodegradation by an algal-bacterial consortium (Muñoz et al., 2003). The pH has also a major impact on nutrients abiotic removal since a significant amount of the inlet N and P is removed via NH_3 volatilization and orthophosphate precipitation at a high pH (9-11) (Craggs et al., 1996; Garcia et al., 2000b). The pH also influences the activity of the bacterial community, and complete bacterial inhibition at pH above 10 is commonly observed in stabilization ponds in a sunny day at noon. This pH-induced bacterial decay, although important for the disinfection of pathogens in sewage treatment, can also cause a decrease in the pollutant removal efficiency (Oswald, 1988; Schumacher et al., 2003). Fortunately, pH control is relatively easy to implement in case of pH-mediated poor process performance.

Light

Algal activity increases with light intensity up to a certain level where the photosynthetic apparatus becomes saturated (usually at 200-400 $\mu E \cdot m^{-2} \cdot s^{-1}$), with a subsequent decrease in the photosynthetic activity at higher light intensities (photoinhibition) (Ogbonna and Tanaka, 2000b; Sorokin and Krauss, 1958). This relationship is however only valid at very low cellular densities where all the cells received the same amount of energy. Hence, it does not represent the reality in most photobioreactors where high cell densities are present. (Contreras-Flores et al., 2003; Richmond, 2000). In such systems, mutual shading occurs and the actual light intensity to which the microalgae are exposed within the reactor becomes also a function of the biomass concentration (Evers, 1991).

In our studies on continuous pollutant biodegradation using *C. sorokiniana*, the total biomass concentration was a function of the amount of pollutant consumed, which itself depended on the O_2 supplied by microalgae. Microalgal activity, and therefore O_2 production, was often limited by light supply (Muñoz et al., 2004). Thus, in light limited cultures, an increase in light input at the reactor's wall increased both the concentration of microalgae and the oxygenation capacity of the system. Thus, in a light limited chemostat photobioreactor operated at a biomass concentration X_1 an increase in the external light input initially increases the light intensity within the culture. This rises the average specific growth rate of microalgae (μ) and therefore biomass concentration in the photobioreactor (X_2). As a result of the higher microalgal concentration, mutual shading increases, which simultaneously reduces the light intensity within the reactor. The microalgal specific growth rate will therefore decrease to its initial value (Figure 7). Thus, a new steady state at a higher biomass concentration is attained. As a result of the higher microalgal concentration and the higher light input, the O_2 supply to the heterotrophic bacteria is higher and therefore the removal rate increases (Muñoz et al., 2004).

Light utilization efficiencies are usually limited by the saturation of the photosynthetic apparatus at light intensities of approx. 200-400 $\mu E \cdot m^{-2} \cdot s^{-1}$ and the damage at higher irradiances (Ogbonna and Tanaka, 2000b). Three practical solutions have been proposed to increase the efficiency of light utilization in photobioreactors. Firstly, the use of dense, well mixed cultures causes the cells to receive light intermittently and consequently the light received by each individual cell is diluted as a result of mutual shading. Thus, by carefully controlling the microbial density in a flat-plate photobioreactor, Hu et al., (1996) reported light utilization efficiencies up to 2000 $\mu E \cdot m^{-2} \cdot s^{-1}$. Secondly, the dilution of the impinging light by carefully designing the photobioreactor can also avoid excessive damage of the photosynthetic apparatus. In these systems, the light irradiating a certain land area occupied by the reactor is distributed on a larger surface using a special bioreactor design, thus reducing the light saturation effect. However the cost of these photobioreactors is normally very high due to their complex designs (Torzillo et al., 2003). Thirdly, the reduction of the size of the antenna of the individual photosynthetic cells usually involves higher photosynthesis rates under high light intensities. Indeed, by reducing the size of the antenna using molecular tools the light adsorption is minimized and less energy is dissipated, which increases the photosynthetic productivity of the cultivation (Melis et al., 1999).

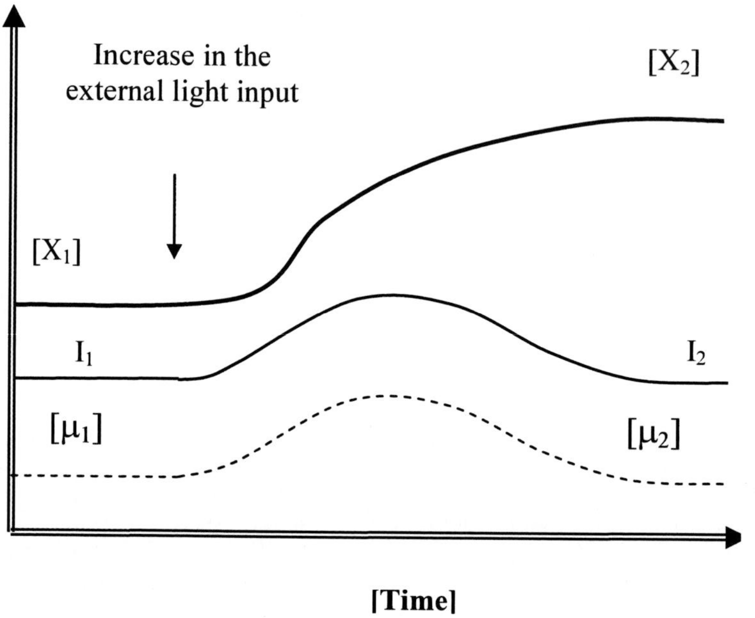

Figure 7. Transient state when increasing the light input in a light limited continuous photosynthetically oxygenated process.

Temperature

The efficiency of microalgae-based treatments normally decreases at low temperatures (Abeliovich, 1986). However, the effect of temperature is intimately related to the composition of the microcosms. If local conditions (or the process control) impose a consortium formed only by one dominating microalga and one bacterial strain, the influence of the temperature will depend on the respective optimum temperatures for the growth of each microorganism. Muñoz et al., (2004) observed that the removal efficiency doubled when the temperature increased by 5 °C using a symbiotic microcosm formed by a *C. sorokiniana* and a *R. basilensis* strain (the activities of both microorganisms increased with the temperature in the tested range). However, if the consortium is formed by various microalgae and bacteria species, each with different optimum temperature for growth, the effect of the temperature on the process could be negligible due to compensation effects (Abeliovich, 1986).

Dissolved Oxygen Concentration (DOC)

The monitoring and control of DOC deserve special attention since high DOC levels can generate photo-oxidative damage on microalgal cells and therefore decrease the treatment plant efficiency (Oswald, 1988; Suh and Lee, 2003). For instance, Matsumoto et al., (1996) reported a decrease in the photosynthetic O_2 production rate from 0.125 to 0.02 $Nm^3 \cdot kg^{-1} \cdot h^{-1}$ when the O_2 partial pressure increased from 0 to 70000 Pa. O_2 supersaturation in enclosed photobioreactors designed for mass algal cultivation can reach up to 400 % and severely inhibits microalgal growth. However, O_2 supersaturation does not constitute such a severe problem in biodegradation processes due to the continuous O_2 consumption by the heterotrophic bacterial cells. For instance, the DOC was always very low (≈ 0 mg·l^{-1}) during the biodegradation of acetonitrile and salicylate in batch mode when the pollutants were present and being degraded. It however rapidly increased after complete pollutant depletion (Guieysse et al., 2002; Muñoz et al., 2005a). A high O_2 concentration is therefore a good indication of complete pollutant depletion during continuous operation (although the reciprocal can not be stated) (Muñoz et al., 2004).

5.2 Predation

Infections by parasitic fungi like *Chytridium* sp. or the development of complicated food chains in the photobioreactor can cause unexpected process failure (Abeliovich and Dikbuck, 1977). However the development of food chains can be easily avoided by daily operating the process at low O_2 levels for a short duration of time (1 hour) in order to suppress the growth of aerobic higher organisms (Abeliovich, 1986).

5.3. Operational Factors

Mixing

Mixing is necessary to 1) prevent algal sedimentation (which could lead to the accumulation and anaerobic decomposition of the biomass inside the reactor); 2) avoid thermal and nutritional gradients in the cultivation broth; 3) enhance transfer of nutrients and metabolites between the cells and the culture medium; and 4) move the microalgal cells from the dark to the illuminated zone in the reactor (Grobbelaar, 2000). This last point is especially important since mixing, together with cell density and light intensity, determine the frequency and duration of the light/dark cycles. It has been proven that high frequency light/dark cycles, attained by intense mixing, enhance photosynthetic rates (Richmond, 2004).

The device used for mixing is also very important because of the potential shear stress imposed on the microalgal cells, which are usually more fragile than bacteria (Barbosa et al., 2004; Mitsuhashi et al., 1995). For instance, Pirt and co-workers (1983) observed a decrease in the maximum specific growth rate of a *Chlorella* strain cultivated in a tubular loop reactor when a centrifugal or rotary positive displacement pump was used for biomass recirculation. In our laboratory, the high shear stress imposed on the microalgae by a centrifugal pump used for medium homogenization was suspected to have caused process failure during the continuous degradation of acetonitrile in a 50-L column loop reactor.

Hydraulic Retention Time

In light limited cultivations, the biomass concentration in the reactor is a function of the light supply and the hydraulic retention time (HRT). Thus, according to the model proposed by Evers (1991), if the HRT in a light limited photobioreactor decreases, part of the biomass will be washed-out from the photobioreactor. This decrease in the microalgal density increases the light intensity inside the culture medium and therefore the specific growth rate of the microalgae (Figure 8). A new steady state at a lower microalgal concentration is therefore reached. Although the specific O_2 production rate is now higher due to the increase in the light intensity, the microalgae concentration is lower, and the total O_2 production could be either lower or higher, depending on the extent of the variations of both parameters.

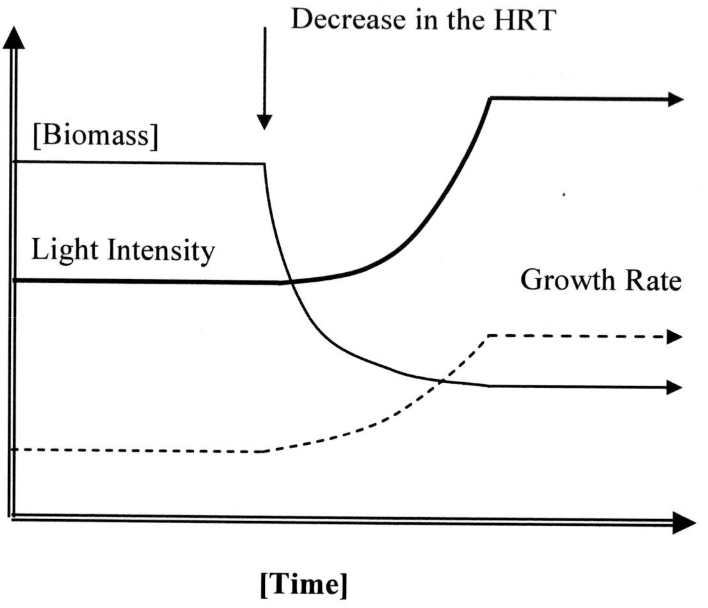

Figure 8. Transient state when decreasing the HRT in a light limited continuous photosynthetically oxygenated process.

Muñoz et al., (2004) observed that when a chemostat photobioreactor was operated at long HRTs (> 2.7 days), the process was characterised by complete pollutant removal, with high values of absorbency and high dissolved oxygen concentrations. Thus, at high HRT the oxygenation capacity of the system was higher than the BOD requirements of the industrial effluent, leading to a surplus of O_2. When the HRT was further decreased both the removal efficiency and the biomass concentration decreased. The DOC also decreased to values of approx. 0.5 $mg \cdot l^{-1}$, indicating that the process was controlled by O_2 supply and therefore by algal activity. When degrading toxic pollutants, the decrease in the removal efficiency brings also a simultaneous reduction in the photosynthetic activity due to the inhibition caused by the rise in the pollutant concentration in the medium.

Biomass Retention

Biomass retention is a very important parameter that determines the light utilization efficiency and prevents from biomass washout. Low HRT is normally associated with low biomass concentrations. In this case, the biomass concentration inside the reactor could not be sufficient to utilize all the light supplied to the system leading to a low O_2 production and therefore to a poor process performance (Muñoz et al., 2004). Biomass retention offers the possibility to maintain a high microalgal concentration. Biomass retention can be implemented by microbial immobilization or by biomass settling and further recirculation into the photobioreactor (Evans and Furlong, 2003; Richmond, 1983). Microalgal recirculation is nowadays the most common strategy as it allows a reasonable control of the biomass concentration without clogging problems or need for a proper immobilization support. Muñoz et al. (2004) thus reported an increase of 44 % in the process efficiency when the biomass concentration increased from 0.4 to 0.6 $g \cdot l^{-1}$ by recirculation. Biomass recirculation also enhanced nutrient removal in a semicontinuous culture of *Scenedesmus* treating a secondary wastewater (De la Noüe and Eidhin, 1988). Too high biomass concentrations are however detrimental as a result of the increase in microalgal dark respiration, which reduces the net amount of O_2 available to bacteria.

Inoculation Strategy

A correct inoculation strategy ensures a fast and reliable process start-up. At the early stages of the biodegradation process, the initial removal rates are correlated to the amount of microalgae initially inoculated (Guieysse et al., 2002). In these cases, microalgal density limits the biodegradation process because O_2 supply is intrinsically linked to microalgal growth and because microalgae usually grow at much lower rates than bacteria. Therefore, the O_2 is consumed as soon as it is released. Then, once a critical microalgal density is reached, the process becomes limited by the light supply to the individual photosynthetic cells (all the light is used). Other authors have also highlighted the importance of the initial cell population density, reporting that photodamage may cause the culture to collapse at very low microalgal densities, even when exposed to low irradiances (Göksan et al., 2003). Therefore, the inoculation of the systems should provide a sufficiently dense microalgal culture in order to avoid culture collapse and to ensure high initial oxygenation rates.

6. Conclusions

Due to their ability to produce O_2, green microalgae and cyanobacteria are capable of supporting aerobic wastewater treatment processes. Photosynthetic oxygenation is more energy efficient (since there is no need for external aeration) and safer (no risk of pollutant volatilization) than mechanical aeration, which make the process very attractive from the economical and technical point of view. However, this is just an example of the potential contribution of microalgae to environmental biotechnology. Nutrient and heavy metal removal can be coupled with organic pollutant mineralization in a single stage process, with the subsequent reduction in plant construction and operation costs. Photosynthetic microorganisms are also capable of assimilating or biotransforming toxic organic compounds like hormones, dyes or phenolic compounds.

Microalgae selection is a very important step that impacts the efficiency and stability of the process. First, the complex interactions between microalgae and bacteria should be considered due to the potential antagonistic/synergistic effects within the microcosm. Secondly, since microalgae generally exhibit lower growth rates and lower tolerance towards toxicants compared to the pollutant degrading bacteria, they are likely to limit the biodegradation process. Microalgae should be therefore selected in regards to their specific growth rate, O_2 production capacity, and tolerance towards the target toxicants. In addition, a good sedimentation capacity is also desired as microalgae harvesting could eventually limit the economic viability of the treatment process

Traditionally, microalgae based treatment processes have been implemented in HRAPs, devised for sewage treatment. Recent studies have also shown the potential of microalgae in the degradation of toxic industrial wastewaters. In this regard, enclosed photobioreactors are preferred as they allow for a better control, lower water losses, lower land requirements, and a reduced risk of hazardous pollutant volatilizations. Among enclosed bioreactors, tubular photobioreactors offer the best light utilization efficiency and process scalability compared to tank or flat plate photobioreactors. The actual high-cost associated with large-scale closed photobioreactors has limited their widespread application. However, the increasing demand for microalgae-derived pharmaceuticals, health and care products or aquaculture will eventually reduce the construction cost of closed photobioreactors, favoring their implementation for environmental applications.

Microalgae based processes are governed by environmental (pH, light intensity, temperature and O_2 concentration), biotic (predation by zooplankton, infections) and operational factors (mixing, HRT, biomass recirculation, pollutant concentration and inoculation strategy). Therefore, a deep understanding of the influence of these factors on the biodegradation performance is required for successful process optimization. Recent studies have shown that mixing and biomass retention are crucial for maintaining high pollutant removal rates.

The water and energy sectors are intrinsically linked as for instance highly-efficient water treatment processes (i.e. reverse osmosis) are not suitable on a global scale due to their high energy demand. Such constrain is likely to increase in a close future due to increasing energy costs and CO_2 taxations. Hence, one must encourage the development of processes which are, if not energy producing, energy and resource efficient. In this regards, sun-powered algae-based processes have a tremendous potential for converting organic matter into a valuable biomass suitable for fertilization and bioenergy production with minimal global impact. Further work should focus on the scale-up and optimization of microalgae-based processes in order to transform this promising bioremediation approach into a well-established and robust reality. In this regard, an adequate piloting is needed when scaling-up the process due to the lack of experience in design and data from full-scale enclosed photobioreactors treating residual wastewater.

References

Abeliovich, A., Azov, Y., 1976. Toxicity of ammonia to algae in sewage oxidation ponds. *Appl. Environ. Microb.* 31, 801-806.

Abeliovich, A., Dikbuck, S., 1977. Factors Affecting Infection of *Scenedesmus obliquus* by a *Chytridium* sp. in Sewage Oxidation Ponds. *Appl. Environ. Microb.* 34, 832-836.

Abeliovich, A., Weisman, D., 1978. Role of Heterotrophic Nutrition in Growth of Alga *Scenedesmus-Obliquus* in High-Rate Oxidation Ponds. *Appl. Environ. Microb.* 35, 32-37.

Abeliovich, A., 1986. Algae in wastewater oxidation ponds: In: Richmond, A. (Eds.), *Handbook of Microalgal Mass culture.* CRC Press, pp. 331-338.

Aiba, S., 1982. Growth kinetics of photosynthetic microorganisms. *Adv. Biochem. Eng.* 23, 85-156.

Akhtar, N., Saeed, A., Iqbal, M., 2003. *Chlorella sorokiniana* immobilized on the biomatrix of vegetable sponge of *Luffa cylindrica*: a new system to remove cadmium from contaminated aqueous medium. *Bioresource Technol.* 88, 163-165.

Aksmann, A., Tukaj, Z., 2004. The effect of Anthracene and Phenanthrene on the Growth, Photosynthesis, and SOD Activity of the Green Alga *Scenedesmus armatus* depends on the PAR Irradiance and CO_2 Level. Arch. *Environ. Contam. Toxicol.* 47, 177-184.

Aziz, M.A., Ng, W.J., 1993. Industrial wastewater treatment using an activated algae-reactor. *Water Sci. Technol.* 28, 71-76.

Azov, Y., Goldman, J.C., 1982. Free Ammonia Inhibition of Algal Photosynthesis in Intensive Cultures. *Appl. Environ. Microb.* 43, 735-739.

Barbosa, M.J., Hadiyanto, R., Wijffels, H., 2004. Overcoming Shear Stress of Microalgae Cultures in Sparged Photobioreactors. *Biotechnol. Bioeng.* 85, 78-85.

Benemann, J.R., Koopman, B., Weissman, J., Goebel, R., 1980. Development of microalgae harvesting and high rate pond technologies in California. In: Shelef, G., Soeder, C.J., (Eds.), *Algae Biomass Production and Use.* Elsevier/North-Holland Biomedical Press, Amsterdam., pp. 457-495.

Boon, A.G. 1983. Aeration methods. In: Barnes, D., Foster, C.F., Johnstone, D.W.M. (Eds.) *Oxidation Ditches in Wastewater Treatment.* Pitman. London. pp. 173-187.

Borde, X., Guieysse, B., Delgado, O., Munoz, R., Hatti-Kaul, R., Nugier-Chauvin, C., Patin, H., Mattiasson, B., 2003. Synergistic relationships in algal-bacterial microcosms for the treatment of aromatic pollutants. *Bioresource Technol.* 86, 293-300.

Borowitzka, M.A., 1997. Microalgae for aquaculture: Opportunities and constraints. *J. Appl. Phycol.* 9, 393-401.

Borowitzka, M.A., 1999. Commercial production of microalgae: ponds, tanks, tubes and fermenters. *J. Biotechnol.* 70, 313-321.

Cañizares-Villanueva, R.O., 2000. Biosorción de metales pesados mediante el uso de biomasa microbiana. *Revista Latinoamericana de Microbiologia* 131-143.

Chevalier, P., De la Noüe, J., 1985. Efficiency of Immobilized Hyperconcentrated Algae for Ammonium and Ortho-Phosphate Removal from Wastewaters. *Biotechnol. Lett.* 7, 395-400.

Chojnacka, K., Chojnacki, A., Gorecka, H., 2005. Biosorption of Cr^{3+}, Cd^{2+} and Cu^{2+} ions by blue-green algae *Spirulina* sp.: kinetics, equilibrium and the mechanism of the process. *Chemosphere.* 59, 75-84.

Clijsters, H., Vanassche, F., 1985. Inhibition of photosynthesis by heavy metals. *Photosynth. Res.* 7, 31-40.

Contreras-Flores, C., Pena-Castro, J.M., Flores-Cotera, L.B., Cañizares-Villanueva, R.O., 2003. Advances in conceptual design of photobioreactors for microalgal culture. *Interciencia.* 28, 450-456.

Costa, A.C.A., Leite, S.G.F., 1990. Cadmium and Zinc Biosorption by *Chlorella-Homosphaera*. Biotechnol. Lett. 12, 941-944.

Craggs, R.J., Adey, W.H., Jenson, K.R., St John, M.S., Green, F.B., Oswald, W.J., 1996. Phosphorus removal from wastewater using an algal turf scrubber. *Water. Sci. Technol.* 33, 191-198.

Day, J.G., Benson, E.E., Fleck, R.A., 1999. In vitro culture and conservation of microalgae: Applications for aquaculture, biotechnology and environmental research. In *Vitro Cell. Dev.-Pl.* 35, 127-136.

De-Bashan, L.E., Moreno, M., Hernandez, J.P., Bashan, Y., 2002. Removal of ammonium and phosphorus ions from synthetic wastewater by the microalgae *Chlorella vulgaris* coimmobilized in alginate beads with the microalgae growth-promoting bacterium *Azospirillum brasilense*. *Water Res.* 36, 2941-2948.

De la Noüe, J., Eidhin, D.N., 1988. Improved Performance of Intensive Semicontinuous Cultures of *Scenedesmus* by Biomass Recirculation. *Biotechnol. Bioeng.* 31, 397-406.

De la Noüe, J., Laliberte, G., Proulx, D., 1992. Algae and Waste-Water. *J. Appl. Phycol.* 4, 247-254.

De la Noue, J., Proux, D., 1988. Biological tertiary treatment of urban wastewater with chitosan-immobilised Phormidium. *Appl. Microbiol. Biotechnol.* 29 (2-3), 292-297.

Divakaran, R., Sivasankara, P., 2002. Flocculation of algae using chitosan. *J. Appl. Phycol.* 14, 419-422.

Eisenberg, D.M., Oswald, W.J., Benemann, J.R., Goebel, R.P., Tiburzi, T.T., 1981. Methane fermentation of microalgae. *Anaerobic Dig., Proc. Int. Symp.* 1st. 99-111.

Essam, T., Magdy, A.A., El Tayeb, O., Mattiasson, B., Guieysse, B., 2005. Biological treatment of industrial wastes in a photobioreactor. *Water Sci. Technol.* (submitted for publication).

Evans, G.M., Furlong, J.C., 2003. Phytotechnology and Photosynthesis. In: *Environmental Biotechnology: Theory and Application. edn First.* Edited by: John Wiley Sons. pp. 163-169.

Evers, E.G., 1991. A Model for Light-Limited Continuous Cultures - Growth, Shading, and Maintenance. *Biotechnol. Bioeng.* 38, 254-259.

Fukami, M., Ohbayashi, H., Yoshioka, Y., kuroishi, M., Yamazaki, S., Toda, S., 1988. A Zinc tolerant chlorotic mutant strain of *Euglena-Gracilis-z* induced by Zinc. *Agr. Biol. Chem. Tok.* 52, 601-604.

Fukami, K., Nishijima, T., Ishida, Y., 1997. Stimulative and inhibitory effects of bacteria on the growth of microalgae. *Hydrobiologia.* 358, 185-191.

Garcia, J., Hernandez-Marine, M., Mujeriego, R. 1998. Tratamiento de aguas residuales urbanas mediante lagunas de alta carga: evaluación experimental. *Ingenieria del Agua.* 5, 35-50.

Garcia, J., Mujeriego, R., Hernandez-Marine, M., 2000a. High rate algal pond operating strategies for urban wastewater nitrogen removal. *J. Appl. Phycol.* 12, 331-339.

Garcia, J., Hernández-Marine, M., Mujeriego, R., 2000b. Influence of phytoplankton composition on Biomass Removal from High-Rate Oxidation Lagoons by Means of Sedimentation and Spontaneous Flocculation. *Water Environ. Res.* 72, 230-237.

Ghirardi, M.L., Zhang, L.L., Flynn, J.W., Seibert, T., Greenbaum, M., Melis, A.E., 2000. Microalgae: a green source of renewable H_2. *Trends Biotechnol.* 18, 506-511.

Goksan, T., Dumaz, Y., Gokpinar, S., 2003. Effect of light paths lengths and initial culture density on the cultivation of *Chaetoceros muelleri* (Lemmermann, 1898). *Aquaculture.* 217, 431-436.

Gonzalez, L.E., Bashan, Y., 2000. Increased growth of the microalga *Chlorella vulgaris* when coimmobilized and cocultured in alginate beads with the plant- growth-promoting bacterium *Azospirillum brasilense*. *Appl. Environ. Microb.* 66, 1527-1531.

Greca, M.D., Fiorentino, A., Guerriero, I., Pinto, G., Pollio, A., Previtera, L., 1997. Biotransformation of adrenosterone into 11-ketotestosterone by *Scenedesmus quadricauda* grown in myxotrophic conditions. *Biotechnol. Lett.* 19, 1123-1124.

Green, M., Mels, A., Lahav, O., Tarre, S., 1996. Biological-ion exchange process for ammonium removal from secondary effluent. *Water. Sci. Technol.* 34, 449-459.

Grobbelaar, J.U., Soeder, C.J., Groeneweg, E.S., Hartig, P., 1988. Rates of Biogenic Oxygen Production in Mass-Cultures of Microalgae, Absorption of Atmospheric Oxygen and Oxygen Availability for Waste-Water Treatment. *Water Res.* 22, 1459-1464.

Grobbelaar, J.U., 2000. Physiological and technological considerations for optimising mass algal cultures. *J. Appl. Phycol.* 12, 201-206.

Guieysse, B., Borde, X., Munoz, R., Hatti-Kaul, R., Nugier-Chauvin, C., Patin, H., Mattiasson, B., 2002. Influence of the initial composition of algal-bacterial microcosms on the degradation of salicylate in a fed-batch culture. *Biotechnol. Lett.* 24, 531-538.

Hirooka, T., Akiyama, Y., Tsuji, N., Nakamura, T., Nagase, H., Hirata, K., Miyamoto, K., 2003. Removal of hazardous phenols by microalgae under photoautotrophic conditions. *J. Biosci. Bioeng.* 95, 200-203.

Hoffman, J.P., 1998. Wastewater treatment with suspended and nonsuspended algae. *J. Phycol.* 34, 757-763.

Hu, Q., Guterman, H., Richmond, A., 1996. A flat inclined modular photobioreactor (FIMP) for outdoor mass cultivation of photoautotrophs. *Biotechnol. Bioeng.* 51, 51-60.

Janssen, M., Tramper, J., Mur, L.R., Wijffels, R.H., 2003. Enclosed Outdoor Photobioreactors: Light Regime, Photosynthetic Efficiency, Scale-Up, and Future Prospects. *Biotechnol. Bioeng.* 81, 193-210.

Kaplan, D., Christiaen, D., Arad, S.M., 1987. Chelating Properties of Extracellular Polysaccharides from Chlorella Spp. *Appl. Environ. Microb.* 53, 2953-2956.

Kaplan, D., Heimer, Y.M., Abeliovich, A., Goldsbrough, P.B., 1995. Cadmium toxicity and resistance in *Chlorella* sp. *Plant Sci.* 109, 129-137.

Khoshmanesh, A., Lawson, F., Prince, I.G., 1996. Cadmium uptake by unicellular green microalgae. *Chem. Eng. J. Bioch. Eng.* 62, 81-88.

Kratochvil, D., Volesky, B., 1998. Advances in the biosorption of heavy metals. *Trends Biotechnol.* 16, 291-300.

Kuba, T., Smolders, G., Van Loosdrecht, M.C.M., Heijnen, J.J., 1993. Biological Phosphorus Removal from Waste-Water by Anaerobic-Anoxic Sequencing Batch Reactor. *Water Sci. Technol.* 27, 241-252.

Laliberté, G., Proulx, G., Pauw, N., De la Noüe, J., 1994. Algal Technology in wastewater treatment. *Ergenisse Limnologie.* 42, 283-302.

Lau, P.S., Tam, N.F.Y., Wong, Y.S., 1995. Effect of Algal Density on Nutrient Removal from Primary Settled Waste-Water. *Environ. Pollut.* 89, 59-66.

Lee, Y.K., Low, C.S., 1991. Effect of photobioreactor inclination on the biomass productivity of an outdoor algal culture. *Biotechnol. Bioeng.* 38, 995-1000.

Lee, Y.K., 2001. Microalgal mass culture systems and methods: Their limitation and potential. *J. Appl. Phycol.* 13, 307-315.

Lee, J.S., Lee, J.P., 2003. Review of Advances in Biological CO_2 Mitigation Technology. Biotechnol. Bioproc. E. 8, 354-259.

Liang, S., Liu, X., Chen, F., Chen, Z., 2004. Current microalgal health food R & D activities in China. *Hydrobiologia.* 512, 45-48.

Lima, S.A.C., Castro, P.M.L., Morais, R., 2003. Biodegradation of p-nitrophenol by microalgae. *J. Appl. Phycol.* 15, 137-142.

Madigan, M.T., Martinko, J.M., Parker, J. 1997. Metabolic Diversity among the Microorganisms. In: *Brock Biology of Microorganisms.* edn eighth. Edited by: Prentice Hall International, Inc.

Malik, A., 2004. Metal bioremediation through growing cells. *Environ. Int.* 30, 261-278.

Mallick, N., 2002. Biotechnological potential of immobilized algae for wastewater N, P and metal removal: A review. *Biometals.* 15, 377-390.

Mara, D.D., Pearson, H., 1986. Artificial freshwater environment: Waste stabilization ponds. In: Rehm, H.J., Reed, G. (Eds.), *Biotechnology. Velagsgesellschaft,* pp.177-206.

Martinez Sancho, M.E., Jimenez Castillo, J.M., Espinola Lozano, J.B., El Yousfi, F., 1993 Sistemas algas-bacterias para tratamiento de residuos liquidos. *Ing. Quim.* 25, 131-135.

Matsumoto, H., Hamasaki, A., Shioji, N., Ikuta, Y., 1996. Influence of dissolved oxygen on photosynthetic rate of microalgae. *J. Chem. Eng. Jpn.* 29, 711-714.

Mattuschka, B., Junghaus, K., Straube, B., 1993. Biosorption of Metals by waste biomass. In: Torma, A.E., Apel, M.L., Brierley, C.L., (Eds.), Biohydrometallurgical Technologies. *The Minerals, Metals and Materials Society:* Warrendale.

McGriff, E.C. Jr., McKinney, R.E., 1972. The removal of nutrients and organics by activated algae. *Water Res.* 6, 1155-1164.

Melis, A., Neidhardt, J., Benemann, J.R., 1999. *Dunaliella salina* (Chlorophyta) with small chlorophyll antenna size exhibit higher photosynthetic productivities and photon use efficiencies than normally pigmented cells. *J. Appl. Phycol.* 10, 515-525.

Mezrioui, N., Oudra, B., Oufdou, K., Hassani, L., Loudiki, M., Darley, J., 1994. Effect of Microalgae Growing on Waste-Water Batch Culture on *Escherichia-Coli* and *Vibrio-Cholerae* Survival. *Water Sci. Technol.* 30, 295-302.

Miron, A.S., Gomez, A.C., Camacho, F.G., Molina Grima, E., Chisti, M.Y., 1999. Comparative evaluation of compact photobioreactors for large-scale monoculture of microalgae. *J. Biotechnol.* 70, 249-270.

Mitsuhashi, S., Hosaka, K., Tomonaga, E., Muramatsu, H., Tanishita, K. 1995. Effects of shear flow on photosynthesis in a dilute suspension of microalgae. *Appl. Microbiol. Biot.* 42, 744-749.

Molina-Grima, E. 1999. Microalgae mass culture methods. In: Flickinger, M.C.D., (Eds.), Encyclopedia of Bioprocess Technology: Fermentation, *Biocatalysis and Bioseparation.* S.W. John Wiley & Sons.

Molina-Grima, E., Belarbi, E-H., Acién Fernandéz, F.G., Robles-Medina, A., Chisti, Y., 2003. Recovery of microalgal biomass and metabolites: process options and economics. *Biotechnol. Adv.* 20, 491-515.

Morita, M., Watanabe, Y., Saiki, H., 2001. Instructions of microalgal biomass production for practically higher photosynthetic performance using a photobioreactor. *Trans IchemE,* 79, part C, 176-183.

Mouget, J.L., Dakhama, A., Lavoie, M.C., De la Noue, J., 1995. Algal growth enhancement by bacteria: Is consumption of photosynthetic oxygen involved? *FEMS Microbiol. Ecol.* 18, 35-43.

Muñoz, R., Köllner, C., Guieysse, B., Mattiasson, B., 2003,. Salicylate biodegradation by various algal-bacterial consortia under photosynthetic oxygenation. *Biotechnol. Lett* 25, 1905-1911.

Muñoz, R., Köllner, C., Guieysse, B., Mattiasson, B., 2004. Photosynthetically oxygenated salicylate biodegradation in a continuous stirred tank photobioreactor. *Biotechnol. Bioeng.* 87(6), 797-803.

Muñoz, R., Jacinto, M.S.A., Guieysse, B., Mattiasson, B., 2005a. Combined Carbon and Nitrogen removal from acetonitrile using algal-bacterial reactors. *Appl. Microbiol. Biot.* 67(5) 699-707.

Muñoz, R., Rolvering, C., Guieysse, B., Mattiasson, B., 2005b. Photosynthetically oxygenated acetonitrile biodegradation by an algal-bacterial microcosm: a pilot scale study. *Water Sci. Technol.*51(12) 261-265.

Muñoz, R., Alvarez, T., Muñoz, A., Terrazas, E., Guieysse, B., Mattiasson, B., 2006. *Sequential removal of heavy metals ions and organic pollutants using an algal-bacterial consortium*. Chemosphere (accepted for publication).

Narro, M.L., 1987. Petroleum Toxicity and the oxidation of aromatic hydrocarbons. In: Fay, P., Van Baalen, C., (Eds.), *The cyanobacteria. Elsevier,* pp. 491-511.

Nicolella, C., van Loosdrecht, M.C.M., Heijnen, J.J., 2000. Wastewater treatment with particulate biofilm reactors. *J. Biotechnol.* 80, 1-33.

Nielsen, J., Villadsen, J., Lidén, G., 2003. Scale-up of bioprocesses. In: *Bioreaction Engineering Principles*. Edn Second. Edited by: Kluwer Academic/Plenum Publishers. pp. 477-517.

Nurdogan, Y., Oswald, W.J., 1995. Enhanced Nutrient Removal in High-Rate Ponds. *Water Sci. Technol.* 31, 33-43.

Nurdogan, Y., Oswald, W.J., 1996. Tube settling of high-rate pond algae. *Water Sci. Technol.* 33, 229-241.

Ogbonna, J.C., Tanaka, H., 2000b. Light requirement and photosynthetic cell cultivation - Development of processes for efficient light utilization in photobioreactors. *J. Appl. Phycol.* 12, 207-218.

Ogbonna, J.C., Yoshizowa, H., Tanaka, H., 2000.Treatment of a high strenght organic wastewater by a mixed culture of photosynthetic microorganisms. *J. Appl. Phycol.* 12 (3-5), 277-284.

Oh, H-M., Lee, S.J., Park, M-H., Kim, H-S., Yoon, J-H., Kwon, G-S., Yoon, B-D., 2001. Harvesting of *Chlorella vulgaris* using a bioflocculant from *Paenibacillus sp.* AM49. *Biotechnol. Lett.* 23, 1229-1234.

Olguin, E.J., 2003. Phycoremediation: key issues for cost-effective nutrient removal processes. *Biotechnol. Adv.* 22, 81-91.

Omar, H.H., 2002. Bioremoval of zinc ions by *Scenedesmus obliquus* and *Scenedesmus quadricauda* and its effect on growth and metabolism. *Int. Biodeter. Biodegr.* 50, 95-100.

Osada, T., Kuroda, K., Yonaga, M., 1995. Reducing Nitrous-Oxide Gas Emissions from Fill-and-Draw Type Activated-Sludge Process. *Water Res.* 29, 1607-1608.

Oswald, W.J., 1976. Gas production from micro algae. Clean Fuels Biomass, Sewage, Urban Refuse, Agric. Wastes, *Symp. Pap.* 311-324.

Oswald, W.J. 1988. Micro-algae and waste-water treatment. In: Borowitzka, M.B.L. (Eds.), *Micro-algal biotechnology.* Cambridge University Press;:305-328.

Oswald, W.J., 1995. Ponds in the twenty-first century. *Water Sci. Technol.* 31, 1-8.

Oswald, W.J., 2003. My sixty years in applied algology. *J. Appl. Phycol.* 15, 99-106.

Paerl, H.W., Kellar, P.E., 1978. Significance of Bacterial Anabaena (Cyanophyceae) Associations with Respect to N_2 Fixation in Freshwater. *J. Phycol.* 14, 254-260.

Palmer, C.M., 1969. A composite rating of algae tolerating organic pollution. *J. Phycol.* 5, 78-82.

Peña-Castro, J.M., Martinez-Jeronimo, F., Esparza-Garcia, F., Canizares-Villanueva, R.O., 2004. Phenotypic plasticity in Scenedesmus incrassatulus (Chlorophyceae) in response to heavy metals stress. *Chemosphere.* 57, 1629-1636.

Peric, J., Trgo, M., Vukojevic Medvidovic, N., 2004. Removal of zinc, copper and lead by natural zeolite-a comparison of adsorption isotherms. *Water Res.* 38, 1893-1899.

Pinto, G., Pollio, A., Previtera, L., Temussi, F., 2002. Biodegradation of phenols by microalgae. *Biotechnol. Lett.* 24, 2047-2051

Pinto, G., Pollio, A., Previtera, L., Stanzione, M., Temussi, F., 2003. Removal of low molecular weight phenols from olive oil mill wastewater using microalgae. *Biotechnol. Lett.* 25, 1657-1659.

Pirt, S.J., Lee, Y.K., Walach, M.R., Pirt, M.W., Balyuzi, H.H.M., Bazin, M.J., 1983. *J. Chem. Tech. Biotechnol.* 33B, 35-58.

Pulz, O. , 2001. Photobioreactors: production systems for phototrophic microorganisms. *Appl. Microbiol. Biot.* 57, 287-293.

Richmond, A., 1983. Phototrophic microalgae. In: Rehm, H.J, Reed, G., (Eds.), *Biotechnology.* Edn First. Verlagsgesellschaft. Vol 3. pp. 109-114.

Richmond, A., 2000. Microalgal biotechnology at the turn of the millennium: A personal view. *J. Appl. Phycol.* 12, 441-451.

Richmond, A., 2004. Principles for attaining maximal microalgal productivity in photobioreactors: an overview. *Hydrobiologia.* 512, 33-37.

Robinson, P.K., Reeve, J.O., Goulding, K.H., 1988. Kinetics of Phosphorus Uptake by Immobilized *Chlorella. Biotechnol. Lett.* 10, 17-20.

Robinson, P.K., 1998. Immobilized Algal Technology for wastewater treatment purposes. In: Wong, Y-S., Tam, Nora, F.Y. (Eds), Wastewater treatment with Algae. *Springer-Verlag and Landes Bioscience,* pp. 1-16.

Rose, P.D., Boshoff, G.A., van Hille, R.P., Wallace, L.C.M., Dunn, K.M., Duncan, J.R. 1998. An integrated algal sulphate reducing high rate ponding process for the treatment of acid mine drainage wastewaters. *Biodegradation.* 9, 247-257.

Safonova, E., Dmitrieva, I.A., Kvitko, K.V., 1999. The interaction of Algae with alcanotrophic bacteria in black oil decomposition. *Resources, conservation and Recycling.* 27, 193-201.

Safonova, E., Kvitko, K.V., Iankevitch, M.I., Surgko, L.F., Afri, I.A., Reisser, W., 2004. Biotreatment of Industrial Wastewater by Selected Algal-Bacterial Consortia. *Eng. Life Sci.* 4, 347-353.

Sawayama, S., Rao, K.K., Hall, D.O., 1998. Nitrate and phosphate ion removal from water by *Phormidium laminosum* immobilized on hollow fibres in a photobioreactor. *Appl. Microbiol. Biot.* 49, 463-468.

Sawayama, S., Minowa, T., Yokoyama, S.Y., 1999. Possibility of renewable energy production and CO2 mitigation by thermochemical liquefaction of microalgae. *Biomass and Bioenerg.* 17(1), 33-39.

Semple, K.T., Cain, R.B., Schmidt, S., 1999. Biodegradation of aromatic compounds by microalgae. *Fems Microbiol. Lett.* 170, 291-300.

Schmitt, D., Muller, A., Csogor, Z., Frimmel, F.H., Posten, C., 2001. The adsorption kinetics of metal ions onto different microalgae and siliceous earth. *Water Res.* 35, 779-785.

Schumacher, G., Blume, T., Sekoulov, I., 2003. Bacteria reduction and nutrient removal in small wastewater treatment plants by an algal biofilm. *Water Sci. Technol.* 47, 195-202.

Sikora, F.J., Tong, Z., Behrends, L.L., Steinberg, S.L., Coonrod, H.S., Ammonium removal in constructed wetlands with recirculating subsurface flows: removal rates and mechanisms. *Wat Sci. Technol.* 32(3) 193-202.

Sorokin, C., Krauss, R.W., 1958. The Effects of Light Intensity on the Growth Rates of Green Algae. *Plant Physiol.* 33, 109-113.

Su, J.J., Liu, B.Y., Chang, Y.C., 2003. Emission of greenhouse gas from livestock waste and wastewater treatment in Taiwan. *Agr. Ecosyst. Environ.* 95, 253-263.

Subaramaniana, G., Uma, L., 1997. Role of cyanobacteria in pollution abatement. In: Sinha, M.P., (Eds.), *Recent Advances in Ecobilogical Research.* APH Publishing Corporation, Vol. 1. pp. 435-443.

Suh, I.S., Lee, C.G., 2003. Photobioreactor Engineering: Design and Performance. *Biotechnology and Bioprocess Engineering.* 8, 313-321.

Tam, N.F.Y., Lau, P.S., Wong, Y.S., 1994. Waste-Water Inorganic N-Removal and P-Removal by Immobilized *Chlorella Vulgaris*. *Water Sci. Technol.* 30, 369-374.

Tam, N.F.Y., Wong, Y.S., 2000. Effect of immobilized microalgal bead concentrations on wastewater nutrient removal. *Environ. Pollut.* 107, 145-151.

Tarlan, E., Dilek, F.B., Yetis, U., 2002. Effectiveness of algae in the treatment of a wood-based pulp and paper industry wastewater. *Bioresource Technol.* 84, 1-5.

Tchobanoglous, G., Burton, F.L., Stensel, D.H., 2003. Metcalf and Eddy. *Wastewater Engineering: Treatment and reuse.* Fourth. Edn. Edited by: Mc Graw Hill.

Tison, D.L., Lingg, A.J., 1979. Dissolved Organic-Matter Utilization and Oxygen-Uptake in Algal-Bacterial Microcosms. *Can. J. Microbiol.* 25, 1315-1320.

Torzillo, G., Pushparaj, B., Masojidek, J., Vonshak, A., 2003. Biological constraints in algal biotechnology. *Biotechnology and Bioprocess Engineering* 8, 339-348.

Travieso, L., Benitez, F., Weiland, P., Sanchez, E., Dupeyron, R., Dominguez, A.R., 1996. Experiments on immobilization of microalgae for nutrient removal in wastewater treatments. *Bioresource Technol.* 55, 181-186.

Travieso, L., Cañizares, R.O., Borja, R., Benitez, F., Dominguez, A.R., Dupeyron, R., Valiente, V., 1999. Heavy metal removal by microalgae. B. *Environ. Contam. Tox*. 62, 144-151.

Travieso, L., Pellon, A., Benitez, F., Sanchez, E., Borja, R., O'Farrill, N., Weiland, P., 2002. BIOALGA reactor: preliminary studies for heavy metals removal. *Biochem. Eng. J.* 12, 87-91.

Tredici, M.R., Carlozzi, P., Zittelli, G.C., Materassi, R., 1991. A Vertical Alveolar Panel (Vap) for Outdoor Mass Cultivation of Microalgae and Cyanobacteria. *Bioresource Technol.* 38, 153-159.

Tredici, M.R., Zittelli, G.C., 1998. Efficiency of Sunlight Utilization: Tubular Versus Flat Photobioreactors. *Biotechnol. Bioeng.* 57, 187-197.

Tredici, M.R., 1999. Bioreactors, photo. In: Flickinger, M.C.D., (Eds.), *Encyclopedia of Bioprocess Technology: Fermentation, Biocatalysis and Bioseparation.* S.W. John Wiley & Sons.

Tsygankov, A.A., 2001. Laboratory scale photobioreactors. *Appl. Biochem. Microbiol.* 37, 333-341.

Van Hille, R.P., Boshoff, G.A., Rose, P.D., Duncan, J.R., 1999. A continuous process for the biological treatment of heavy metal contaminated acid mine water. *Resour. Conserv. Recy.* 27, 157-167.

Van Loosdrecht, M.C.M., Hooijmans, C.M., Brdjanovic, D., Heijnen, J.J., 1997. Biological phosphate removal processes. *Appl. Microbiol. Biot.* 48, 289-296.

Van Loosdrecht, M.C.M., Jetten, M.S.M., 1998. Microbiological conversions in nitrogen removal. *Water Sci. Technol.* 38, 1-7.

Villaescusa, I., Fiol, N., Martinez, M., Miralles, N., Poch, J., Serarols, J., 2004. Removal of copper and nickel ions from aqueous solutions by grape stalks wastes. *Water Res.* 38, 992-1002.

Volesky, B., 2001. Detoxification of metal-bearing effluents: biosorption for the next century. *Hydrometallurgy.* 59, 203-216.

Vollenweider, R.A., 1985. Elemental and Biochemical-Composition of Plankton Biomass - Some Comments and Explorations. *Arch. Hydrobiol.* 105, 11-29.

Wilde, E.W., Benemann, J.R., 1993. Bioremoval of Heavy Metals by the Use of Microalgae. *Biotechnol. Adv.* 11, 781-812.

Wolfaardt, G.M., Lawrence, J.R., Robarts, R.D., Caldwell, D.E., 1994. The Role of Interactions, Sessile Growth, and Nutrient Amendments on the Degradative Efficiency of a Microbial Consortium. *Can. J. Microbiol.* 40, 331-340.

Wood, A., 1987. Simple Wastewater treatment system incorporating the selective cultivation of a filamentous algae. *Water. Sci. Technol.* 19, 1251-1254.

Yan, G., Yu, J.Y., Wang, Y.X., 1996. The effects of pH and temperature on orthophosphate removal by immobilized *Chlorella vulgaris*. *Biotechnol. Lett.* 18, 893-896.

Yu, R-Q., Wang, W-X., 2004. Biokinetics of cadmium, selenium, and zinc in freshwater alga *Scenedesmus obliquus* under different phosphorus and nitrogen conditions and metal transfer to *Daphnia magna*. *Environ. Pollut.* 129, 443-456.

In: Leading-Edge Environmental Biodegradation
Editor: L. E. Pawley, pp. 199-222

ISBN: 978-1-60021-903-0

Chapter 6

MICROORGANISMS BIOPROSPECTION: A NEW TENDENCY IN MICROBIAL ECOLOGY

Elcia Margareth Souza Brito*
Instituto de Investigación en Biología Experimental,
Universidad de Guanajuato
***Leandro Helgueira Andrade*†**
Departamento de Química Fundamental
Universidade de São Paulo
***César Augusto Caretta*‡**
Departamento de Astronomía
Universidad de Guanajuato
***Robert Duran*§**
Institut Pluridisciplinaire de Recherche sur l'Environnement et les Matériaux
Université de Pau et des Pays de l'Adour

ABSTRACT

Since long ago the men have been using plants, animals and also microorganisms as source of chemicals and products for his interest. With the recent development in the exploration of the microbiological and genetical world, this resource became still more promising, and stimulated a new area of research, the bioprospection. Roughly speaking it comprises the search of new economically viable compounds, produced via catabolic reactions, which can be efficiently manipulated and applied to some chemical process of interest in the industry, health or environment. As an example of new trends on Ecology Microbiology, this chapter is concerned on the bioprospection of bacteria, looking for

* emsbrito@gmail.com
† leandroh@iq. usp.br
‡ caretta@astro.ugto.mx
§ robert.duran@univ-pau.fr

biocatalysts to prepare pure enantiomerically compounds, to be applied on pharmacology, and hydrocarbonoclastic consortia, to be applied on bioremediation of oil contaminated environments. 106 bacteria strains, with potential for synthesis of chiral molecules, were obtained using 8 different carbon sources (4 nitriles, 2 alcohols and 2 ketones). Moreover, 5 bacteria strains, capable of degrading hydrocarbon compounds (between 16 - 45%) in laboratory conditions, were also isolated from a hydrocarbonoclastic consortium. Unexpectedly, when this consortium was applied to some *in situ* microcosms (planted on a mangrove ecosystem) the oil compounds were not reduced significantly. This result shows the necessity of making *in situ* experiments to test the real potentiality of the microorganisms for remediation of contaminated sites. It also reinforces that natural ecological systems are very complex and hard to be controlled, and that our knowledge about the biotechnological potential of microorganisms has much to be developed. In the meanwhile, we discuss the importance of selecting a suitable culture media (carbon source) and the modern tools for studying the biodiversity of the samples and the chemical potential of the isolates.

1. Introduction

It is long known that men employ plants or parts of them to cure diseases (for instance, the Ebers papyrus of Egipt, dated of about 1500 b.c., and the chinese medicine records from the imperor of Huang Ti, at 200 b.c.), but only at the end of XIX century appeared for the first time a term to describe the research or the study of the active effect of substances from the plants, animals or microorganisms, the pharmacognosy, that is, nowadays, a sub-area of pharmacy. The term bioprospection has been used specifically to express the collecting and testing of biological samples (from plants, animals or microorganisms) or the indigenous knowledge which helps the discovering and exploitation of genetic or biochemical resources [6]. Although the biodiversity of macroorganisms is also a very important biological reserve, here we will focus on the microbial diversity and their more recently discovered potential for bioprospection.

The knowledge of the "microbial world" increased largely with (a) the advent of the microscopy, (b) the discovery of the genetic material and (c) the use of DNA code for taxonomical studies. With microscopes the microoganisms were classified by their morphologic and physiologic characteristics, as is done for macroorganisms. However, by contrast to animals and plants, the microbial morphology is not complex enough to create a consistent phylogenic classification. Fortunately, the knowledge that DNA molecules hold of genetic information, together with the possibility of sequencing them, allowed the study of the microbial evolution using molecular approaches. The statement that DNA is a molecular chronometer for taxonomy studies happened on the 70's. In the 80's strong claims emerged for the use of the larger rRNA molecules on the microbial ecology studies [39; 40; 41; 59; 60; 64; 81]. Woese [82] used 16S rRNA coding sequences to propose the separation of the living beings in three great domains: Archeobacteria, Bacteria and Eukaria. Nowadays the microbiology biodiversity study is relatively easy to do as compared to the early 80's, since new technologies for DNA sequencing and data processing are available.

For defining structure and diversity of microbial communities the environmental microbiologists have been identifying and analyzing their microbial populations by their richness and everness [54]. Richness is defined as the number of species within a community and everness is the amount of each population that constitutes the community. Unquestionably, the microbial world is much more complex that it was thought before. Most of the bacterial studies have been done, and are still being carried out, by conventional culture-dependent methods. Unfortunately, since the majority of the microorganisms are not cultivable, by using these methods only less than 1% of existing bacteria on Earth have been identified [1]. It is also estimated that nearly 80% of Archea and Bacteria of the biosphere may be found at large depths, high temperatures and extreme pressions within the interior of the planet [55; 79]. Additionally, there are many environments still unexplored on the planet surface. As an example of these environments the most extreme are: (a) high temperature and acid or basic sites (e.g. hot springs or geysers); (b) elevated temperature, pression, and metal and dissolved minerals concentration sites (e.g. hydrothermal vents like those of Marianas Trench); (c) extremely cold sites (e.g. permafrosts, glaciers y subglaciar lakes); (d) high salt concentraction sites (e.g. the Dead Sea and the Great Salt Lake); (e) extremely dry and hot sites (e.g. the Atakama Desert and the Valley of the Death); (f) hyper-alkaline sites (e.g. the Magadi Lake); and (g) hyper-acid sites (e.g. the Tinto River). In addition to these naturally extreme environments, there are the anthropic polluted areas, which also expose the microorganisms to extreme condictions of pH, temperature, radioactivity, anoxic and/or reducing condictions and chemical products (e.g. heavy metals and hydrocarbons). Although it is dificult to simulate and manipulate extreme environments, some research has been carried out, making use of new technologies to create in laboratories favorable conditions that allow the growth of these extremophiles [61].

Despite of the general concern on the biodiversity preservation, the bioprospection is essentially an economic exploration of the natural resources before they disappear [50; 56]. As a consequence, topics as the bioprospection and the patent law have became very popular in the scientific and no scientific media. Another outcome is a political conflict: the countries with higher natural conserved biodiversity usually have no means to study this biodiversity, while many countries that have high bioprospection technologies have small natural resources. Then, while the formers try to protect their biodiversity with laws to combat the biopiratery, the latter use the patent law to protect their technology. Moreover, the bioprospection usually aims some direct biotechnological applicability. Chemists and pharmacists are, in general, interested on searching a more efficient, economically more viable, or ecologically more appropriate compound for a well established chemical process. The microbial ecologists, on the other hand, are seeking microorganisms and/or catabolic gene capable of mitigating a pollutant in order to remediate contaminated environments [13].

In this chapter we will discuss some important and pioneering aspects of the methology used on the bioprospection of microorganisms. The questions that will be addressed on this enterprise are the following: Since there is a huge bacterial diversity to be discovered and a vast amount of environments still unexplored, where to begin the bioprospection (which environment is more promising) and which strategy to addopt? Considering that most of the microbial researches have been substantiated on cultivable species, which culture media is better suited for gathering a specific group of microorganisms? How to access the genetic material of uncultivable microorganisms? Considering that it is usual to isolate a large number of strains in a specific culture media, is it worthwhile to work with all of them or is it better to choose some?

2. Planing the Bioprospection of Microorganisms

The main possible steps of the work on bioprospection of microorganisms can broadly be defined. The first step consists on sampling the material (soil, sediments, water or particulated material) that will be used as inocula on culture media to obtain microorganism consortia and/or to isolate the strains. Once pure strains are obtained, their phylogenetic analysis is done. In this stage, a plenty of tests for the bacterial characterization are done: morphological (shape, size, gram stain, mobility), physiological and biochemical (e.g., oxidase, catalase, peroxydase reactions, etc.). Then the DNA is extracted, and the 16S rRNA gene is amplified by PCR using universal primers and it is then sequenced for identification. After a careful verification of the quality of the produced data, a BLAST program is run to compare the obtained sequence to those deposited at an online databank (for instance, the National Center for Biotechnology Information, NCBI website: http://www.ncbi.nlm.nih.gov) and check if it is new. Finally, the evolutionary trees of the strains is performed using other bioinformatic tools: first the sequences are aligned by a *ClustalW* program, the alignments are carefully checked, and then the phylogenetic tree is constructed using one of several available programs (PAUP, PHYLIP, MEGA, GCG, phylodendron, etc.).

In order to select, within all the isolates, the more suitable strains for proceeding the bioprospection work (enzymes and/or catabolic gene), they are submitted to additional qualitative and quantitative experiments for substrate specificity. In the search for biocatalysts to prepare chiral compounds (such as alcohol enantiomerically pure), the isolates are submitted to different carbon sources, such as prochiral or racemic compounds. One way that has been succesfully used to find biocatalysts which exhibit high yield and enantioselectivity along the reaction is the screening test carried out on microplates: different strains are added to each well with the reaction media and, after an evaluation period, their productivity are compared. Another possibility is to take the strain that has a very high production but needs a long time to grow, identify the catabolic gene responsable for the production and clone it on a fast growing bacterium. On the other hand, for searching for biodegraders, *in vitro* and *in situ* tests containing the pollutant that is targeted are carried out, and the ability of the microorganisms to degrade this pollutant is evaluated.

The studies that use microcosms to evaluate the potentiality of consortia usualy employ molecular tools, which allow to work with the total DNA of a sample, to access the fluctuation of microbial diversity (ecological succession). Examples of molecular fingerprinting methods available for monitoring species on environmental samples are: the DGGE (Denaturing Gradient Gel Electrophoresis) and TGGE (Temperature Gradient Gel Electrophoresis), the SSCP (Single-Strand Conformational Polymorphism), RISA (rRNA Intergenic Spacer Analysis) and ARISA (Automated Ribosomal Intergenic Spacer Analysis), and the T-RFLP (Terminal Restriction Fragment Length Polimorphysm). For a review, see Dahllof [24] and Darling and Blum [25], whom evaluate these and other DNA-based approaches currently used, such as: SSP (Species-Specific PCR), FISH (Fluorescent In Situ Hybridization), DNA barcoding, and POAs (Phylogenetic Oligonucleotide Micro Arrays).

The DGGE and TGGE are based on electrophoresis of 16S rDNA fragments in polyacrylamide gels containing a linearly increasing gradient of denaturants, where the DNA fragments of the same length, but with different base-pair sequences, can be separated [54]. Both require a CG clamp (40-45 bases rich CG sequence) attached to the 5' end of the forward primer, to prevent complete melting of the PCR product. This technique has the advantage to allow the recovery of the DNA fragment of interest directly from the gel for the determination of the nucleic acid sequence. The drawback is that it may occur co-migration of different DNA fragments [3; 58; 62; 73], leading to an underestimation of the existing populations.

The SSCP, as the DGGE and TGGE, is based on the relation between the electrophoretic mobility of a single stranded DNA and its conformation, but with the electrophoresis done on a non-denaturing acrylamide gel [74]. The different bands, in this case, result from the characteristic tree-dimensional shape of nucleic acids and their nucleotide sequence. In contrast to DGGE, the SSCP do not require the CG clamp which can be source of artifacts on the annealing step. Traditionally the PCR products are separated using a vertical gel electrophoresis apparatus [52; 72; 74; 75] but an automated sequencer can also be used [23]. On the automated sequencer, fluorescently labeled primers and a sample with the fluorescent inner standard are used, which allow to compare migration patterns between samples [23]. The drawbacks of SSCP method are the possible occurance of co-migration and formation of heteroduplex, while factors such as gel matrix, temperature, and fragment size can influence its sensibility [53].

The RISA technique exploits the variability in the length of the interspecific spacer between the small (16S) and large (23S) rRNA gene subunits [67; 71]. In RISA the PCR product (a mixture of fragments contributed by all community members) is denatured and electrophoresed on the polyacrylamide gel, and similarly to DGGE, the DNA fragments are visualized by silver staining, producing a fingerprint of the sample community [12]. Moreover, techniques that use silver staining tend to have low resolution comparing to capillary electrophoresis system, so flourished the ARISA, an automated RISA. On this method the application of an automated DNA sequencer improves the resolution and efficiency of the analysis of the generated fragments. The ARISA uses fluorescent-tagged oligonucleotide primers for PCR amplification. During the electrophoresis, the fluorescent fragments are detected by an automated system [38]. Differences in intensity of the bands on RISA, or of profiles on ARISA, allow comparing little changes on the community population.

The T-RFLP also utilizes fluorescent-tagged oligonucleotide primers for the PCR amplification, but combines treatment of the amplified DNA with restriction enzymes and separation of the fragments in an automated sequencer [7; 9; 54; 63]. It is a very fast method, allowing the processing of many samples at the same time, and very sensible for the determination of ribotypes [14], but there are also some inconveniences: it uses a sequencer, an expensive equipment, and it is not enough sensible for the detection of small populations. Thus, the choice of the best-suitable molecular technique depends on the goals of the work and also on the laboratory conditions.

Concerning the uncultivable microorganims, the total DNA recovered from the sample can be used to construct environmental libraries by cloning large DNA fragments (>100kb) with bacterial artificial chromosomes (BACs). The libraries are then characterized by 16S rRNA coding sequences analysis. These BAC libraries can also be submitted to probes for evaluating their applicability on the synthesis of chiral substances or on the transformation of a pollutant. From the fragments of DNA clones it is possible to obtain the gene of interest by PCR or sub-cloning.

The last step of the adopted approach is the obtention of the catabolic genes that include all the functional sequences encoding the elements which are responsible for the gene structure, regulation and transcription. Several DNA regions can be used to reach this objective [69], from which the most basic signal sensor is a single consensus of sequences (conserved sequence fragment). Then, when a microorganism that degrade a pollutant or produce an interesting compound is found, the presence of the possible catabolic genes involved on this process can be scanned. For instance, strains that grow on polycyclic aromatic hydrocarbons (PAHs) problably have dioxygenase gene, and those that grow on aliphatics, monoxygenases, and so on. To access novel genes, the DNA (from strains or BACs) may be amplified by PCR using specific already validated primers. Usually these primers are designed by the researchers, based on genic sequences of genes with similar function, already described, which are taken from online databanks (for example, NCBI). Again, the obtained sequences are aligned (using *ClustalW*), and then they are inspected to find conserved fragments. These conserved fragments will be used to design the primers. Type strains (that carry the target or a similar gene) are used to find the best PCR-conditions and for primer validation. Then, the primers are used for PCR amplification of DNA sequences containing the novel gene. PCR amplified fragments are checked on agarose electrophoresis gels, cut from the gel, sequenced and then the sequences analysed by the bioinformatic tools cited above.

Doubtless the meta-(genomic, proteomic and bolomic) studies generate great amounts of data which require powerful computational tools for storing, manipulating and analysing them. There are many examples of the use of bioinformatics on the analysis presented above: the integration and comparison of genetic sequences to those deposited on genebank databases, the alignment of sequences, the construction of phylogenetic trees, etc. Moreover, *in silico* resources are used for accessing all the available information associated, not only the sequences, with the microorganisms already studied (isolate history, growth requeriments, physiological proteomic information, and so on) [70]. They are also used for applying statistical tests to evaluate and check the reliability of the results, and for simulating certain intermediary processes in order to optimize some choices. In the latter case, amplification and restriction of genetic sequences can be, for example, simulated to produce T-RFLP fingerprints and, by comparing the results for different restriction enzymes, the more suitable enzyme for dintinguishing the specie populations can then be choosen (e.g. [34]).

Definitelly, to accomplish a complete bioprospection work, the research groups must have well equipped laboratories. The basic microbiology needs, for example, microscopes and incubators. For the biomolecular analysis, PCR-termocyclers and sofisticated apparatus like capillary electrophoresis are necessary to perform DNA sequencing or fingerprinting. Most often, gas-chromatography (GC), liquid-chromatography and/or chromatography-mass-spectometry instruments are used in order to measure degradation rates of a pollutant or products produced by microorganisms. In order to carry out microbial prospection, besides sophisticated and expensive instruments, high level expertise on different disciplines are required. Thus, the best way is to establish collaborative links between chemists and microbiologists.

3. Manipulation of the Microbial Material

A little about carbon chemistry

Chiral substances are defined as molecules for which their mirror images cannot be superimposed (Figure 1). This happen when the central atom (called chiral or assimetric atom), usually a chiral carbon (but N, P and Si also can form chiral centers), is bounded to differents chemical species. The presence of a chiral center not necessarily gives to the molecule the chiral properties, similarly if a molecule has more than one chiral center, it is not necessarily a chiral substance. The two molecules that are specular images of a chiral substance are named as enantiomers, and a mixture of equal amounts of the two enantiomers is named a recemic mixture.

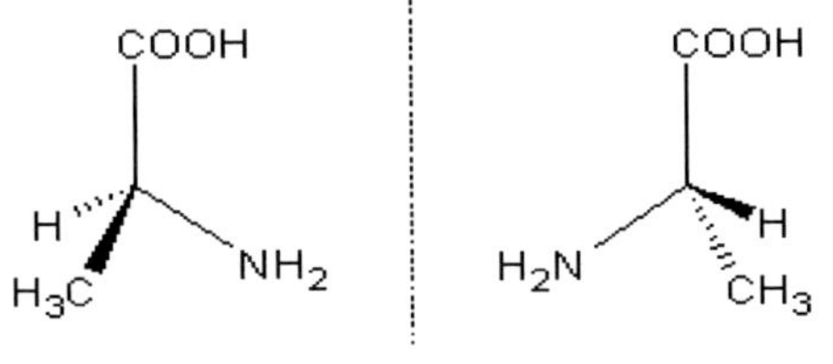

Figure 1. Chiral molecules (Alanine).

When the chiral characteristic is not based on the presence of an assimetric carbon, it occurs due to the presence of geometric degrees of freedom: parts of the molecule can change the orientation respect to a symmetry plane or an axis, yielding distinct chemical properties to them (Figure 2).

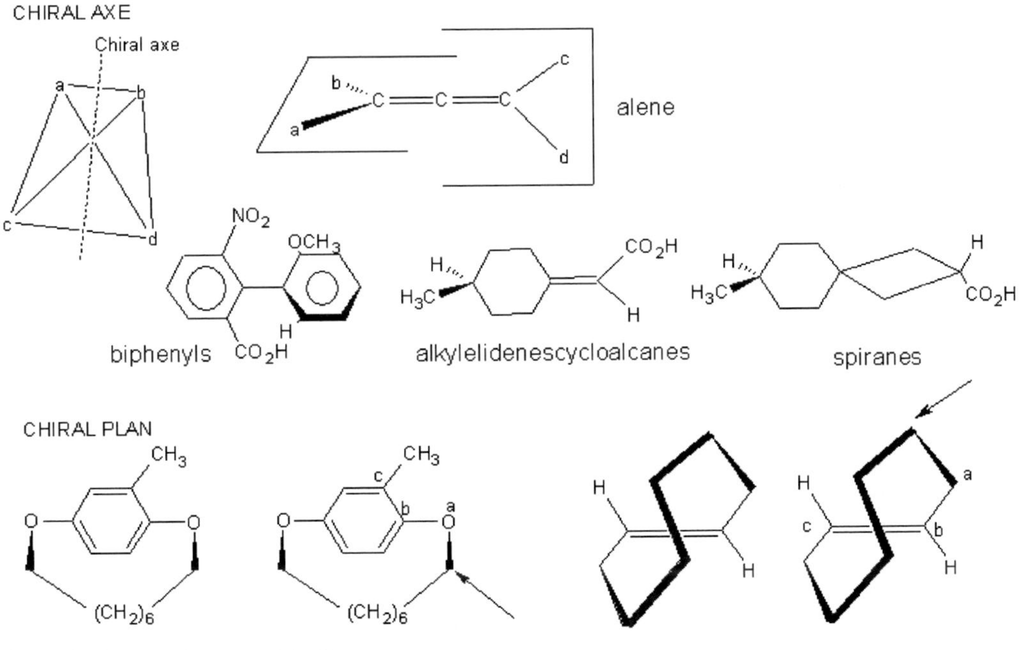

Figure 2. Plane and axis of chirality.

In the chemical synthesis, mixtures of enantiomeric substances are more frequently produced, in contrast to the live organisms that tend to produce enantiomerically pure substances. Moreover, racemic mixtures can act distinctly within the biologic activity: the thalidomide, for instance, has sedative and hypnotic activities on the dextrorotatory form, while on the levorotatory form it has a teratogenic activity. Similarly, the (R,R)-Ethambutol can produce vision disturb, while on the (S,S) form it is bacteriostatict and antimycobacterial [9]. These examples strongly support the necessity to bioprospect enzymes which can catalyze selective reaction to produce enantiomerically pure chiral substances, especially for the production of drugs.

Interesting Enzymes for Bioprospection

There is a vast number (over 3,000) of enzymes already identified [49]. The enzymes are proteins capable to catalyse biochemical reactions in a high velocity and specificity that is essential for environment adaptations. There are also enzymes with structural and regulatory functions, and also some with catalytical promiscuity, a property that has received much attention recently [22].

Concerning the enzymatic bioprospection, the capacity of a microorganism to survive on a specific media is usually explored. This media contains the product that is responsible for stimulating a catabolic route which involves one or several enzyme classes. The organic compounds most commonly used on enzymatic bioprospection are aliphatic and cyclic hydrocarbons, PAHs, alcohols, nitriles, and so on, which stimulate the production of oxidases and hydrolases. On the other hand, for the bioprospection applied to pharmaceutics and fine chemistry, it is the chiral specificity of the enzymes that flourish them. So, reactions that can produce chiral substances mediated by enzymes, like esterification, transesterification, and aminolysis or amidation are of special interest since they are important in the formation of stereo- and regiochemical products.

Hydrocarbons can be degraded by several microorganisms [5], using different enzymatic metabolims. The hydrocarbons structurally more simple, like the aliphatic hydrocarbons, are the easiest to degrade. The alkanes, for instance, can be degraded by three differents metabolic attacks: (a) monoterminal oxydation, (b) biterminal oxydation (by some bacteria and fungi), in which both terminal methyl group of n-alkanes are sequentially hydroxylated, and (c) subterminal oxydation. In all these cases hydrolases or monooxygenases are involved [21]. Similarly, the biochemical convertion of cyclic alkane and alcohols involve dehydrogenases, monoxygenases and hydrolases. The microbial degradation of aromatic hydrocarbons usually is initiated by oxygenases, followed by stereoselective reactions of hydroxylation [15]. For example, for the degradation of toluene and xylenes, the xylene monooxygenase (XMO) settles the carboxylic acids formation. Other enzymes may be involved, like the benzyl alcohol dehydrogenase and benzaldehyde dehydrogenase, which catalyze for the formation of benzoic acids. Regarding to the degradation of PAHs and Poly-Chlorinated Biphenyl compounds (PCBs), beyond the dehydrogenases, the dioxygenases also can be involved. Briefly, in the degradation reaction mechanisms of these compounds, the molecular oxygen is added on a cis-position of the aromatic ring by a dioxygenase, and subsequently the catechol dehydrogenation occurs, followed by the cleavage of the ring [2; 37].

For industrial applications, the most frequently used enzymes are those that belong to the class of hydrolases, more specifically those which cleave carboxilic ester bonds and carboxyl esterases, since they can produce enantioselective substances. Important examples of this class are the nitrilases involved on nitrile biotransformation. The nitrilases are enzymes which catalyze the direct hydrolysis of organonitrile derivatives to the corresponding carboxylic acid and ammonia. An alternative route for the nitrile catabolism is by a combination of nitrile hydratase and amidase, which follow a dienzymatic pathway for the convertion to acids: the nitrile hydratase form amides from nitrile, which can be subsequently hydrolysed by amidases [18; 28; 47; 65; 66]. The major disadvantage of using nitrilases and nitrile hydratases in industrial applications is their relative poor stability, probably due to thiol group at their active sites [48]. In order to overcome this disadvantage, nitrile hydratases have been extensively studied [18; 28; 47; 66], their genes characterized [30], several cloning and expression systems have been developed [18; 19; 29; 31] and molecular tools have been developed for the detection of new genes [65]. In the same way transglutaminases, enzymes that catalyze the formation of isopeptide bonds (ε-(γ-glutamyl) lysine bonds), show enantio-selective transamidase activities that are of special interest for the production of modified proteins [32; 46].

The alcohol dehydrogenases is another class of enzymes that have been extensivelly studied because of their conserved structure. Based on the relative activity towards alcohol catalysis they have been classified as primary-alcohol and secondary-alcohol desydrogenases. The secundary-alcohol desydrogenase are attractive to research into chiral chemical production due to their broad specificity and their high enantiospecific conversion of prochiral ketones to alcohols [35]. The lipases are also other examples of enzymes utilized in the modern world on some industrial catalytic processes, like addictive detergents and the production of optically pure enantiomers [4; 45; 83].

Culture Media

The microbiological bioprospection is a novel and scarcely explored area, and so is the determination of the ideal culture medium (composition, temperature, pH, etc) to make grow and isolate the microorganisms under study. The work carried out by Domingues [27], with the *Thermus* strains, and that by Onillo et al. [61], with a psycobacteria that produces an α-L-rhamnonosidase (one enzyme that showed high flexibility at low temperatures), are examples of this laborious effort. In this section we will develop the reasoning on the choice of the culture media for the bacteria that produce chiral substances and for bacteria showing hydrocarbonoclastic activities, some recent results are included.

The culture media may be chosen as a function of the microorganism group that is targeted. The organisms must find in this synthetic environment all the substances required for energy generation and cellular biosynthesis: C, H, O, N, S, P, K, Mg, Fe, Ca, Mn, and trace metals as Zn, Co, Cu, and Mo. Which of these nutritional elements that might be included on the culture medium depends on the microoganism class desired. Certain microorganisms may also require small amounts of determined organic compounds for growth, the growth factors: vitamins, amino acids and/or nucleid acids. Moreover, it is also necessary to maintain an acceptable range of physical conditions (O_2 concentration, temperature, pH, etc). The phototroph microorganisms need light, the anaerobes only grow on the abscence of oxygen, the thermophiles demand high temperatures, and so on. On Table 1 some compounds usually used on culture media are presented together with their function.

Culture media may be classified as chemically-defined or complex (undefined) media. A chemically-defined medium is one in which the exact chemical composition is known, as the minimal media, whereas in a complex medium the exact chemical constitution is unknown. When the researcher needs to know the exact nutritional requirements for the growth of a certain microorganism, it is recommended to use minimal media. On the other hand, complex media (like the Luria-Bertani media, dubbed LB) is appropriate when a full range of growth factors is required, such as for the growth of a large amount of a microorganism or an assembly of microorganisms. Interestingly, there are some microorganisms that do not grow on complex media.

Table 1. Usual components of media culture and their physiologycal function

Component	Function
Beef extract	Source of vitamins and other growth factors
Yeast extract	Source of vitamins and other growth factors
Peptone	Source of amino acids, N, S, and P
Casamino acids	Source of amino acids, N, S and P
Glucose; Sucrose; Trisodium citrate	C and energy source
K_2HPO_4 ; KH_2PO_4	pH buffer; P and K source
$(NH_4)_2HPO_4$	pH buffer; N and P source
$MgSO_4$ $7H_2O$	S and Mg^{++} source
$FeSO_4$ $7H_2O$; $FeCl_2$	Fe^{++} source
$MnSO_4$ $7H_2O$	Mn^{++} Source
NH_4Cl	N source
$CaCl_2$ $2H_2O$	Ca^{++} source
Elemental Sulfur	Energy source
CO_2	C source
KCl	K^+ source
NaCl	Na^+ source for halophiles and inhibitory to nonhalophiles
Agar	Inert solidifying agent

When a specific bacterial group is targeted such as a tolerant microorganism to certain compounds, or a microorganism capable to synthetize one enzyme, not only the media is important but also the carbon source that will be used. In this case, minimal media enriched by only one carbon source is more indicated because exceedingly stressing conditions are strongly selective, ideal to search highly specialized microorganisms. For example, to bioprospect hydrocarbonoclastic bacteria it is more suitable to use as carbon source only one compound, such as naphthalene or xylene. For isolating biphenyl degraders, one may use biphenyl, for degraders of nitrile, acetonitril, and so on. Sometimes a combination of an enriched complex media (e.g. sucrose) and a stressing carbon source (e.g. petroleum) is used. This strategy allows getting a large numbers of microorganisms, but the diversity will probably be poor – one may even loose the gene or characteristic that is being searched. Moreover, the concentration is also important, mainly when the carbon source is a toxic compound: an excessive concentration can inhibit so strongly the growth that nothing is obtained. The carbon source may also be chosen based on the enzyme mediated reactions that are looked for, or on the formed sub-products.

In what follows, some results on bioprospection research will be presented, divided on the two main purpose of this investigation: (i) economic (bacteria capable to produce chiral compounds) and (ii) environmental (bacteria that can be applied on oil remediation).

Bacteria Potentially Producing Chiral Substances by Biocatalysis

The appropriate carbon sources for bioprospection of bacteria that catalyze enantioselective reactions were selected taking into account the presence of functional groups in their structures which are transformed by the desired enzymes. We intend to bioprospect the following enzymes: nitrilase, nitrile hydratase, amidase and alcohol dehydrosenases. After the bioprospection assay, we can evaluate the enantioselectivity of the reactions and the broad substrate spectrum of the isolated bacteria. Below we report an example related to this approach, in which nitriles, alcohols and ketones are used as carbon sources. Two sites were sampled, one where a high biodiversity is expected (the Amazonia Rainforest, on the state of Rondonia, Brazil) and another with extreme conditions (the Antarctic, on the Commander Ferraz Brazilian Station). From both sites, soil and sediment samples were taken.

The main goal of this study was to bioprospect bacteria to act as biocatalysts in the nitrile and amide hydrolysis, ketone reduction and alcohol oxidation/deracemization. In this way, we selected different carbon sources for the bioprospection assay.

The compounds selected as carbon sources for nitrile and amide hydrolysis were benzonitrile, 2-cyanoacetamide, succinonitrile and benzamide; for alcohol oxidation/ deracemization [1-(4-chlorophenyl)ethanol and 1-(4-methylphenyl)ethanol)]; and for ketone reduction [1-(4-chlorophenyl)ethanone and 1-(4-methylphenyl)ethanone].

For the first culture, we used minimum liquid mineral media enriched with each one of these compounds, all of them at a final concentration of 50 $mg.L^{-1}$. Then, for strain isolation, we applied the same conditions but using for culture solid media, in which the compounds were spread over the agar. The incubation was done at 20°C for Antarctic samples and 30°C for Amazon samples. Under these conditions, 106 strains were obtained, which supports the efficiency of the used criteria on choosing the culture media.

It is worth noting that the strains isolated on media containing nitriles were forced to use these compounds also as nitrogen source, since no other source of this element was available. Comparing the distinct nitriles used, the succinonitrile and benzonitrile (14 and 15 strains obtained, respectively) showed better results than 2-cyanoacetamide and benzamide (3 and 5 strains, respectively).

Twenty-five strains were isolated on 1-(4-chlorophenyl)ethanol and only 7 strains on 1-(4-methylphenyl)ethanol, suggesting a bigger efficiency of the first compound as carbon source to bioprospect the samples. From the 7 strains isolated on 1-(4-methylphenyl)ethanol, only one came from the Antarctic sample.

And finally, concerning the ketones, both showed similar efficiency (18 and 19 strains, respectively from 1-(4-chlorophenyl)ethanone and 1-(4-methylphenyl)ethanone.

A sharp evaluation of the best carbon sources includes not only the larger number of isolated strains but also the final compounds they are able to synthesize.

The next step in this study is the evaluation of the enantioselectivity of the enzymatic reactions. For this purpose, we employed the new bacteria strains in bioreduction of prochiral ketones (Figure 3, scheme 1) and oxidation/deracemization of racemic alcohols (Figure 3, scheme 2). For the study of bioreduction, we selected six prochiral aromatic ketones [1-(phenyl)ethanone, 1-(4-methylphenyl)ethanone, 1-(4-methoxyphenyl)ethanone, 1-(4-chlorophenyl)ethanone and 1-(4-nitrophenyl)ethanone] and for alcohol oxidation, we selected six racemic alcohols [1-(phenyl)ethanol, 1-(4-methylphenyl)ethanol, 1-(4-methoxyphenyl)ethanol, 1-(4-chlorophenyl)ethanol and 1-(4-nitrophenyl)ethanol].

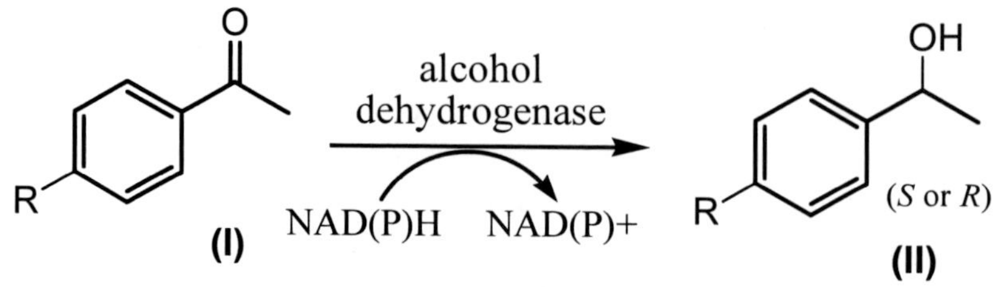

R = H; Me; MeO; Cl; NO_2
Scheme 1.

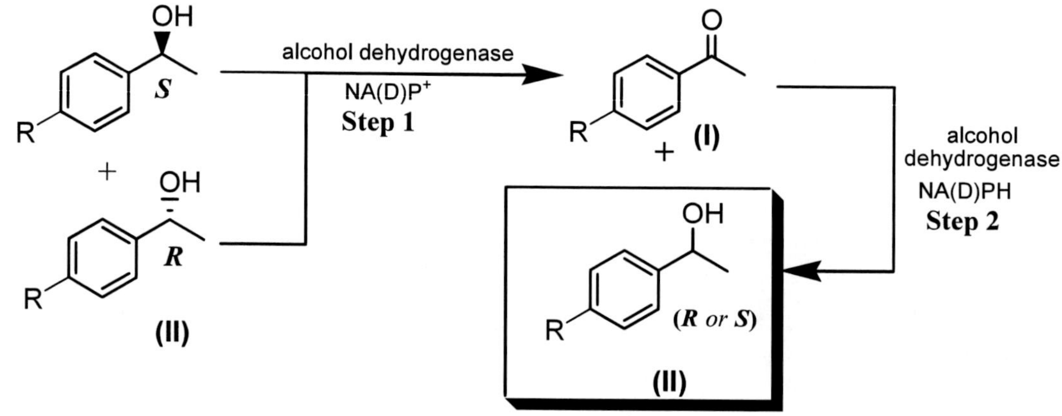

R = H; Me; MeO; Cl; NO_2
Scheme 2.

Figure 3. (1) Bioreduction of prochiral ketones, (2) Enantioselective oxidation/deracemization of racemic alcohols.

The reactions were carried out by incubation of the different bacteria with its corresponding inducer in microtiter plate. After the growing step, the substrate (ketones or alcohols) were added to each bacteria culture. The enantioselectivity of the enzymatic transformation was measured by chiral GC analysis (total time of each analysis was between 4 to 8 minutes). This methodology was employed to determine, quickly, the enantiomeric excess of the enzymatic reactions, since our main goal is the bioprospection of biocatalysts which exhibit high enantioselectivity along the enzymatic transformations.

The best results aiming the synthesis of chiral alcohols with high yield and enatioselectivity were obtained when we employed 1-(4-chlorophenyl)ethanol as inducer. The enantioselective oxidation of racemic alcohols II gave the corresponding enriched (*R*)-alcohols in good yield (up to 64%) and enantiomeric excess (up to 97%) using the bacteria, designated as RC1AC23, as enzymatic source of alcohol dehydrogenases, ADH/NAD(P)H. In addition, we employed 1-(4-methylphenyl)ethanol as inducer and a bacteria designated as RC1AF57 gave the (*S*)-alcohols in good yield (up to 89%) and enantiomeric excess (up to 99%).

As shown in the following chromatograms (Figure 4), both enantiomers of the chiral alcohols can be obtained by the use of different bacteria (RC1AC23 and RC1AF57). This enantiocomplementarity in the synthesis of both enantiomers of the desired chiral alcohols were easily achieved using this bioprospection methodology.

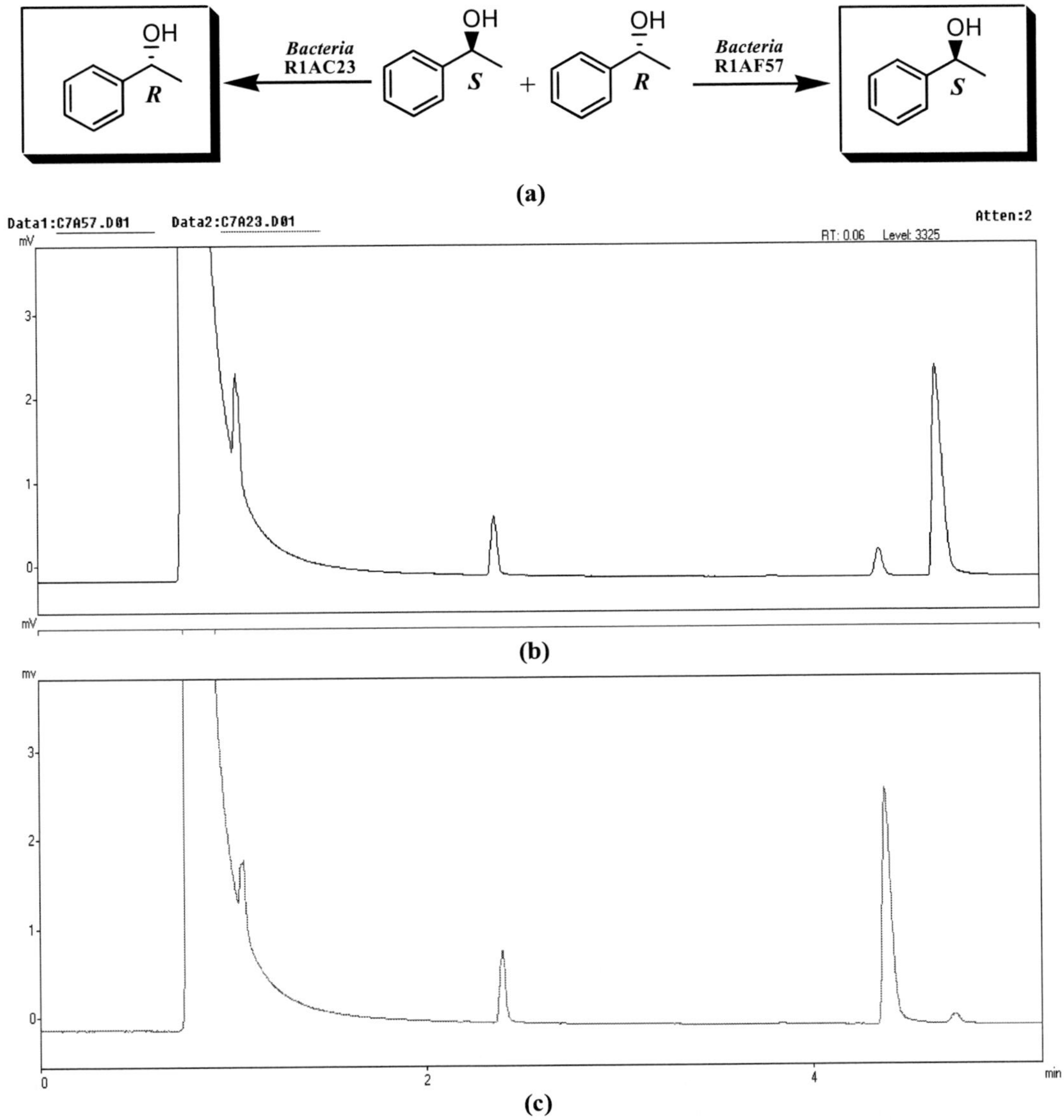

Figure 4. (a) Enantiocomplentary in the synthesis of 1-phenyl-ethanol in enantiomeric enriched form; (b) Chromatogram obtained from reaction of (*RS*)-1-phenyl-ethanol whit R1AF57 after 72 hours reaction; (c) Chromatogram obtained from reaction of (*RS*)-1-phenyl-ethanol whit R1AC23 after 72 hours reaction.

In summary, this bioprospection methodology followed by the evaluation of the enantioselectivity of the enzymatic reactions is a fast way to isolate microorganisms able to catalyze enantioselective chemical transformations.

Hydrocarbonoclastic Bacteria

When isolation targets hydrocarbonoclastic bacteria, the ideal strategy is to induce a metabolism that is intermediated by oxydases (mono- and dioxygenases) and hydrolases. In this case, it is important to take into acount not only the hydrocarbons to be used as carbon source but also the exact concentration of these compounds because they are all toxic (to different levels). For example, Furukawa et al. [42] suggest concentrations as small as 0.005% w/v for PCBs resistent microorganisms. The more frequently used PAHs are naphtalene and phenanthrene since they are the less toxic and relatively easy to degrade. The main encompassed enzymes in the related degradation pathway are the dioxygenases. Neverthless, it is recommended to probe more complex PAH models, which catabolism may involve distinct enzymatic routes. In this case, the compound concentration must be even smaller in order to avoid completely inhospitable conditions that will kill all the microorganisms. The pyrene, a PAH with 4 joined rings, is the least toxic of the more complex PAHs, and therefore well indicated for the use as a carbon source. Another group of dangerous and persistent organic pollutants, frequently investigated on bioprospection works, is the PCBs. The microbiologic degradation pathways of these compounds are similar to thoses of toluene and naphtalene, and are also mediated by dioxygenases. Fungi produce well adapted enzymes such as lignin and manganese peroxydases for dechlorination reactions [26; 33].

An example of a bioprospection study of hydrocabonoclastic bacteria is presented in what follows. The chosen sample site was a mangroove area impacted by oil spill – the Guapimirim Mangroove (Guanabara Bay), located close to Rio de Janeiro city (Brazil) [13]. Sediment samples were collected and used as inocula in a minimum liquid mineral media, supplemented with five distinct model hydrocarbons: octadecane, pristane, naphthalene, fluoranthene and pyrene. The aliphatic alkane (octadecane) may select microorganisms that produce hydroxylases, hydratases and monoxygenases. The same happen with the ramified-alkane (pristane), but using distinct metabolic routes. The PAHs (naphtalene, fluoranthene and pyrene) use dioxygenases mediated metabolisms as explained before. Three distinct concentrations were used: 250 $mg.l^{-1}$ for pristane or octadecane, 100 $mg.l^{-1}$ for naphthalene or fluoranthene and 50 $mg.l^{-1}$ for pyrene. The incubations were done at 37°C. As a result of this treatment, 64 strains were isolated: 14 on octadecane, 8 on pristane, 18 on naphthalene, 13 on fluoranthene and 11 on pyrene.

In the first proof, the strains were cultivated on the complex LB media enriched with the respective hydrocarbon. Forty percent of them showed no growth on this complex media, suggesting that they could only sprout on extremely stressing conditions similar to those on which they were originally living. These strains are very promising for remediation applications: once seeded on an impacted environment they will metabolize the pollutant till its exhaustion and then they will die as a consequence of their inhability to grow in naturally complex media. The other 60% may also be applied on remediation, especially if they present high degradation rates, but only in well controled experiments.

Using again the minimum media enriched with the hydrocarbon from which they had been isolated, we measured the degradation potential of each strain. Only 58% of them could be classified as hydrocarbonoclastic (degradation rates between 1 and 100%). Among them, three strains showed ability to degrade 100% of their respective PAH: GPM2603 (pristane), GPM2544 (pristane) and GPM2545 (naphthalene).

We also tested the capacity of some selected strains to degrade other hydrocarbons further than the one from which they had been isolated [13]. Four pyrene degraders, representative of distinct genera and with relatively high degradation rates (from 16 to 55%), were selected, since this is the most toxic and most persistent hydrocarbon among the probed ones (Table 2). The strain GPM2609, a gram-positive bacteria identified as *Cellulomonas* sp., grew and degraded phenanthrene and fluoranthene apart from pyrene. Since these are PAHs, the associated enzymes were probably dioxygenases. However, it was not capable to degrade octadecane, although this is an hydrocarbon relatively easy to degrade, possibly because its enzymatic route involves monoxygenases. The GPM2509, similar to an *Alcanivorax* sp., showed capacity to degrade completely the aliphatic hydrocarbon, but the most interesting property of this strain was its ability to also degrade some PAHs (pyrene and fluoranthene) – this was the first time that this capacity is reported for *Alcalinivorax* sp. The strains GPM2002 and GPM2541 were able to degrade phenanthrene and octadecane, beyond pyrene, but degraded nothing of fluoranthene. The fluoranthene is composed by three aromatic rings, joined by a non aromatic pentacycle, which probably difficults its biodegradation or create a distinct enzymatic route from those employed by the other probed PAHs. Based on a long study with the strain *Mycobacterium* sp. [17], a complex metabolic route was suggested for the fluoranthene degradation, which do not take to a total mineralization.

From the four bacteria analysed in this experiment, three are able to degrade both aliphatic and aromatic hydrocarbons, even though their reaction mechanisms involve distinct enzyme classes. These results suggest that there may have occured on the sites, before the sampling, a transfer of catabolic genes (by transposition or conjugation events), which could be facilitated by the natural conditions of mangroves (tides and wetlands). Since there are not many microbial bioprospection studies on mangroove ecosystems, this work also revealed the much promising that they are: a rich microbial diversity and microorganisms with high potential for degrading oil hydrocarbons were found.

Table 2. Degradation rate of different hydrocarbons (Py, pyrene, Oc octadecane, Ph, phenanthene and Fl, fluoranthene) of four strains isolated on pyrene as only carbon source

Strain	Identification[a]	Py	Oc	Ph	Fl
GPM2509	*Alcanivorax venusti* ISO4[b]	16%	100%	0%	11%
GPM2609	*Cellulomonas sp.* IFO16066[c]	55%	0%	100%	15%
GPM2002	Hydrothermal vent strain AG33[d]	21%	5%	18%	0%
GPM2541	*Marinobacter sp.* 407.13[e]	27%	10%	50%	0%

(a) Closest relative after 16S rRNA encoding gene analysis. Strains isolated from: (b) Mediterranean seawater, Coast of Santa Pola, Spain [36]; (c) mangrove rhyzosphere, Japan [76]; (d) Galapagos hydrothermal vent samples [78]; (e) cultures of *Alexandrium sp.* associated with the production of paralityc shellfish toxins [Gallander, S., unpublished].

Microcosms

The degradation experiments are usually carried out *in vitro* microcosms (such as, penicillin bottles, erlenmeyer, aquarium, bioreactor etc.). In these partially controled micro-environments only a small number of natural conditions can be reproduced since the complexity of the analysis increase exponentially with the number of parameters. The *in situ* microcosms are the extremes of this complexity and, therefore, they are rarely executed. Also, both the legal questions involved on these experiments and the impossibility to control the natural conditions tend to be restrictive. Several studies showed that microcosms are useful to follow the modification of the structure of bacterial communities after fuel or crude oil contamination [10; 11].

The application of selected or adapted microorganisms to the natural enviroment, for remediation, agricultural goals or even for cientific research, has been severely debated: since we can not completely control them, we do not know how the environment will react. In other words, we ignore if the effects of the introduction of a selected or genetically manipulated microorganism can induce an "ecological disequilibrium" even more impactant than a pollution discharge. By contrast, the optimists defend that, in the case of an antropogenic environmental accident, the introduction of these microorganisms can not impact more than the accident itself. They claim that the introduced microorganisms will survive only while they are subjected to stressing conditions (when the competition is low), and, as soon as the natural conditions are re-established, they will decline by competition. Another interesting possibility is that the catabolic genes involved on the degradation of the pollutant can be present on mobile elements. This will accelerate the recuperation of the environment because the introduced microorganisms will spread these genes to the indigenous populations by horizontal transfer. Since the native microorganisms are naturaly addapted, they will compete with those introduced even during the stressing conditions. An alternative is the landfarming: part of the contamined material is transfered to some place were the physico-chemical conditions (pH, temperature, nutrients, carbons source) and the addition of microorganisms are controled. Anyhow, in general it is not possible to withdraw all the dirt material, particularly when the accident assumes huge proportions.

Habitually, *in situ* tests are carried out on already contaminated sites. This was the case of the Guapimirim Mangroove experiment presented above. This site, a mangroove with a remaining area of 135 km^2, was subjected to an oil spill accident on January 2000. A collapsed pipeline spread approximately 1,3 million tones of oil contamining the surrounding beaches and the protected area of Guapimirim Mangrove. The prevailing situation of the region is characterized by the reception of urban and industrial discharges (nearly 17 tons per day), appart from atmospheric imputs, runoff water and petroleum from tanker trafic [44]. Sediments collected from this area were used as inocula to a petroleum enrichment culture. The hydrocarbonoclastic bacterial consortium obtained was used *in situ* microcosms to verify its potenciality to remediate this area (details will be published elsewhere).

From the same consortium, 6 hydrocarbonoclastic strains were isolated (Table 3). Although these strains showed the ability to degrade, *in vitro*, aliphatic and aromatic hydrocarbons, the consortium applyed to the *in situ* microcosms treated with petroleum did not reduce the hydrocarbons significantly (Figure 5).

There are many natural factors that might have contribute to this result: (a) the abundance of organic material probably supplied an easier carbon source; (b) the oil may have been adsorbed to the pores of the sediment matrix (typically clay) and became unavailable; (c) the action of the tides.

It is also important to emphasize that only a week after seeding the hydrobarbonoclatic consortia on the microcosms, some oil emulsification was observed. These emulsions, that are little oil micelles, normally result from the action of exo-enzymes produced by bacteria as part of the degradation process. Remarkably, after that the oil did not diminish as expected. In the *in vitro* experiments the microorganisms are on continuous stress and have only one carbon source available, and so their exclusive energy source, in contrast to the situation on *in situ* experiments. Probably the microorganisms used the exo-enzymes only to reduce the oil toxicity, a reaction that also usually happen *in vitro* microcosms. The microorganisms have spent energy only to turn their sourrounding habitable, but chose the less expensive and more abundant natural carbon sources instead of the oil hydrocarbons as energy source.

Table 3. Origin and identification of hydrocarbon-degrading bacterial isolates from Mangrove sediments (source Brito et al., [13])

Strain	Identification[a]	HC source[b]	HC deg[c] (%)
GPM2501	*Alcanivorax venusti* ISO4 [d]	Oc	23
GPM2502	*Micrococcus luteus* - Ballarat [e]	Oc	7
GPM2503	*Alcanivorax venusti* ISO4 [d]	Oc	45
GPM2509	*Alcanivorax venusti* ISO4 [d]	Py	16
GPM2511	Hydrothermal vent strain AG33 [f]	Py	11
GPM2524	*Alcanivorax venusti* ISO1 [g]	Oc	20

(a) Closest relative after 16S rRNA encoding gene analysis. Strains isolated from: (d) Mediterranean seawater, Coast of Santa Pola, Spain [36]; (e) Activated-sludge plant in Ballarat, in Victoria, Australia [80]; (f) Galapagos hydrothermal vent samples [78]; (g) Mediterranean seawater, Coast of Alicante, Spain [36];

(b) Hydrocarbon source (HC): Fluoranthene "Fl", Octadecane "Oc" and Pyrene "Py";

(c) Hydrocarbon degradation was measured after 21days of incubation.

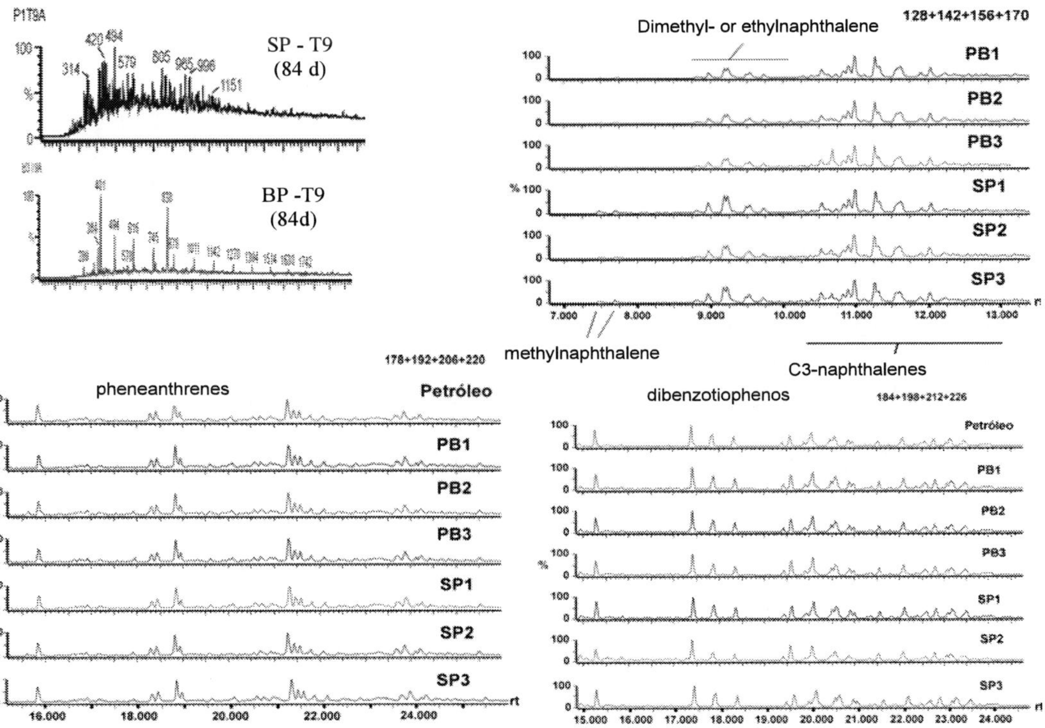

Figure 5. *In situ* degradation tests 84 days after the treatment with petroleum (SP microcosms) or with petroleun and hydrocarbonoclastic bacteria consortia (BP) isolated from a Guapimirim mangrove showing the persistency of the petroleum on the microcosms even the fractions less persistent (naphthalenes, phenanthrenes and dibenzotiophenes).

This and other similar works on mangroves contaminated by oil spills [16; 43; 51; 68; 77] unveil the fragility of such ecosystems and the dificulty to remediate them. It also reinforces the importance of realizing *in situ* microcosmos experiments to really test the applicability of the chosen bioremediation technology.

4. CONCLUSION

In this section we discussed the avant-garde methodology for the bioprospection of microorganisms, particularly the cultivation of bacteria and the testing of their potentiality to produce enantiomeric substances and degrade oil hydrocarbons. The results of two case studies were presented, comprehending strains selected from three distinct ecosystems: one extremely cold (Antartic), one impacted by anthropogenic pollution (artificially extreme) and one with high biodiversity (Amazon rainforest). All of them revealed to be much favorable for bioprospection of microorganisms. This helps us to answer the first question presented on the introduction: extreme and biodiverse environments are the most promising for bioprospection approaches. Organisms that survive on extreme conditions described on the introduction may present biochemical and physiological adaptations, allowed by atypical enzymes, with large potential for biospropection. Other environments with great genetic potential are the unexplored or little explored forests, since these ecosystems present the greatest biodiversities of the planet. On the other hand, the strategy may be to choose the environment based on the product that we are looking for. For example, if microorganisms with potential to degrade polymers are targeted, samples from places that naturally have polymers, such as rubber-trees forests would be chosen for bacterial isolation. Environments contaminated with oil or petroleum are potential places for finding hydrocarbon resistent microorganisms and/or with the capacity to degrade hydrocarbon compounds. If thermoresistant enzymes are expected, hyperthermal environments may be explored, and so on.

Concerning the more suitable culture medium to access the potential microorganisms of these environments, much has been said on the third section. Since on bioprospection one is open to find new and unexpected organisms, using a specific media will limit his search. Thus, it is recommended to probe only a minimum mineral medium enriched with a carbon source suitable for the prospection goals. In order to find this suitable carbon source, a previous sample is taken only for testing different carbon sources. For the uncultivable microorganisms, the strategy is to study directly the total DNA from the sample, using metagenomic tools. An example of these tools is the development of a PCR based method for nitrile hydratase genes screening [65].

The last question on the selection was "which is the most appropriate microorganisms to work with?" The answer to this question must consider both the growth time and the compound production of each strain. If only the interest on a high productivity of the compound is considered, the option to isolate bacterial strains that have the best effective overcome combining the two rates must be chosen. It is possible that an isolate with low production of one catabolic but that grows fastly is more beneficial than one that has a very high production but needs much time for growth. In the case study on the bioprospection of hydrocarbonoclastic bacteria presented here, we showed that the selection we did of 4 strains that grown and degraded pyrene was succesful because all of them revealed the ability to consume other hydrocarbons (both aliphatic and aromatic compounds). This is a strong evidence of the presence of catabolic genes on plasmidic transfers (spread before the sampling), since the degradation of these compounds use distinct enzymatic routes. The exploration of this mechanism has great potential in the recuperation of impacted environments.

REFERENCES

[1] Amann, RI; Ludwing, W; Schleifer, JH. Identification and In Situ detection of individual microbial cells without cultivation. *Microbiol. Rev.*, 1995, 59(1):143-169.

[2] Arensdorf, JJ; Focht, DD. A meta cleavage pathway for 4-chlorobenzoate, an intermediate in the metabolism of 4-chlorobiphenyl by *Pseudomonas cepacia* P166. *Appl. Environ. Microbiol.*,1995, 61(2) 443-447.

[3] Arraya, R; Yamaguchi, N; Tani, K; Nasu, M. Change in the bacterial community of natural river biofilm during biodegradation of aniline derivad compounds determined by denaturing gradient gel electrophoresis. *J. Health Sci.*, 2003, 49(5): 379-385.

[4] Aryee, ANA; Simpson, BK; Villalonga, R. Lipase fraction from the viscera of grey Mollet (Mugil cephalus): Isolation, partial purification and some biochemical characteristics. *Enz.Microbial Technol.*, 2007, 40: 394-402.

[5] Atlas, RM. Microbial degradation of petroleum hydrocarbons and environmental perspectives. *Microbiol. Rev.*, 1981, 45(1):180-209.

[6] Azevedo, *CMA. Biopropeccao: Coleta de material biologico com finalidade de exploracao de recursos genéticos.* Sao Paulo, ed. Conselho Nacional da Reserva Biologica da Mata Atlantica; Cadernos FAPESP n° 17, 2003.

[7] Bardgett, RD; McAllister, E. The measurement of soil fungal: bacterial biomass ratios as an indicator of ecosystem self-regulation in temperate meadow grasslands. *Biol Fertil Soils,* 1999, 29:282–290

[8] Bermudez, JAZ; Barragat, P. Chiral Drugs: from the Chemical dimension to political discussion. *Cadernos Saúde Pública,* 1996, 12(1):47-51.

[9] Blackwood, CB; Marsh, T; Kim, SH; Paul, EA. Terminal restriction fragment length polymorphism data analysis for quantitative comparison of microbial communities. *Appl. Environ. Microbiol.*, 2003, 69(2):926-932.

[10] Bordenave, S; Jézéquel, R; Fourçans, A; Budzinski, H; Merlin, FX; Fourel, T; Goñi, MS; Guyoneaud, R; Grimaud, R; Caumette, P; Duran, R. Degradation of Erika oil. *Aquatic Living Resources*, 2004a, 17:261-268.

[11] Bordenave, S; Fourçans, A; Blanchard, S; GOÑI, MS; Caumette, P; Duran, R. Structure and functional analyses of bacterial communities changes in microbial mats following petroleum exposure. *Ophelia*, 2004b, 58:195-203.

[12] Borneman, J; Triplett, EW. Molecular microbial diversity in soils from Eastern Amazonia: evidence for unusual microorganisms and microbial population shifts associated with deforestation. Appl. Environ. Microbiol., 1997, 63(7): 2647-2653.

[13] Brito, EMS; Guyoneaud, R; Goni-Urriza, M; Ranchou-Peyrouse, A; Verbaere, A; Crapez, MAC; Wasserman, JCA; Duran, R. Characterization of Hydrocarbonoclastic

Bacterial Communities from Mangrove Sediments of Guanabara Bay, Brazil. *Research in Microbiology*, 2006, 157: 752-762.

[14] Brodie, E; Edwards, S; Clipson N. Soil Fungal community structure in a temperature upland grassland soil. *FEMS Microbiol. Ecol.*, 2003, 45: 105-114.

[15] Buhler, B; Witholt, B; Hauer, B; Schimidt, A. Characterization and application of xylene monoxygenases for multistep biocatalysis. *Appl. Environ. Microbiol.,* 2002, 68(2):560-568.

[16] Burns, KA; Codi, S; Duke, NC. Gladstone, Australia Field Studies: weathering and degradation of hydrocarbons in oil mangrove and salt marsh sediments with and without the application of experimental bioremediation protocol. *Mar. Pol. Bull.*, 2000, 41 (7-12): 392-402.

[17] Cerniglia, CE. Biodegradation of Polycyclic Aromatic Hydrocarbons. *Biodegradation,* 1992, 3: 351-368.

[18] Chan Kwo Chion CKN; Duran, R; Arnaud, A; Galzy, P. N-terminal amino acid sequence of *Brevibacterium sp.* R312 wide spectrum amidase. *Appl. Microbiol. Biotechnol.*, 1991, 36:205-207.

[19] Chan Kwo Chion CKN; Duran, R; Arnaud, A; Galzy, P. Electroporation of whole cells of *Brevibacterium sp.* R312 nitrile hydratase producing strain. *FEMS Microbiol. Lett.*, 1991, 81:177-184.

[20] Chan Kwo Chion CKN; Duran, R; Arnaud, A; Galzy, P. Cloning vectors and antibiotic resistance markers for *Brevibacterium sp.* R312. *Gene*, 1991c, 105 :119-124.

[21] Cheng, Q; Thomas, SM; Rouviere, P. Biological conversion of cyclic alkanes and cyclic alcohols into dicarboxylic acids: biochemical and molecular bases. *Appl. Microbiol. Biotechnol.*, 2002, 58:704-711.

[22] Copley, SD. Enzymes with extra-talents: moonlighting functions and catalytic promiscuity. *Curr. Op. Chemical Biology*, 2003, 7:265-272.

[23] Dabert, P; Siave, B; Delgenes, J-P; Moletta, R; Godon, J-J. Characterization of the microbial 16s rDNA diversity of an aerobic phosphorus-removal ecosystem and monitoring of its transition to nitrate respiration. *Appl. Microbiol. Biotechnol.*, 2001, 55: 500-509.

[24] Dahllof, I. Molecular community analysis of microbial diversity. *Curr. Op. Biotechnol.*, 2002, 13: 213-217.

[25] Darling, JA; Blum, MJ. DNA-based methods for monitoring evasive species : a review and prospectus. Biol. Invarions, 2007, 1387-1407.

[26] Deschler, C; Duran, R; Junqua, M; Landou, C; Salvado, JC; Goulas, P. Involvement of 3,4-dichlorophenol hydroxylase activity in degradation of 3,4-dichlorophenol by the white-rot fungus *Phanerochaete chrysosporium. J. Molecular Catalysis, B : enzymatic.*, 1998, 5:423-428.

[27] Domínguez A; Fuciños,P; Rúa, ML; Pastrana, L; Longo, MA; Sanromán, MA; Stimulation of anovel thermostable extracellular lipolytic enzyme in cultures of *Thermus* sp. *Enzyme Microbial Technol.*, 2007, 40 187-194.

[28] Duran, R; Chan Kwo Chion CKN; Biljet, F., Arnaud, A; Galzy, P. The N-terminal amino acid sequences of *Brevibacterium* sp. R312 nitrile hydratase. *J. Basic Microbiol.*, 1992, 32:13-19.

[29] Duran, R; Chan Kwo Chion CKN; Arnaud, A; Galzy, P. Isolation of a strong promoter from *Brevibacterium sp.* R312. *Zentral. Microbiol.*, 1992, 147:499-502.

[30] Duran, R; Nishgiyama, M; Horinouchi, S; Beppu, T. Characterization of nitrile hydratase genes cloned by DNA screening from *Rhodococcus erythropolis*. *Biosci. Biotech. Biochem.*, 1993, 57:1323-1328.

[31] Duran, R. New shuttle vectors for *Rhodococcus sp.* R312 (formerly *Brevibacterium sp.* R312) a nitrile hydratase producing strain. *J. Basic Microbiol.*, 1998, 38:101-106.

[32] Duran, R; Junqua, M; Gancet, C; SCHMITER J. M., Goulas, P. Purification, characterization, and gene cloning of transglutaminase from *Streptoverticillium cinnamoneum* CBS 683.687. *Biochimie,* 1998, 80:313-319.

[33] Duran, R; Deschler, C; Precigou, S; Goulas, P. Degradation of chlorophenols by *Phanerochaete chrysosporium*: effect of 3,4-dichlorophenol on extracellular peroxidases activities. *Appl. Microbiol. Biotechnol.*, 2002, 59:284-288.

[34] Edwards IP; Turco, RF. Inter- and intraespecific resolution of nrDNA TRFLP assessed by computer – simulated restriction analysis of a diverse collection of ectomycorrhyzal fungi. Mycol. Res., 2005, 109(2):212-226.

[35] Faber, K. Biotransformations in Organic Chemistry 2000, 4th edition, Springer, Berlin.

[36] Fernández-Martínez, J; Pujalte, MJ; García-Martínez, J; Mata, M; Garay, E; Rodríguez-Valera, F. Description of *Alcanivorax* sp. nov. and Reclassification of *Fundibacter jadensis* DSM 12178T (Bruns and Berthe-Corti, 1999) as *Alcanivorax jadensis* comb. nov., Members of the Emended Genus *Alcanivorax*. Int J. Syst. Evolutionary Microbiol. 2003, 53: 331-338.

[37] Finette, BA; Subramanian, V; Gibson, TD. Isolation and characterization of *Pseudomonas putida* PpF1 mutants defective in the toluene dioxygenase enzyme system. *J Bacteriol.,* 1984, 160(3):1003-1009.

[38] Fisher, MM; Triplett, EW. Automated approach for ribosomal intergenic spacer analysis of microbial diversity and its application to freshwater bacterial communities. *Appl. Environ. Microbiol.*, 1999, 65(10): 4630-4636.

[39] Fox, GE; Magrum, LJ; Balch, WE; Wolfe, RS; Woese,CR. Classification of methanogenic bacteria by 16S ribosomal RNA characterization. *Proc. Natl. Acad. Sci. USA*, 1977, 4537-4541.

[40] Fox, GE; Peckman, KJ; Woese, CR. Comparative cataloging of 16S Ribosomal Nucleic Acid: molecular approach to prokaryotic systematics. *Int. J. Syst. Bacteriol.*, 1977, 27:44-57.

[41] Fox, GE; Stackebrandt, E; Hespell, RB; Gilson, J; Maniloff, J; Dyer, TA; Wolfe, RS; Balch, WE; Tanner, R; Magrum, LJ; Zablen, LB; Blakemore, R; Gupta, R; Bonen, L; Lewis, BJ; Stahl, DA; Luehrsen, KR; Chen, KN; Lane, DJ; Pace, B; Olsen GJ; Stahl, DA; Sogin, ML; Pace NR. Rapid determination of 16S ribosomal RNA sequences for phylogenetic analysis. *Proc. Natl. Acad. Sci. USA*, 1985, 82:6955-6959.

[42] Furukawa, K; Tonomura, K; Tanibayeshi, A. Effect of chlorine substitution on the biodegrability of Polychlorinated Biphenyls. *Env. Appl. Microbiol.,* 1978, 35(2): 223-227.

[43] Garrity, DS; Levings, SD; Burns, KA. The Galeta Oil Spill I: Long-term effect on the physical structure of the mangrove fringe. *Estuarine Coastal Shelf Sci.*, 1994, 38: 327-248.

[44] Godoy, JM; Moreira, I; Mendes, LB. The Study of ^{210}Pb Method Applicability to Guanabara Sedimentation Rates. Encontro Nacional de Aplicações Nucleares, 1. Recife, 1991. Anais. Recife, UFPE, 2 (1991) 285-304.

[45] Hamachander, C; Puvanakrishnan, R. Lipase from *Ralstonia picketii* as an addictive in laundry detergent formulations. *Proc. Biochem.*, 2000, 35: 809-814.

[46] Junqua, M; Duran, R; Gancet, C; Goulas, P. Optimization of microbial transglutaminase production. *Appl. Microbiol. Biotechnol.*, 1997, 48:730-734.

[47] Kaareh, MR; Legras, JL; Duran, R; Chan Kwo Chion CKN; Arnaud, A; Galzy, P. Purification and properties of the nitrile hydratase of a new strain of *Rhodococcus sp. Zentral. Microbiol.*, 1999, 146:89-98.

[48] Kaul, P; Banerjee, A; Mayilraj,J; Banerjee, UC; Screening for enantioselective nitrilases: Kinetic resolution of racemic mandelonitrile to (R)-(-)-mandelic acid by new bacterial isolates. *Tetraedron: Asymetry,* 2004, 15: 207-211.

[49] Koeler, MK;Wong, CH. Enzymes for chemical synthesis. *Nature.* 2001, 409:232-240.

[50] Koopman, J. Reconciliation of proprietary interests in genetics and knowledge resources: Hurry cauteriously! *Ecol. Economics,* 2005, 53:523-541.

[51] Levings, SD; Garrity, DS; Burns, KA. The Galeta Oil Spill III: Chronic realing, long-term toxicity of hydrocarbon residue and effects on epibiota on the mangrove fringe. *Estuarine Coastal Shelf Sci.*, 1994, 38: 365-395.

[52] Lee, D-H; Zo, Y-G; Kim, S-J. Nonradioactive methods to study genetic profiles of natural bacteria communities by PCR-Single-Strand-Conformation Polymorphism. *Appl. Environ. Microbiol.*, 1996, 62(9): 3112-3120.

[53] Liu, Q; Sommer, SS. Parameters affecting the sensitivities of didioxy fingerprinting and SSCP. *PCR Methods Applic.*, 1994, 4: 97-108.

[54] Liu, WT; Marsh, TL; Cheng, H; Forney, LJ. Characterization of microbial diversity by determining Terminal Restriction Fragment Length Polymorphism of genes encoding 16S rRNA. *Appl. Env. Microbiol.*, 1997, 63(11):4516-4522.

[55] Mandelsman, J. Metagenomics : application af genomics to uncultured microorganisms. *Microbiol. Mol. Biol. Rev.*, 2004, 68(4) :669-685.

[56] Morin, FJ. Les accords de bioprospection favorisent-ils la conservation des resources génétiques? *Revue de droit de l'Université de Sherbrooke*, 2003, 34(1): 307-343.

[57] Muyzer, G; Wall, EC; Uitterlinden, AG. Profiling of a complex microbial population by denaturing gradient gel electrophoresis analysis of polymerase chain reaction-amplifield genes coding for 16S rRNA. *Appl. Env. Microbiol.*, 1993, 59(3): 695-700.

[58] Muyzer, G. DGGE/TGGE a method for identifying genes from natural ecosystems. *Curr. Op. Microbiol.*, 1999, 2:317-322.

[59] Olsen, GJ; Pace, NR; Nuell, M; Keine, BP; Gupta, R; Woese, CR. Sequence of the 16S gene from the thermoacidophilic archaebacterium *Sulfolobus solfataricus* and its evolution implication. *J.Mol. Evol.*, 1985, 22: 301-307.

[60] Olsen, CR; Lane, DJ; Giovannoni, SJ; Pace, NR. Microbial ecology and evolution: a ribosomal RNA approach. *Annu. Rev. Microbiol.*, 1986, 40: 377-355.

[61] Orillo, AG; Ledesma, P; Delgado, OD; Spagna, G; Breccia JD. Cold-active □-Lrhaminosidade from psychrotoletant bacteria isolated from a sub-antartic ecosystem. *Enzyme Microbiol.Technol.*, 2007, 40:236-241.

[62] Orita, M; Ywahana, H; kanazawa, H; Hayashi, K; Seriya, T. Detection of polymorphism of human DNA by Gel Electrophoresis as a Single-Strand Conformation Polymorphysm. *Proc. Natl. Acad. Sci. USA*, 1989, 86(8): 2766-2670.

[63] Overmann, J; Coolen, MJL; Tuschak, C. Specific detection of different phylogenetic groups of chemocline bacteria based on PCR and denaturing gradient gel electrophoresis of 16S rRNA gene fragments. *Arch. Microbiol.*, 1999, 172:83-94.

[64] Pace, NR; Stahl, DA; Lane, DJ; Olsen, GJ. The analysis of natural microbial population by ribosomal RNA sequence. *Adv. Microbiol. Ecol.*,1986, 9:1-55.

[65] Precigou, S; Goulas, P; Duran, R. Rapid and specific identification of nitrile hydratase (NHase) encoding genes in soil samples by polymerase chain reaction. *FEMS Microbiol. Letters*, 2001, 204:155-161.

[66] Precigou, S; Wieser, M; Pommares, P; Goulas,P; Duran, R. *Rhodococcus pyridinovorans* mw3, a nitrile hydratase producing bacterium. *Biotechnology Letters*, 2004, 26:1379-1384.

[67] Ranjard, L; Poly, F; Lata, J-C; Mougel, C; Thioulouse, J; Nazareth, S. Characterization of bacterial and fungal soil communities by automated ribosomal intergenic spacer analysis fingerprint: biological and methodological variability. *Appl. Environ. Microbiol.*, 2001, 67(10): 4479-4487.

[68] Ransay, MA. Effect of bioremediation on the microbial community in oil mangrove sediments. *Mar. Poll. Bull.*, 2000, 41(7-12): 413-419.

[69] Rehn, BHA. Bioinformatic tolls for DNA/protein sequences analysis functional assignment of genes and protein classification. *Appl. Microbiol. Biotechnol.*, 2001, 57:579-592.

[70] Ryan MJ; Smith, D. Fungal genetic resource centers and the genomic challenge. *Mycol. Res.,* 2004, 108(12):1351-1362.

[71] Sanz, JL; Kochling, T. Molecular biology techniques used in wastewater treatment: an overview. *Proc. Biochem.*, 2007, 42: 119-133.

[72] Schwieger, F; Tebbe, C. A new approach to utilize PCR-Single-Strand-Conformation Polymorphism for 16S rRNA gene-based microbial community analysis. *Appl. Environ. Microbiol.*, 1998, 64(12): 4870-4876.

[73] Singler, WV; Turco, RF. The impact of chlorothalonil application on soil bacteria and fungal populations as assessed by denaturing gradient gel electrophoresis. *Appl. Soil Ecol.*, 2002, 21:107-118.

[74] Spinard, L; Mazars, R; Theillet, C. Protocols for an improved detection of point mutation by SSCP. *Nucl. Acid. Res.*, 1991, 19(4): 4009.

[75] Stach, JEM; Bathe, S; Clapp, JP; Burns, RG. PCR-SSCP comparison of 16S rDNA sequence diversity in soil DNA obtained using different isolation and purification methods. *FEMS Microbiol. Ecol.*, 2001, 36: 139-151.

[76] Takeuchi, M; Hatano, K. Phylogenetic analysis of actinobacteria in the mangrove rhizosphere. *IFO Res. Commun.*, 1999, 19: 47-62.

[77] Tam, NFY; Wong, TWY; Wong, YS. A case study on fuel oil contamination in a mangrove swamp in Hong Kong. *Mar. Poll. Bull.*, 2005, 51: 1092-1100.

[78] Teske, A; Brinkhoff, T; Muyzer, G; Moser, DP; Jannasch, HW. Diversity of Thiosulphate-Oxidizing Bacteria from Marine Sediments and Hydrothermal Vents. *Appl. Environ. Microbiol.,* 2000, 66: 3125-3133.

[79] Torsvik, V; Goksoyr,R; Daal, FL. High diversity in DNA of soil bacteria. *Appl. Environ. Microbiol.*, 1990, 56(3): 782-787.

[80] Wieser, M; Denner, EB; Kampfer, P; Schumann, P; Tindall, B; Steiner, U; Vybiral, D; Lubitz, W; Maszenan, AM; Patel, BJ; Seviour, RJ; Radax, C; Busse, HJ; Emended

descriptions of the genus *Micrococcus, Micrococcus luteus* (Cohn 1872) and *Micrococcus lyslae* (Kloos et al. 1974). *Int. J. Syst. Evol. Microbiol.*, 2002, 52: 629-637.

[81] Woese, CR. The phylogeny of prokaryotes. *Science*, 1980, 209:457-463.

[82] Woese, CR. Bacterial evolution. *Microbiol. Rev.*,1987, 51(2):221-271.

[83] Xin, JY; Li, SB; Xu, Y; Chui,JR; Xia, CG. Dynamic Enzymatic resolution of Naproxin methyl ester in a membrane bioreactor. *J. Chem. Technol.,* 2001, 76: 579-585..

In: Leading-Edge Environmental Biodegradation
Editor: L. E. Pawley, pp. 223-247
ISBN: 978-1-60021-903-0

Chapter 7

LIGNOCELLULOLYTIC MICROORGANISMS FROM A COMPOSTING HABITAT: CURRENT AND FUTURE PROSPECTS

M. C. Vargas-García, M. J. López, F. Suárez-Estrella and J. Moreno
Departamento de Biología Aplicada, Universidad de Almería, La Cañada de San Urbano, Almería, Spain

ABSTRACT

Composting is one of the most successful tools for recycling organic wastes, but it is also a process with a broad diversity from a microbial point of view. The nutritional nature of the materials liable to suffer this kind of biotransformation, as well as the environmental conditions that prevail throughout the process, favour the presence of microbial populations adapted to the specific circumstances of every stage. From thermophilic to xenobiotic-degrading microorganisms, there are very different microorganisms in compost, many of which exhibit interesting capabilities, not only for the composting process, but for other environmental applications. Among these communities, one of the most attention-grabbing is that of lignocellulolytic microorganisms, which play a starring role in the global carbon cycle since lignocellulose is the major biopolymer in nature. Nowadays, lignocellulolytic microorganisms are becoming more and more important on account of their implication in industrial processes. Thus, in the paper industry there are an increasing number of factories that have substituted the chemical processes by biological treatments, which are friendlier from an environmental perspective. On the other side, the exhaustion of fossil fuels has turned the eyes to alternative sources, biomass bioconversion to ethanol being one of the most promising. And it is expected in the future these activities to be engaged to a greater extent since both citizens and governments demand for a bigger environmental commitment by industries. Now it is time for scientists to answer this demand and make all these processes both environmental and economically viable.

1. INTRODUCTION

Lignocellulose is the most important biopolymer in nature, not only on account of its abundance but for its characteristics. Nevertheless, this biomass has been considered for a long time just as a waste or used for cattle feeding at best. This consideration has changed in the last years and nowadays different industries have focused their attention on lignocellulose as alternative feedstock. Thus, traditional applications like paper manufacture, fuels from renewable sources or agricultural uses (animal feed and organic amendment) have increased the demand for lignocellulosic substrates [1], as well as some new utilization that appears, mainly in the chemical industry for the production of commodity and specialty products [2]. The broad distribution along the whole world and the availability at very low prices [3] are ample incentives to explain the interest for lignocellulosic biomass. Moreover, it has to be considered the environmentally beneficial since many of these raw materials can act as pollutants if they are not processed [4].

Most of the processes which make possible the transformation of the lignocellulosic raw materials into value-added products have a common characteristic: microorganisms. The metabolic capabilities of different microbial strains promote the hydrolysis of the macromolecules that constitute the lignocellulosic material and the resulting hydrolysates are converted into products by these or other microorganisms. Thus, it is clear that microorganisms play a key role in lignocellulose technology, both in relation to the direct utilization of the biopolymer and the utilization of the lignocellulolytic enzymes for industrial processes. These enzymes are strongly demanded for different industrial applications, such as the decolourization of dyes [5], the release of toners from office wastes [6], the clarification of fruit juices [7] or to improve of alcoholic drinks such as beer and wine [8]. The benefit of using this kind of biological treatment instead of chemical or physical processes lies on the reduction in energy costs [1, 6] and the lesser environmental damage that these enzymatic treatments cause [4].

The biotransformation of the lignocellulosic fraction is a complex process that needs the participation of a great number of enzymes, mainly with hydrolytic and oxidative activities. Some of them show a high specificity that limit their intervention in processes different from their own, but others have an ample range of performance which qualify them to promote modifications in molecules with structural similarities to the target molecule. Thus, this versatility makes lignocellulolytic microorganisms of interest for applications related to lignocellulose as well as for their catabolic potential on other recalcitrant compounds [9]. These biochemical abilities have been successfully used for the bioremediation of both xenobiotics and naturally occurring molecules such as aromatic hydrocarbons, pesticides or explosives [10], and together with the applications in industries previously commented as chemicals, food, pulp and paper, textile and laundry or agriculture, justify the attention that in recent years these enzymes have received. Actually, several reviews on lignocellulolytic microorganisms and their enzymes, have been published [11-14], with special emphasis on mechanisms and applications. Nevertheless, most of these reviews contain information dealing with only one lignocellulolytic enzyme (xylanase, laccase, cellulase…). This chapter is focused on the whole potential of lignocellulolytic microorganisms and the role of composting as one of the most suitable environment for these microorganisms. Composting provides an excellent habitat for the isolation of lignocellulolytic microorganisms due to both the nature of the composted materials and the environmental conditions. On one hand, most of the organic substrates used in composting are from vegetal origin [15]; what causes that lignocellulose is the main substrate. On the other hand, the high temperatures prevailing during the thermophilic phase of the process lead to the degradation of lignocellulose [16].

2. Composting

Composting is considered one of the most efficient treatments used in managing organic wastes, since the recovery of the resources contained in wastes is more complete than in other processes like landfilling or incineration [17]. It is an aerobic process by which biodegradable organic matter is decomposed under controlled conditions by a complex and changing microbial population. Throughout the composting process there are many fluctuations in both chemical and physical parameters, and therefore, also in microbial communities. Successful composting is then determined by the presence of the suitable microbial population at every stage and the existence of the right physico-chemical conditions to support this population, mainly temperature, oxygen, moisture, pH and C:N ratio.

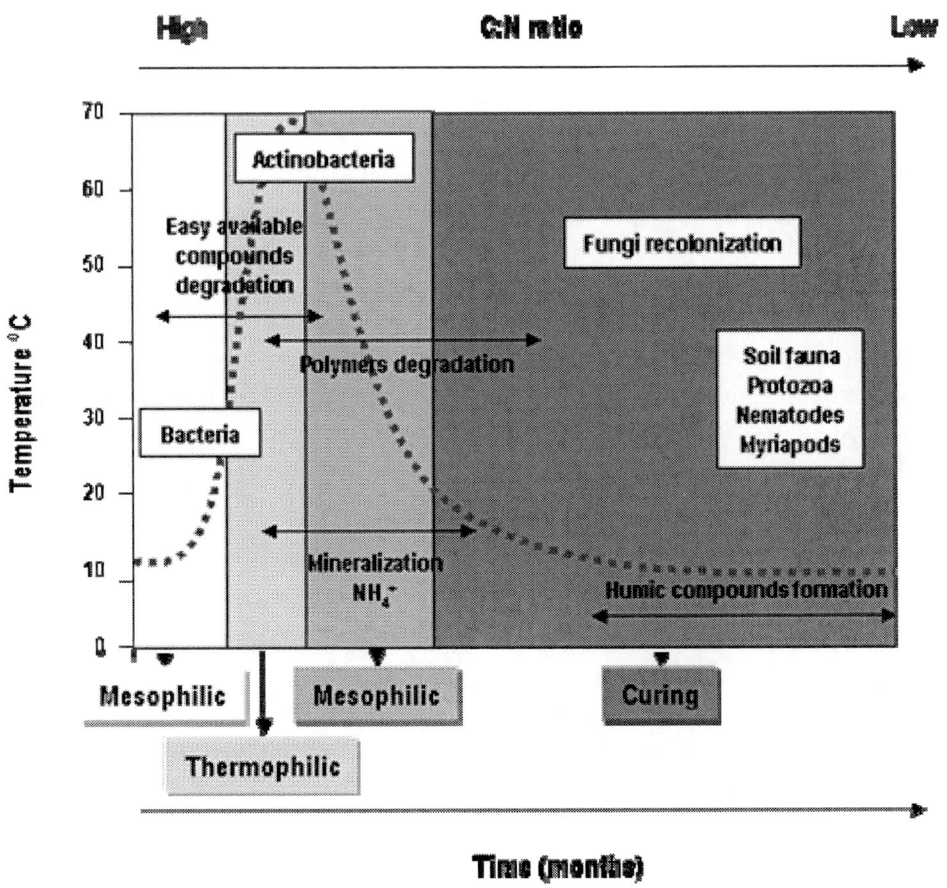

Figure 1. Microorganisms and events throughout the composting process.

2.1. Temperature

The microbial activity on the nutritional substrates causes an increase of temperature. The extent in which this increase takes place depends on the degradability and energy content of the substrates, the dispersion of the generated energy and other environmental conditions such as aeration and moisture [15]. Nevertheless, under controlled conditions the composting process can be divided into three phases according to thermal values: a first mesophilic phase, a termophilic phase and a second mesophilic phase or cooling.

The initial mesophilic stage goes on for 1-2 days during which microbial metabolism on easily decomposable compounds causes the accumulation of heat and the rise of the temperature from ambient to 40-45°C, when the thermophilic phase begins. This stage lasts as long as there are substrates that support microbial activity and temperature can reach up to 70-80°C. These high temperatures are major responsible for the sanitation capacity of composting as well as they promote the biotransformation of lignocellulosic polymers, although for the later activity a better degradation pattern can be obtained if temperature do not go beyond 60°C [18] because of the higher microbial diversity [15, 19]. When nutrients start to be scarce, the decomposition rates slow and temperature decreases to ambient. It gives way to the cooling phase and finally, after this bio-oxidative stage, the curing or maturation phase characterized by the formation of stable compounds, mainly humic substances [20].

2.2. Aeration

Aeration provides oxygen concentration enough for the efficient activity of aerobic microorganisms. If oxygen becomes a limiting factor, there is a reduction in temperature values and the rate of decomposition which means longer composting processes and lower sanitation capacity. [21, 22]. On the contrary, excessive aeration also decreases temperature and also may cause an undesirable loss of water [23].

The appropriate air flow must be selected according to the type of waste, the stage of the process and the environmental conditions [17] and it can be supplied by mechanical turning (windrow processes), by forced aeration (static piles) or by a combination of this two methods (aerated/turned systems).

2.3. Moisture

Adequate water availability is necessary for composting on account of the microbial demand for a medium in which nutrients that support metabolic activity can be dissolved. Moisture contents far from the optimal conditions stop the biological process, either by dehydration (deficient water availability) or by anaerobic conditions (excessive water availability) [24].

Although there is not a common optimal moisture content for composting, a range between 40 and 70% has been referenced as more suitable for most processes [25, 26]. Nevertheless, the characteristics of the raw materials and the system used to supply aeration determine the optimal moisture content for composting organic substrates. In this sense, the use of forced aeration systems makes necessary a strict control of moisture in order to avoid desiccation.

2.4. pH

The pH values throughout the composting process are closely related to microbial activity. As the process begins, the degradation of easily degradable substrates promotes the initial accumulation of organic acids and the acidification of the compost environment [27]. The subsequent biotransformation of these acidic compounds and the formation of ammonia cause an increase in pH. In both cases, the pH range should not exceed the limits for microbial activity. Microorganisms that are important in composting develop optimal activities at pH between 7.0 and 8.0, although carbohydrate degradation takes place in a wider range, from 6.0 to 9.0 [25].

2.5. Carbon : Nitrogen Ratio (C:N)

The nutrient balance is one of the most important factors in composting. Among the different components that contribute to this balance, the C:N ratio probably is the most influent. The optimal C:N ratio in the starting material is around 25, although a range between 20 and 40 leads to an adequate process depending on the properties of the raw substrates [28]. When the concentration of carbon is not enough, the low C:N ratio slows down decomposition and nitrogen losses may increase because of volatilization as ammonia [29]. This phenomenon becomes important when both temperature and pH are high [30, 31]. On the contrary, if excessive carbon amounts cause a high C:N ratio (>40), the biotransformation is delayed [32], and microorganisms must oxidize the excess of carbon to get a more adequate C:N ratio for their biological activities.

3. Microbiology of Composting

The importance of nutritional and environmental factors in composting processes has been stated in previous sections. Satisfactory composting process only takes place if conditions at every stage are adequate to support the right microbial populations and their activities. Throughout the process, different communities succeed one to another according to the particular environmental conditions of the specific phase. This interaction between microorganisms with different metabolic activities makes possible the mineralization of the organic substrates [33]. As composting starts, primary decomposers act upon easily available compounds, mainly sugars, causing an increase on temperature that leads to the thermophilic phase. During this initial stage, the high nutrient level mostly favours the growth of bacteria because of their shorter generation time [15], although fungi contribute to this initial activity too.

The increase of temperature gives raise to the inactivation of mesophilic microorganisms, being thermophilic and thermotolerant species those that take the leading part. Among them, actinomycetes play an outstanding role, not only for their degrading ability but for their implication in the sanitation capacity of the composting process [34, 35]. Actinomycetes have been reported as degraders of cellulose, hemicellulose, lignin and even xenobiotics compounds [9, 36]. Other prokaryotes that are active at the thermophilic range include both endospore-forming bacteria like *Bacillus* spp. and related species and non-spore forming bacteria as those belonging to genus *Thermus* and *Hydrogenobacter* [17]. If temperature does not exceed 60°C, thermophilic fungi are primary contributors to biological activity, since they are considered as the most important lignocellulose degraders [37]. *Thermomyces*, *Penicillium*, *Aspergillus*, *Geotrichum*, *Mucor* and especially *Phanerochaete* and *Cladosporium* have been referenced in the thermophilic phase of composting [16, 17]. During this phase, the scarcity of easily degradable carbon sources forces the microorganisms to growth from lignocellulosic materials. This is a very complex process that needs the synergistic action of species with different metabolic activities, since the main constituents of lignocellulose (cellulose, hemicellulose and lignin) are bound by covalent and non-covalent linkages that produce an intricate structure which is hard to access [1]. The specific nature of the substrate being composted and the environmental conditions, mainly aeration, determine the time microorganisms take to complete this phase (from days to months).

As substrates become exhausted, microbial activity starts to be limited and, consequently, temperature decreases. Thus, a new change in populations takes place. Mesophilic microorganisms recolonize the material, either by reactivation of dormant species or by inoculation from environment, causing an increase in biodiversity, especially in fungi [38], that will be important in the curing stage due to their higher metabolic versatility [15, 39]. From a microbiological point of view, this maturation stage may be dominated by fungi and actinomycetes [40]. Although chemical and physical processes are predominant during this period, these microorganisms, as well as some bacterial species, take part in the stabilization process since they are involved in the synthesis of polysaccharides, in the nitrogen-fixation and in the redox transformation of hydrogenated, nitrogenated and sulphur compounds [15] as well as in the mineralization of some recalcitrant fractions [41].

3.1. Cellulolytic Microorganisms in Composting Processes

Cellulose is a linear homopolymer of glucose units linked by ß-glycosidic bonds, although cellobiose is the repeating unit on account of the appearance of alternate reversed glucoses. In nature, the cellulose molecule shows up to 15000 monomeric units, although it has been reported higher degrees of polymerization in some algae [42]. These linear chains join together through hydrogen bonds, which give rise to larger units named elementary fibrils. These are in turn assembled in microfibrils and, finally, in cellulose fibers [3]. The cellulose fibers are associated with other polymers, mainly hemicellulose and lignin, and this association, which is different depending on plant cell type and maturity degree, makes difficult its biodegradation [6]. Hydrolysis is also hindered by its crystalline structure, since the degree of packing for molecules in this matrix is so tight that enzymes access is rather difficult. Nevertheless, not all cellulose regions are structured in this way. There are a variable percentage of non-organized chains known as amorphous regions that are more easily degraded.

Despite of recalcitrant nature of cellulose, cellulolytic capabilities are widely distributed among microorganisms. Both bacteria and fungi include members that are able to degrade cellulose. Within bacteria most of them belong to orders *Actinomycetales* and *Clostridiales*, which are usually found in composting processes [17, 43, 44]. Species from *Bacillus* have been also referenced in relation to cellulose decomposition [45], as well as common members of the compost microbiota [46]. Regarding fungi, decomposition of cellulosic substrates is a well-known capacity for many species belonging to Ascomycete, Basidiomycete and Deuteromycete groups [3]. Although some members of two first groups have been detected in composting, like *Chaetomium*, *Coriolus* or *Phanerohaete*, most of the fungi reported in this environment are included in Deuteromycetes. Thus, *Acremonium*, *Alternaria*, *Aspergillus*, *Cladosporium*, *Fusarium*, *Geotrichum*, *Humicola*, *Paecilomyces*, *Penicillium*, *Thermomyces* or *Trichoderma* are usually cited as normal inhabitants in composting processes [15].

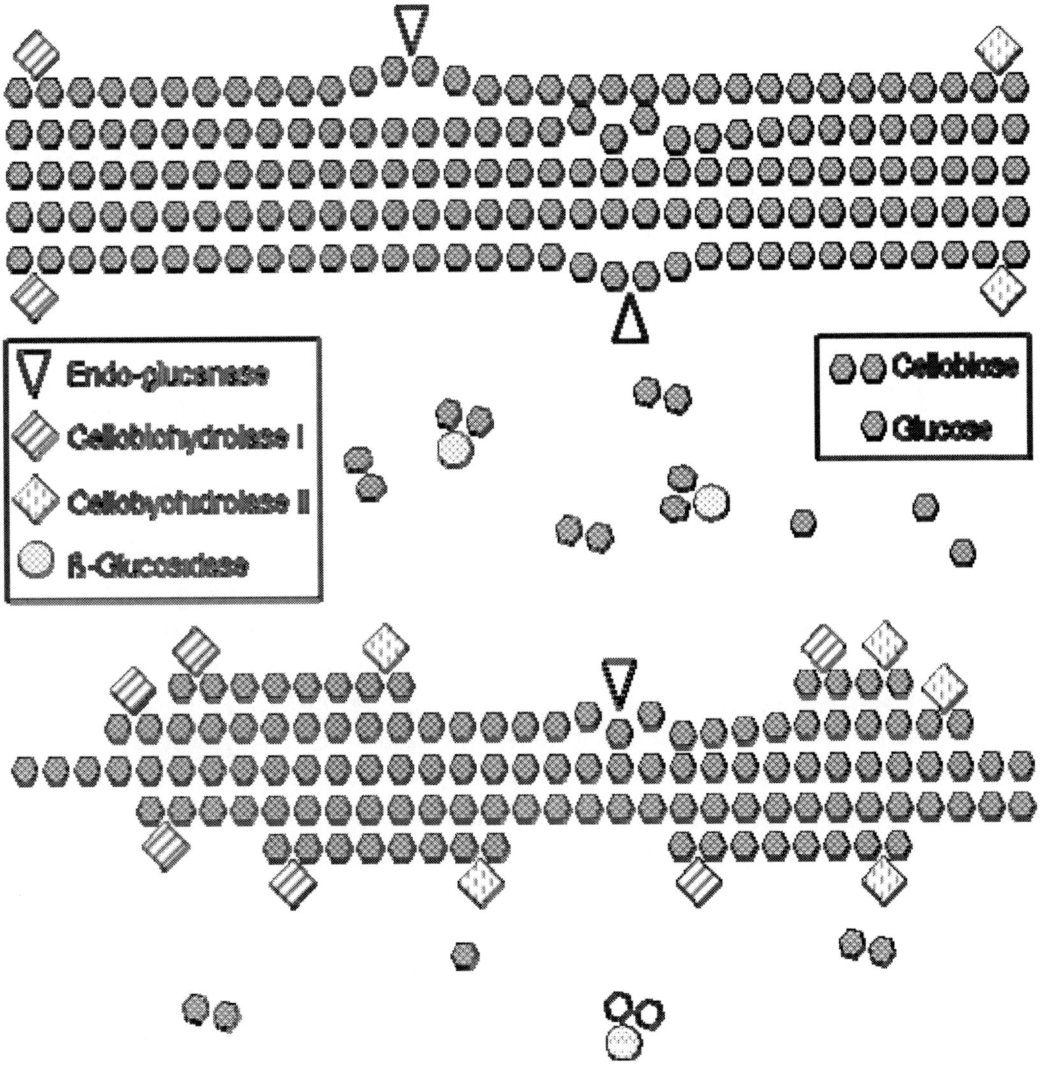

Figure 2. Biodegradation of cellulose [1].

Mechanisms for cellulose decomposition in anaerobic microorganisms are mostly associated to cell-bound enzymatic complexes named cellulosomes, while the aerobic ones act through enzymes that are excreted into the growth medium [1, 47]. Whether cellulose is hydrolyzed by aerobic or anaerobic processes, complete degradation requires the synergistic and coordinated action [3] of three major types of enzymes: endoglucanases, exoglucanases and ß-glucosidases. Endoglucanases generate oligosaccharides of variable length by randomly cutting intramolecular glycosidic bonds in amorphous regions. Exoglucanases, also called cellobiohydrolases, release glucose or cellobiose from chain ends. The final hydrolysis from cellobiose to glucose is carried out by ß-glucosidases, although some of them can act upon cellodextrins [48].

3.2. Hemicellulolytic Microorganisms in Composting Processes

Unlike cellulose, hemicelluloses are referred to polymeric compounds with different composition and structure. Nevertheless, all of them are made up of short heteropolysaccharidic chains with both pentoses and hexoses. Among pentosans (xylans), D-xylose and L-arabinose are the major constituents, while D-glucose, D-galactose and D-mannose compose hexosans (mannans). Sugars are joined by ß-1,4-glycosidic bonds, although galactosyl residues link through positions 1 and 3.

Xylans are the most common hemicelluloses. Although there are modifications depending on the origin, all xylans present a D-xylose backbone. Differences appear in substituents: L-arabinose as sole substituent, which is common in grasses and cereals, or 4-O-methyl-D-glucuronic acid and acetyl groups together with L-arabinose in woods [42]. Through L-arabinosyl and glucuronosyl residues, xylans may establish covalent bonds with phenolic compounds of lignin [49], while hydrogen bonds favour the union with other components, both cellulosic and hemicellulosic [50]. Thus, the intricate matrix between xylan and cellulose probably prevents enzymatic hydrolysis of the last one [51].

Bacteria and fungi are considered as the main hemicellulose degraders, although some yeasts species like *Aureobasidum*, *Cryptococcus* and *Trichosporon* have been reported as xylanase producers [52, 53]. Among bacteria, the ability to degrade hemicellulose has been described both in aerobic and anaerobic species. Aerobic bacterial genera detected in composting like *Bacillus*, *Cellulomonas*, *Pseudomonas*, *Streptomyces*, or *Themonospora* [15] include microorganisms able to hydrolyze different hemicelluloses [54-57], while *Clostridium* is the most important anaerobic genus, among those described in composting processes, that produces hemicellulose degrading enzymes [58]. In relation to fungi, many cellulolytic species also produce hemicellulolytic enzymes. Thus, *Acremonium*, *Alternaria*, *Aspergillus*, *Chaetomium*, *Fusarium*, *Paecilomyces*, *Penicillium*, *Thermomyces* and *Trichoderma* add to the ability of degrading cellulose, as it has been cited above, the production of some enzymes related to hemicellulose hydrolysis [7, 8, 37]. Some of them can even act upon different kind of hemicelluloses, like *Alternaria*, *Aspergillus*, *Fusarium* or *Trichoderma* [15].

As it has been commented for cellulose, the efficient degradation of hemicellulose needs the concerted action of many enzymes [59]. Moreover, different strategies are detected between aerobes and anaerobes microorganisms. Aerobic species act through extracellular enzymes, although in some cases the complete degradation to monosaccharides is carried out by cell-associated enzymes, what prevents the utilization of these nutrients by non-hemicellulolytic surrounding microorganisms [60]. Hemicellulases from anaerobic bacteria are included in the cellulosomes, together with cellulolytic enzymes, since they seem to favour the unwrapping of cellulose from the lignocellulose matrix rather than promote the degradation of hemicellulose [61]. Different types of hemicellulases exist depending on the substrate they hydrolyze, the bonds they break and the products they generate [62]. Nevertheless, it can be established two major groups according to the location of the hydrolyzed residues: endo-enzymes, if they act upon the polymeric backbone to produce oligomers, and exo-enzymes, if they release monomeric units by the attack of chain ends. Among endo-enzymes, endo-ß-xylanase is the most common [63], although endo-mannanases, endo-galactanases and endo-arabinanase have been also described [63-65]. ß-Xylosidases and ß-mannobiosidases, which are the two main exo-enzymes characterized, release respectively xylose and mannose from oligomers, while α-galactosidases and ß-glucosidases produce galactose and glucose. Moreover, a variety of accessory enzymes are needed for complete hemicellulose degradation, like acetyl, ferulic and p-coumaric esterases, α-arabinofuranosidases or α-glucuronidases [59].

3.3. Ligninolytic Microorganisms in Composting Processes

Lignin is a highly branched aromatic and amorphous polymer composed of three principal building blocks with phenyl-propane structure: p-coumaryl alcohol, coniferyl alcohol and sinapyl alcohol. These three cynnamyl alcohol precursors join together in variable proportions depending on plant species and tissue, as well as location within cell wall (42). Another source of diversity is the type of bond between phenyl-propanoid units, since more than ten different types of aryl ether and carbon-carbon linkages have been referenced [66]. Through secondary reactions of polymerization, lignin associates to cellulose and, specially, hemicellulose, producing a complex structure which is quite difficult to break down [67].

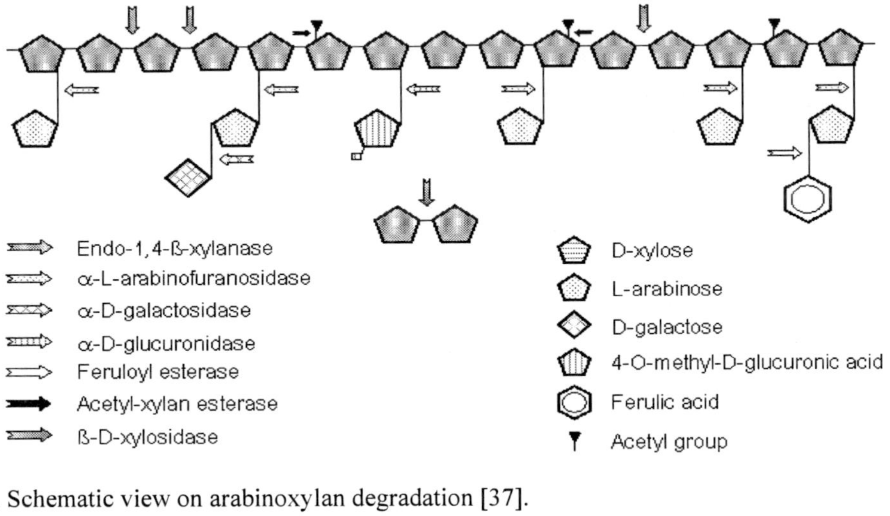

Figure 3. Schematic view on arabinoxylan degradation [37].

The decomposition of lignin is carried out more efficiently by fungi [68]. White-, brown- and soft-rot fungi can degrade lignin to different extents, some specifically and some simultaneously with cellulose and hemicellulose [42]. The most extensive degradation of lignin is that produced by white-rot fungi [69], like *Ceriporiopsis, Cladosporium, Dichomitus, Fomes, Ganoderma, Heterobasidion, Phanerohaete, Phellinus, Phlebia*, or *Trametes*. Among them *Cladosporium, Phanerochaete*, and *Trametes* have been reported in composting processes [15, 70]. Brown-rot fungi *Polyporus, Serpula* and *Gloephyllum* demethylate degrade lignin, although to a lesser extent that white-rot fungi [68]. Nevertheless, none of them have been referenced as common species in compost. On the contrary, soft-rot fungi as *Alternaria, Aspergillus, Fusarium, Paecilomyces, Penicillium* or *Trichoderma* are usually detected in composting processes [15], and all of them have proved their ability to degrade lignin, although in a slow and incomplete way [42]. Although they have received much lesser attention than fungi, some prokaryotes are able to modify lignin being of ecological significance, especially when extreme environmental conditions limit the growth of fungi [68]. Actinomycetes like *Streptomyces* and *Nocardia* are the most efficient, although they do not show as much capacity as fungi [68]. Other lignin-degrading bacteria have been identified in the genera *Alcaligenes, Acinetobacter, Arthrobacter, Erwinia, Pseudomonas* or *Xanthomonas*, most of them being compost-inhabitants [15, 16].

As correspond to its structural complexity, lignin degradation is carried out by a large enzymatic system consisting of three major types of enzymes: peroxidases, laccases and H_2O_2-producing peroxidases. All of them act extracellularly and show low specificity. Two different peroxidases have been described: lignin (LiP) and manganese (MnP) peroxidases. LiP is more efficient since it oxidizes both phenolic and non-phenolic compounds [71]. This reaction is promoted by H_2O_2, which is able to gain electrons from the enzyme. This action turns LiP into a highly reactive intermediate with capacity to oxidize aromatic compounds, forming cation radicals which undergo different reactions with no enzymatic catalysis [72]. MnP show similar catalytic cycle, although the electrons for the reduction come from Mn^{2+} instead of aromatic compounds. The Mn^{3+} generated in this action is responsible for the formation of highly reactive aromatic radicals. Some microorganisms, like *Pleurotus* and *Bjerkandera* produce a versatile peroxidase (VP) that is able to oxidize both LiP and MnP substrates [73].

Since peroxidase activity is H_2O_2-dependent, ligninolytic microorganisms also produce enzymes that generate this compound. Intracellular and extracellular oxidases acting upon different substrates have been described, like glucose-oxidase, methanol-oxidase, fatty acyl-CoA oxidase or aryl-alcohol oxidase. These enzymes, despite of being indirectly associated to lignin depolymerization, are very important for a successful process.

Laccases are blue-copper phenoloxidases with broad specificity towards aromatic compounds containing hydroxyl and amine groups. These enzymes oxidize both phenolic and non-phenolic substrates by gaining one electron which is further transferred to an O_2 molecule. The aromatic free radical generated promoted the polymer cleavage.

Finally, other enzymes like quinone oxidoreductases and dehydrogenases are essential in lignin degradation, since they prevent re-polymerization phenomena among some compounds generated as lignin degradation progress.

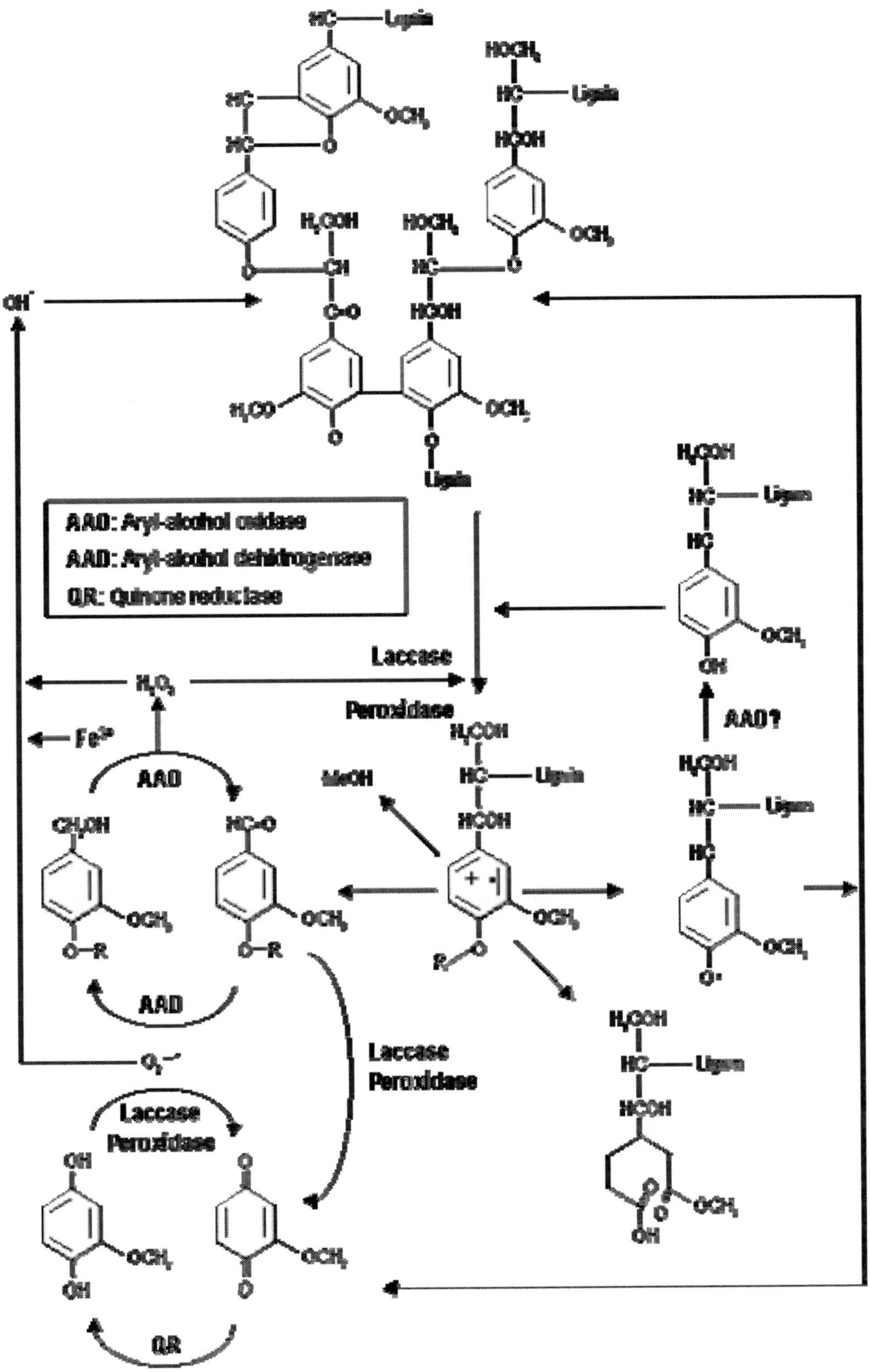

Figure 4. Enzymatic and non-enzymatic reactions during lignin biodegradation [116].

Table 1. Main lignocellulolytic activities detected in microbial species isolated in composting processes. (C: Cellulolytic activity; HC: Hemicellulolytic activity; L: Ligninolytic activity)

Bacteria	Actinobacteria	Fungi
Acinetobacter (L)	*Micronospora* (C)	*Acremonium* (C, HC)
Alcaligenes (L)	*Nocardia* (L)	*Alternaria* (C,HC)
Arthrobacter (C, L)	*Streptomyces* (C, HC, L)	*Aspergillus* (C, HC, L)
Bacillus (C, HC, L)	*Thermoactinomyces* (C, HC)	*Chaetomium* (C, HC)
Cellulomonas (C)	*Thermonospora* (C, HC)	*Cladosporium* (C)
Clostridium (C, HC)		*Fusarium* (C, HC, L)
Erwinia (L)		*Geotrichum* (C)
Pseudomonas (C, HC, L)		*Humicola* (C)
Xanthomonas (C, HC, L)		*Pacilomyces* (C, HC, L)
		Penicillium (C, HC, L)
		Phanerochaete (C, HC, L)
		Thermomyces (C, HC)
		Trametes (C, HC, L)
		Trichoderma (C, HC, L)

4. Industrial and Biotechnological Applications of Lignocellulolytic Microorganisms

Lignocellulose is the major component of biomass, the most abundant source of renewable organic matter. In nature, plant materials are the common origin for lignocellulose, but during last decades different human activities have strongly increased the world production up to levels which turns the situation in a serious problem. This scene, together with the exhaustion of the traditional fossil energy sources and the growing demand of society for a more sustainable way of life, has promoted a change in the way biomass is considered. Nowadays, lignocellulosic materials not only can be directly used for the production of animal feeds or organic amendments for agriculture, as it has been traditionally, but they have potential to be the raw material for a new biobased industry [2]. Liquid fuels and chemicals are some of the products that can be generated from lignocellulose, while different industries as chemical, food, brewery and wine, textile and laundry or pulp and paper take advantage of lignocellulolytic enzymes. On the other hand, the low specificity of some of these enzymes makes possible their application in bioremediation processes.

Although biobased products and processes still have some downsides, among them availability or excessive high costs, a variety of benefits are derived from their use. Many of these benefits are associated to environment, since they reduce the emission of carbon dioxide and other gases responsible for climate change. Moreover, the impact on health of workers and people who lives near factories is lower because of fewer toxic fumes emissions. From a social point of view, the implementation of systems based on biomass utilization may enhance stability of rural communities with no other possibilities of subsistence. These upsides, as well as the possibility of reducing costs by improving both properties of raw materials and microorganisms/enzymes used [4], might make real a biobased economy in the next future.

4.1. Pulp and Paper Industry

Pulping and bleaching are two major phases in the production of paper. Pulping makes possible the extraction of cellulose fibres from plant material, while bleaching favours the decolourization of fibres by the removal of residual lignin and chromophore substances formed in the breakdown of lignin. Conventional methods used have some disadvantages from both economical and environmental point of view [74]. Thus, mechanical pulping processes are extremely energy intensive and produce paper with less strength [75], while chemical processes, both for pulping and bleaching, apply hazardous compounds for workers and environment in large amounts [76].

In order to avoid these drawbacks, new biological procedures are being introduced for either the partial or complete substitution of the traditional methods. From the initial investigations by Lawson & Still [77] to nowadays, great advances in knowledge of how biopulping and biobleaching proceed have been achieved. Despite of some studies that refer energy and pollutants savings up to 30% and 25%, respectively [78], conventional methods are still dominant at industrial scale, although combined processes with biological pretreatments are being applied in many factories [78].

Although processes with whole microbial cells are feasible [75], the use of enzymes seems to produce better results, since microorganisms may affect cellulose in some extent, even those which depolymerize lignin selectively. The environmental conditions seem to be responsible for this change in the metabolic activity [76]. The action of ligninases and hemicellulases promote the removal of lignin and hemicellulose with minimal affectation of cellulose [79]. Laccases delignified woody and non-woody fibres [80], and together with MnP and LiP, have been used in biopulping and biobleaching [81]. This last process is carried out by the additional action of xylanases, which are able to relax the strong matrix that xylan and lignin form. A synergic effect is produced when xylanase is combined with other hemillulolytic enzymes such as α-arabinofuranosidase, mannanase and α-galactosidase [82, 83]. Thus, lignin is more easily removed by further treatments [8], which results in higher final brightness and pulp strength, and lower energy consumption and bleaching chemicals demand [55]. These effects can be achieved only if enzymes are completely free of cellulase activity and show maximal levels of activity when temperature is high and pH is alkaline, since these are the conditions for pulp when bleaching is carried out [7]

4.2. Production of Biological Fuels

The next future exhaustion of fossil fuels and the necessity to reduce CO_2 emissions in order to prevent the climate change has turned the eyes of both scientists and politicians to alternative and greener fuels. One of the most promising technologies in this respect is the production of bioethanol from biomass, not only because it is an environmentally friendly fuel but for the evenly distribution of raw material around the world [84]. Sugar cane and starch-containing materials are commonly used in the production of ethanol, despite that they are demanded for other applications. Moreover, the fuel obtained from hydrolysates of these materials seems to reduce CO_2 emissions in lesser extent than desirable [85]. Lignocellulosic biomass can be a better choice due to the lower contribution to the global warming through the emission of gases and the inexpensive cost [6], which can contribute to get bioethanol be more competitive in relation to fossil fuels.

The successful use of the whole energetic potential of lignocellulose makes necessary the efficient depolymerization of the polysaccharidic fractions to soluble sugars. The hydrolytic process can be performed by the action of physico-chemical parameters or enzymes [51]. Although the former promote the release of monomeric sugars in an efficient way, many other degradation products that inhibit microbial fermentation are formed in different concentration depending on the type of pretreatment [86]. Thus, both for environmental and technical reasons, enzymatic hydrolysis are preferred over physico-chemical processes [51]. Nevertheless, the application of physico-chemical treatments previously to the enzymatic process results in higher accessibility for enzymes and greater sugar concentration [84].

Both cellulolytic and hemicellulolytic enzymes are needed for a complete use of sugars present in lignocellulose, since microorganisms used in the fermentation stage can be genetically modified for pentose utilization [87]. Thus, exo/endoglucanases, cellobiohydrolases and cellobiases lead to the release of glucose, while exo/endoxylanases, arabinosidases, esterases and glucuronidases make similar action in relation to xylose and arabinose. Nevertheless, enzymatic preparations containing only cellulolytic enzymes might be used on account of low content of arabinose in most types of lignocellulosic materials [84] and the non-specific nature of some glucanases, which allow them to act upon xylan as well [51].

Lignocellulolytic enzymes can be applied in other bioprocesses for the production of energy. Biofuel cells are an emerging and very promising technology for meeting energy needs in the future, although nowadays their application is limited because of the lack of knowledge about some metabolic aspects of the system [88]. Nevertheless, laccases have been described as one of the few enzymes able to accept electrons from the cathode [89].

4.3. Bioremediation

One of the most negative aspects of the actual way of living is pollution. Many of the processes by which products that sustain progress are produced demand the utilization of great amounts of hazardous compounds, part of which are released to the environment. Because of their high recalcitrant nature, most of them tend to accumulate in soil, water and air [9]. The anthropogenic origin of these pollutants is responsible for their persistence, since nature do not have the appropriate biological mechanism for their transformation. Nevertheless, non-specific enzymes acting upon similar molecules or bonds can promote their biodegradation as it has been stated for lignocellulolytic systems [90].

Among lignocellulolytic enzymes, bioremediation processes are mainly associated to lignin-degrading enzymes, although cellobiose dehydrogenase has been reported in the degradation processes of some xenobiotics as well [91]. Laccases and peroxidases have been described as potential degraders of numerous pollutants, most of them being aromatic compounds. Polycyclic aromatic hydrocarbons, chlorinated phenols, nitro-substituted compounds, dioxins or synthetic dyes that are used as explosives, pesticides or synthetic polymers can be transformed into highly reactive cationic radicals by these enzymes trough a one-electron oxidation [92]. The free radicals react spontaneously and promote the transformation of the contaminants into non-toxic compounds or biologically assimilated forms [90].

One controversial point in bioremediation lies in the utilization of cell-free enzymes *versus* whole cells. Cell-free enzymes are able to act under a wider range of environmental conditions than microbial cells and in the presence of high concentration of toxic substances that are inhibitory to microorganisms [93], as well as they are not affected by predators or inhibitors of microbial metabolism. On the other hand, the complete biodegradation of the pollutant demands the combined action of different metabolic ways that takes place only if the whole cell is present. Another disadvantage for enzymes is the potential loss of activity and even the occurrence of natural immobilization processes [10]. There is a third alternative in relation to lignocellulolytic microorganisms and remediation of contaminated environments: composting. Primary ingredients of composting can be added to the polluted soil to start the process and the subsequent activity of microorganisms within them, which promote the degradation of the pollutants [9]. Some studies have proved the validity of these strategies for decontaminating soils polluted with polycyclic aromatic compounds [94], pentachlorophenol [95], petroleum-based oil wastes [96], as well as the addition of mature compost [97, 98].

4.4. Green Chemistry

Chemical industry nowadays is based on fossil resources. Despite of the fundamental role they have played in human progress over the last century, the drawbacks of this system are becoming evident in such extent that both consumers and administrators demand a more sustainable industry which does not rely on finite resources and decrease the environmental damage [99]. On this scenario, a bio-based chemical industry is gaining momentum due to the use of renewable feedstocks as well as the implementation of safer and more eco-friendly processes [100].

As happens with bioethanol, sugar cane and starch-containing materials have been the preferred substrate for the production of bio-based commodity chemicals [101]. Nevertheless, the high cost of these raw materials affects the competitivity of the bio-based products in relation to those produced from traditional industry. The utilization of lignocellulosic biomass, which is cheaper, can contribute to reduce costs, although full use of nutritional and energetic potential of this feedstock forces microorganisms to ferment not only hexoses but pentoses too. From these nutritional sources both commodity products and specialty chemicals have been obtained. Ethanol is probably the most important among commodities, not only for its application as fuel but for its ability to act as precursor of different organic compounds needed for the production of paints, solvents, plastics, etc. [4]. Other alcohols, as well as aldehydes and organic acids have been produced from sugar of lignocellulosic origin: acetaldehyde, butanediol, glycerol, acetic acid, lactic acid, succinic acid are some of these products [101]. In relation to specialty chemicals, lignocellulosic biomass has been used for the production of amino acids, antioxidants, vitamins, pesticides, thickening agents, flavours, sweeteners, fragrances and, overall, enzymes [4]. Lignocellulolytic enzymes have numerous applications in such different industries as textile and leather manufacturing, wood processing, detergent production, agriculture, biotechnology and food market [2]. Surely the production of enzymes will be the most important area of development for lignocellulolytic microorganisms if the future brings an industry based on lignocellulosic materials [4].

In textile industry, cellulases allow the controlled modification of the cellulosic fibres, being bio-stoning and bio-polishing the most successful applications. Bio-stoning, which have replaced the previous "stone-washing" method to promote the aged effect on blue jeans and denim, increases productivity, reduces risks for workers and environment and favours automation [102]. In non-denim textiles, cellulases prevent phenomena like fuzz formation and pilling which reduce the commercial value of cellulosic fabrics [103]. As a component of detergents, cellulases help fabrics to preserve their new appearance [104] by removing degraded microfibrils and dirt particles and improving colour brightness [102]. More unusual although very promising is the utilization of xylanases for processing different plant fibres [8] or promote the solubilization of stubborn stains from vegetal origin [105]. In agriculture, cellulases can be used to enhance the efficiency of biocontrol agents against weeds [106] and phytophatogens [107], while laccases have been applied with biotechnological purposes as components of biosensors for the determination of lignin or phenolic compounds [108, 109].

4.5. Food and Feed Industry

Enzymes can carry out multiple functions on food and feed industry. Thus, the action of these biocatalysts can positively affect the nutritional quality of the product, increase the level of safety, upgrade the whole production process and generate new products [110]. Many different enzymes are usually applied in the food and feed industry, among which those showing lignocellulolytic capabilities play a prominent role. Laccases, xylanases, arabinofurosidases or cellulases take part in very different processes for the generation of food products such as bread and related products, wine and beer, fruit juices and animal feed [2, 102, 111].

In the bread-making process, xylanases and other hemicellulases promote the break down of hemicellulose resulting in a better redistribution of water and softer dough which makes easier the mixing [8]. They also favour the growth of the dough, increase the volume of bread and delay the staling process [83, 105]. On the other hand, the presence of arabinoxylooligosaccharides in bread is believed to have an improved health effect [8]. In relation to ligninolytic enzymes, it has been proved that laccases increase the strength of crumb structure and stability, while stickiness is reduced [111].

The use of cellulases and hemicellulases during wine elaboration favours the colour extraction by means of a better skin maceration, makes easier the clarification and filtration processes and produces wines of higher quality [102]. One of the most important factors in quality is aromatic compounds, which can be released from major polymeric structures by the action of arabinofuranosidases and glucosidases [83]. Both enzymes also prevent the transformation of these molecules into less-fragrant compounds during wine aging. In turn, laccases promote the stabilization of wine by the oxidation of polyphenolic substances, which are later removed [113]. Similar function is carried out in beer, since the presence of polyphenols favour the precipitation of proteins and the formation of hazes during storage [13]. Glucanases also contribute to avoid the haze formation when they are added at the mashing stage and during primary fermentation, especially if poor quality cereals are used [114].

Together with pectinases, cellulases and hemicellulases extract, clarify and stabilize juice from fruits and vegetables. They facilitate the maceration process, promoting the extraction of fruit ingredients and increasing the yield of juice [102], while in later stages reduce viscosity and hydrolyze substances that can produce turbidity [8]. Laccases have been applied to prevent turbidity too, as well as to promote stabilization. Nevertheless, results in relation to this last factor are contradictory since some reports suggest stability is higher when chemical conventional treatments are used [111], and some others point out that good stability can be reached by combining the action of laccases with ultrafiltration methods, especially in light-colour juices [115].

One of the most traditional applications of lignocellulosic materials is the production of animal feed. For a long time, they have been directly used with no prior treatment but nowadays the utilization of enzymes leads to more digestible feeds which results in a more efficient use of the energetic content [8]. Thus, the enzymatic treatment of substrate mainly with xylanase and glucanase stimulates weight gain both in monogastric and ruminant species, although some controversial results have been reported for the latter [104].

Table 2. Major applications of lignocellulolytic enzymes

<table>
<tr><td>Field</td><td>Cellulases</td><td>Hemicellulases</td><td>Ligninases</td></tr>
<tr><td>Bioconversion</td><td colspan="2">Hydrolysis of polymeric substances for the release of monomeric sugars used as substrates for the production of bioethanol and chemicals</td><td></td></tr>
<tr><td>Textiles</td><td>Bio-stoning and bio-polishing</td><td>Solubilization of stubborn stains</td><td></td></tr>
<tr><td>Laundry</td><td>Improvement of washing efficiency</td><td></td><td></td></tr>
<tr><td>Pulp and paper</td><td>Deinking</td><td colspan="2">Biopulping and biobleaching</td></tr>
<tr><td>Livestock</td><td colspan="2">Improvement of digestibility and nutritional properties of animal feed</td><td></td></tr>
<tr><td>Environment</td><td></td><td></td><td>Bioremediation of habitats polluted with xenobiotics and other recalcitrant compounds</td></tr>
<tr><td>Food</td><td colspan="3">Bread making
Extraction, clarification and stabilization of fruit and vegetable juices
Extraction, clarification and stabilization of wine</td></tr>
<tr><td>Biotechnology</td><td></td><td></td><td>Biosensors</td></tr>
</table>

5. Conclusion

Metabolic microbial versatility is a powerful tool from which mankind can take important advantages. Nevertheless, minimal use of a such potential has been done, although the last years have seen a dramatic increase in biotechnology on account of social pressure for the implementation of a safer and environmentally respectful industrial activity and the next exhaustion of the fossil resources that support our way of living. A predominant biomass-based industry is still far from reality but closer every day. Actually, some conventional products originated from fossil resources are being replaced by homologous biomass-based products as well as different political strategies incentive in a bio-based economy through directives and laws that make bio-products more competitive.

Microorganisms, mainly lignocellulolytic microorganisms, play a key role in this new industry, since they are primarily responsible for the natural transformation of lignocellulose. Major enzymes involved in the degradation processes of the different polymers integrated in the lignocellulosic matrix find numerous applications of commercial interest, in addition to their participation in the treatment of lignocellulosic materials that promote the release of monomeric sugars used as substrates in other bioprocesses. From this perspective, lignocellulolytic microorganisms and, in turn, environments where they are present are of major interest. Composting, the process by which residual organic matter is biotransformed into a value added product, is one of the most appropriate habitats for lignocellulolytic microorganisms because of the high lignocellulose content of most of the wastes. Having this in mind, as well as the special environmental conditions that prevailed throughout the process, composting must be considered as a primary source for lignocellulolytic microorganisms and enzymes showing activity in a wide range on conditions.

References

[1] Malherbe, S. & Cloete, T. E. (2002). Lignocellulose biodegradation: fundamental and applications. *Rev Environ Sci Biotechnol, 1*, 105-114.

[2] Howard, R. L., Abotsi, E., Jansen van Rensburg, E. L. & Howard, S. (2003). Lignocellulose biotechnology: issues of bioconversion and enzyme production. *Afr J Biotechnol, 2*, 602-619.

[3] Lynd, L. R., Weimer, P. J., van Zyl, W. H. & Pretorius, I. S. (2002). Microbial cellulose utilization: fundamentals and biotechnology. *Microbiol Mol Biol R, 66*, 506-577.

[4] Dale, B. E. (2003). "Greening" the chemical industry: research and development priorities for biobased industrial products. *J Chem Technol Biot, 78*, 1093-1103.

[5] Claus, H., Faber, G. & König, H. (2002). Redox-mediated decolorization of synthetic dyes by fungal laccasse. *Appl Microbiol Biotechnol, 59*, 672-678.

[6] Pérez, J., Muñoz-Dorado, J., de la Rubia, T. & Martínez, J. (2002). Biodegradation and biological treatments of cellulose, hemicellulose and lignin: an overview. *Int Microbiol, 5*, 53-63.

[7] Beg, Q. K., Kapoor, M., Mahajan, L. & Hoondal, G.S. (2001). Microbial xylanases and their industrial application: a review. *Appl Microbiol Biotechnol, 56*, 326-338.

[8] Polizeli, M. L. T. M., Rizzati, A. C. S., Monti, R., Terenzi, H. F., Jorge, J. A., Amorim, D. S. (2005). Xylanases from fungi: properties and industrial applications. *Appl Microbiol Biotechnol, 67*, 577-591.

[9] Semple, K. T., Reid, B. J. & Fervor, T. R. (2001). Impact of composting strategies on the treatment of soils contaminated with organic pollutants. *Environ Pollut*, 112, 269-283.

[10] Gianfreda, L. & Rao, M. A. (2004). Potential of extra cellular enzymes in remediation of polluted soils: a review. *Enzyme Microb Tech*, 35, 339-354.

[11] Collins, T., Gerday, C. & Feller, G. (2005). Xylanases, xylanase families and extremophilic xylanases. *FEMS Microbiol Rev*, 29, 3-23.

[12] Sukumaran, R. K., Singhania, R. R. & Pandey, A. (2005). Microbial cellulases – Production, applications and challenges. *J Sci Ind Res India*, 64, 832-844.

[13] Rodríguez Couto, S. & Toca Herrera, J.L. (2006). Industrial and biotechnological applications of laccases: a review. *Biotechnol Adv*, 24, 500-513.

[14] Topakas, E., Vafiadi, C. & Christakopoulos, P. (2007). Microbial production, characterization and applications of feruloyl esterases. *Process Biochem*, 42, 497-509.

[15] Ryckeboer, J., Mergaert, J., Vaes, K., Klammer, S., de Clerq, D., Coosemans, J., Insam, H. & Swings, J. (2003). A survey of bacteria and fungi occurring during composting and self-heating processes. *Ann Microbiol*, 53, 349-410.

[16] Tuomela, M., Vikman, M., Hatakka, A. & Itävaara, M. (2000). Biodegradation of lignin in a compost environment: a review. *Bioresource Technol*, 72, 169-183.

[17] Boulter, J. I., Boland, G. J. & Trevors, J. T. (2000). Compost: a study of the development process and end-product potential for suppression of turfgrass disease. *World J Microb Biot*, 16, 115-134.

[18] Cegarra, J., Alburquerque, J. A., Gonzálvez, J., Tortosa, G. & Chaw, D. (2006). Effects of the torced ventilation on composting of a solid olive-mill by-product ("alpeorujo") managed by mechanical turning. *Waste Manage*, 26, 1377-183.

[19] Cahyani, V. R., Matsuya, K., Asakawa, S. & Kimura, M. (2004). Sucesión and phylogenetic profile of eukaryotic communities in the composting process of rice straw estimated by PCR-DGGE análisis. *Biol Fertil Soils*, 40, 334-344.

[20] Narihiro, T. & Hiraishi, A. (2005). Microbiology of fed-batch composting. *Microbes Environ*, 20, 1-13.

[21] Roberts, G. D. & Roberts, V. J. (1999). Bio-catalytic oxidation reactor. US Patent, N° 5981270.

[22] Suárez-Estrella, F., Vargas-García, M. C., Elorrieta, M. A., López, M. J. & Moreno, J. (2003). Temperature effect on *Fusarium oxysporum* f.sp. *melonis* survival during horticultural waste composting. *J Appl Microbiol*, 94, 475-482.

[23] Yamada, Y. & Kawase, Y. (2006). Aerobic composting of waste activated sludge: kinetic analysis for microbiological reaction and oxygen consumption. *Waste Manage*, 26, 49-61.

[24] Tiquia, S. M., Tam, N. F. Y. & Hodgkiss, I. J. (1998). Changes in chemical properties during composting of spent pig litter at different moisture contents. *Agr Ecosyst Environ,* 67, 79-89.

[25] Nekluydov, A. D., Fedotov, G. N. & Ivankin, A. N. (2006). Aerobic processing of organic waste into composts. *Appl Biochem Micro+*, 42, 341-353.

[26] Nelson, V. L., Crowe, T. G., Shah, M. A. & Watson, L. G. (2006). Temperature and turning energy of composting feedlot manure at different moisture contents in southern Alberta. *Can Biosyst Engine*, 48, 6.31-6.37.

[27] Hellmann, B., Zelles, L., Palöjarvi, A. & Bai, Q. (1997). Emission of climate-relevant trace gases and succession of microbial communities during open-windrow composting. *Appl Environ Microb*, 63, 1011-1018.

[28] Chang, J. I., Tsai, J. J. & Wu, K. H. (2006). Thermophilic composting of food waste. *Bioresource Technol*, 97, 116-122.

[29] Liang, Y., Leonard, J. J., Feddes, J. J. R. & McGill, W. B. (2006). Influence of carbon and buffer amendment on ammonia volatilization in composting. *Bioresource Technol*, 97, 748-761.

[30] Pagans, E., Barrena, R., Font, X. & Sánchez, A. (2006). Ammonia emissions from the composting of different organic waste. Dependency on process temperature. *Chemosphere*, 62, 1534-1542.

[31] Moldes, A., Cendón, Y. & Barral, M. T. (2007). Evaluation of municipal solid waste compost as a plant growing media component, by applying mixture design. *Bioresource Technol*, 98, 3069-3075.

[32] Ros, M., García, C. & Hernández, T. (2006). A full-scale study of treatment of pig slurry by composting: kinetic changes in chemical and microbial properties. *Waste Manage*, 26, 1108-1118.

[33] Steger, K., Eklind, Y., Olsson, J. & Sundh, I. (2005). Microbial community growth ad utilization of carbon constituents during thermophilic composting at different oxygen levels. *Microb Ecol*, 50, 163-171.

[34] Doumbou, C. L., Hamby Slove, M. K., Crawfor, D. L. & Beaulieu, C. (2001). Actinomycetes, promising tools to control for plant diseases and to promote plant growth. Phytoprotection, 82, 85-102.

[35] Suárez-Estrella, F., Vargas-García, M. C., López, M. J. & Moreno, J. (2006). Antagonistic activity of actinomycetes and fungi from horticultural compost against phytopathogen microorganisms. In: Mendez-Vilas, A. (Ed.), *Modern multidisciplinary applied microbiology: Exploiting microbes and their interactions* (pp. 370-374).Weinheim, Wiley-VCH.

[36] Abdula, H. M. & El-Shatoury, S. A. (2007). Actinomycetes in rice straw decomposition. *Waste Manage*, 27, 850-853.

[37] Aro, N., Pakula, T. & Penttilä, M. (2005). Transcriptional regulation of plant cell wall degradation by filamentous fungi. *FEMS Microbiol Rev*, 29, 719-739.

[38] Tiquia, S. M. (2005). Microbial community dynamics in manure composts based on 16S and 18S rDNA T-RFLP profiles. *Environ Technol*, 26, 1101-1114.

[39] Alfreider, A., Peters, S., Tebbe, C. C., Rangger, A. & Insam, H. (2002). Microbial community dynamics during composting of organic matter as determined by 16S ribosomal DNA analysis. *Compost Sci Util*, 10, 303-312.

[40] Hermann, R. F. & Shann, J. F. (1997). Microbial community changes during the composting of municipal solid waste. *Microb Ecol*, 33, 78-85.

[41] Steger, K. (2006). Composition of microbial communities in composts. *PhD Thesis*, Swedish University of Agricultural Sciences, Uppsala, Sweden.

[42] Kuhad, R. C., Singh, A. & Eriksson, K. E. L. (1997). Microorganisms and enzymes involved in the degradation of plant fiber cell walls. *Adv Biochem Eng Biot*, 57, 45-125.

[43] Peters, S., Koschinsky, S., Schwieger, F. & Tebbe, C. C. (2000) Succession of microbial communities during hot composting as detected by PCR-single-strand-conformation polymorphism-based genetic profiles of small-subunit rRNA genes. *Appl Environ Microb*, 66, 930-936.

[44] Pedro, M. S., Haruta, S., Hazaka, M., Shimada, R., Yoshida, C., Hijura, K., Ishii, M. & Igarashi, Y. (2001). Denaturing gradient gel electrophoresis analysis of microbial community from field-scale composter. *J Biosci Bioeng*, 91, 159-165.

[45] Lu, W. J., Wang, H. T., Yang, S. J., Wang, Z. C. & Nie, Y. F. (2006) Isolation and characterization of mesophilic cellulose-degrading bacteria from flower stalks-vegetable waste co-composting system. *J Gen Appl Microbiol*, 51, 353-360.

[46] Takaku, H., Kodaira, S., Kimoto, A., Nashimoto, M. & Takagi, M. (2006). Microbial communities in the garbage composting with rice hull as an amendment revealed by culture-dependent and –independent approaches. *J Biosci Bioeng*, 101, 42-50.

[47] Zhang, Y. H. P., Himmel, M. E. & Mielenz, J. R. (2006). Outlook for cellulose improvement: screening and selection strategies. *Biotechnol Adv*, 24, 452-481.

[48] Zhang, Y. H. P. & Lynd, L.R. (2004). Toward an aggregated understanding of enzymatic hydrolysis of cellulose: noncomplexed cellulase systems. *Biotechnol Bioeng*, 88, 797-824.

[49] Uffen, R. L. (1997). Xylan degradation: a glimpse at microbial diversity. *J Ind Microbiol Biot*, 19, 1-6.

[50] Paananen, A., Oesterberg, M., Rutland, M., Tammelin, T., Saarinen, T., Tappura, K. & Stenius, P. (2004). Interaction between cellulose and xylan: an atomic force microscope and quartz crystal microbalance study. In: Gatenholm, P. & Tenkanen, M. (Eds.), *Hemicelluloses ACS Symposium Series Vol 864* (pp. 167-183). Washington, DC: ACS.

[51] Kabel, M. A., van der Maarel, M. J. E. C., Slip, G., Voragen, A. G. J. & Schols, H. A. (2006). Standard assays do not predict the efficiency of commercial cellulase preparations towards plant material. *Biotechnol Bioeng*, 93, 56-63.

[52] Liu, W., Lu, Y. & Ma, G. (1999). Induction and glucose repression of endo-ß-xylanase in the yeast *Trichosporon cutaneum* SL409. *Process Biochem*, 34, 67-72.

[53] Laitila, A., Wilhelmson, A., Kotaviita, E., Olkku, J., Home, S. & Juvonen, R. (2006). Yeasts in an industrial malting ecosystem. *J Ind Microbiol Biot*, 33, 953-966.

[54] Irwin, D., Jung, E. D. & Wilson, D. B. (1994). Characterization and sequence of a *Thermonospora fusca* xylanase. *Appl Environ Microbiol*, 60, 763-770.

[55] Kansoh, A. L. & Nagieb, Z. A. (2004) Xylanase and mannanase enzymes from Streptomyces galbus NR and their use in biobleaching of softwood kraft pulp. *Anton Leeuw Int J G*, 85, 103-114.

[56] Raposo, M. P., Inácio, J. M., Mota, L. J. & De Sá-Nogueira, I. (2004). Transcriptional regulation of genes encoding arabinan-degrading enzymes in *Bacillus subtilis*. *J. Bacteriol*, 186, 1287-1296.

[57] Amaya-Delgado, L., Vega-Estrada, J., Flores-Cotera, L. B., Dendooven, L., Hidalgo-Lara, M. E. & Montes-Horcasitas, M. C. (2006). Induction of xylanases by sugar cane bagasse at different cell densities of *Cellulomonas flavigena*. *Appl Microbiol Biot*, 70, 477-481.

[58] Han, S. O., Yukawa, H., Inui, M. & Doi, R. H. (2003). Regulation of expression of cellulosomal cellulase and hemicellulose genes in *Clostridium cellulovorans*. *J Bacteriol*, 185, 6067-6075.

[59] Shallom, D. & Shoham, Y. (2003). Microbial hemicellulases. *Curr Opin Microbiol*, 6, 219-228.

[60] Nagy, T., Emami, K., Fontes, C. M. G. A., Ferreira, L. M. A., Humphry, D. R. & Gilbert, H. J. (2002). The membrane-bound α-glucuronidase from *Pseudomonas cellulose* hydrolizes 4-O-methyl.D-glucuronoxylooligosaccharides but not 4-O-methyl –D-glucuronoxylan. *J Bacteriol*, 184, 4925-4929.

[61] Zverlov, V. V., Schantz, N., Schmitt-Kopplin, P. & Schwarz, W. (2005). Two new major subunits in the cellulosome of Clostridium thermocellum: xyloglucanase Xgh74A and endoxylanase Xyn10D. *Microbiology*, 151, 3395-3401.

[62] Jeffries, T. W. (1994). Biodegradation of lignin and hemicellulases. In: Ratledge, C. (Ed.), *Biochemistry of microbial degradation* (pp. 233-277). Dordrecht, The Netherlands: Kluwer Academic Publishers.

[63] Hägglund, P., Eriksson, T., Collén, A., Nerinckx, W., Claeyssens, M. & Stålbrand, H. (2003). A cellulose-binding module of the *Trichoderma reesei* ß-mannanase Man5A increases the mannan-hydrolysis of complex substrates. *J Biotechnol*, 101, 37-48.

[64] Leal, T. F. & Sá-Nogueira, I. D. (2004). Purification, characterization and functional analysis o fan endo-arabinanase (AbnA) from Bacillus subtilis. *FEMS Microbiol Lett*, 241, 41-48.

[65] Ryttersgaard, C., Le Nours, J., Lo leggio, L., Jørgensen, C. T., Christensen, L. L. H., Bjørnvad, M. & Larsen, S. (2004). The structure of endo-ß-galactanase from *Bacillus licheniformis* in complex with two oligosaccharide product. *J Mol Biol*, 341, 107-117.

[66] Brunow, G. (2001). Methods to reveal the structure of lignin. In: Steinbüchel, A. & Hofrichter, M. (Eds.), *Biopolymers: Lignin, humic substances and coal* (pp. 89-116). Weinheim, Germany: Wiley-VCH.

[67] Monties, B. & Fukushima, K. (2001). Occurrence, function and biosynthesis of lignins. In: Steinbüchel, A. & Hofrichter, M. (Eds.), *Biopolymers: Lignin, humic substances and coal* (pp. 1-64). Weinheim, Germany: Wiley-VCH.

[68] Tuomela, M. (2002). Degradation of lignin and other ^{14}C-labelled compounds in compost and soil with an emphasis on white rot-fungi. *PhD Thesis*, University of Helsinki, Helsinki, Finland.

[69] Hatakka, A. (2001). Biodegradation of lignin. In: Steinbüchel, A. & Hofrichter, M. (Eds.), *Biopolymers: Lignin, humic substances and coal* (pp. 129-180). Weinheim, Germany: Wiley-VCH.

[70] Granit, T., Chen, Y. & Hadar, Y. (2007). Humic acid bleaching by white-rot fungi isolated from biosolids compost. *Soil Biol Biochem*, 39, 1040-1046.

[71] Martínez, A. T. (2002). Molecular biology and structure-function of lignin-degrading peroxidases. *Enzyme Microb Tech*, 30, 425-44.

[72] Higuchi, T. 2004. Microbial degradation of lignin: role of lignin peroxidase, manganese peroxidase and laccase. *Proc Jpn Acad, Ser B*, 80, 204-214.

[73] Heinfling, A., Martínez, M. J., Martínez, A. T., Bergbauer, M. & Szewzyk, U. (1998). Transformation of industrial dyes by manganese peroxidases from *Bjerkandera adusta* and *Pleurotus eryngii* in a manganese independent-action. *Appl Environ Microbiol*, 64, 2788-2793.

[74] Shatalov, A. A. & Pereira, H. (2007). Xylanases pre-treatment of giant reed organosolv pulps: direct bleaching effects and bleach boosting. *Ind Crop Prod*, 25, 248-256.

[75] Akhtar, M., Scott, G. M., Swaney, R. E. & Shipley, D. F. (2000). Biomechanical pulping: a mill scale evaluation. *Resour Conserv Recy*, 28, 241-252.

[76] Breen, A. & Singleton, F. L. (1999). Fungi in lignocellulose breakdown and biopulping.

[77] Lawson, L. R. & Still, C. N. (1957). The biological decomposition of lignin – Literature survey. *Tappi J*, 40, 56A-80A.

[78] Atik, C., Imamoglu, S. & Bermek, H. (2006). Impact of xylanase pre-treatment on peroxide bleaching stage of biokraft pulp. *Int Biodeter Biodegr*, 58, 22-26.

[79] Techapun, C., Poosaran, N., Watanabe, M. & Sasaki, K. (2003). Thermostable and alkaline-tolerant microbial cellulose-free xylanases produced from agricultural wastes and the properties required for use in pulp bleaching bioprocesses: a review. *Process Biochem*, 38, 1327-1340.

[80] Camarero, S., García, O., Vidal, T., Colom, J., del Río, J. C., Gutiérrez, A., Gras, J. M., Monje, R., Martínez, M. J. & Martínez, A. T. (2004). Efficient bleaching of non-wood high quality paper pulp using laccase-mediator system. *Enzyme Microb Tech*, 31, 113-120.

[81] Brasil de Souza-Cruz, P., Freer, J., Siika-Aho, M. & Frraz, A. (2004). Extraction and determination of enzymes produced by *Ceriporiopsis subvermispora* during biopulping of *Pinus taeda* wood chips. *Enzyme Microb Tech*, 34, 228-234.

[82] Clarke, J. H., Davidson, K., Rixon, J. E., Halstead, J. R., Fransen, M. P., Gilbert, H. J. & Hazlewood, G. P. (2000). A comparison of enzyme-aided bleaching of softwood paper pulp using combinations of xylanasemannanase and α-galactosidase. Appl Microbiol Biotechnol, 53, 661-667.

[83] Numan, M. T. & Bhosle, N. B. (2006). α-L-Arabinofusidases: the potential applications in biotechnology. *J Ind Microbiol Biotechnol*, 33, 247-260.

[84] Hahn-Hägerdal, B., Galbe, M., Gorwa-Grauslund, M. F., Lidén, G. & Zacchi, G. (2006). Bio-ethanol – the fuel of tomorrow from the residues of today. Trends Biotechnol, 24, 549-556.

[85] Farrell, A. E., Plevin, R. J., Turner, B. T., Jones, A. D., O'Hare, M. & Kammen, D. M. (2006). Ethanol can contribute to energy and environmental goals. *Science*, 311, 506-508.

[86] Klinke, H. B., Thomsen, A. B. & Ahring, B. K. (2004). Inhibition of ethanol-producing yeast and bacteria by degradation products produced during pre-treatment of biomass. *Appl Microbiol Biotechnol*, 66, 10-26.

[87] Hahn-Hägerdal, B., Karhumaa, K., Fonseca, C., Spencer-Martins, I. & Gorwa-Grauslund, M. F. (2007). Towards industrial pentose-fermenting yeast strains. *Appl Microbiol Biotechnol*, 74, 937-953.

[88] Rabaey, K., Boon, N., Siciliano, S. D., Verhaege, M. & Verstraete, W. (2004). Biofuel cells select for microbial consortia that self-mediate electron transfer. *Appl Environ Microb*, 70, 5373-5382.

[89] Alcalde, M., Ferre, M., Plou, F. J. & Ballesteros, A. (2006). Environmental biocatalysis: from remediation with enzymes to novel green processes. *Trends Biotechnol*, 24, 281-287.

[90] Mester, T. & Tien, M. (2000). Oxidation mechanism of ligninolytic enzymes involved in the degradation of environmental pollutants. *Int Biodeter Biodegr*, 46, 51-59.

[91] Cameron, M. D., Timofeevski, S. & Aust, S. D. Enzymology of *Phanerochaete chysosporium* with respect to the degradation of recalcitrant compounds and xenobiotics. *Appl Microbiol Biotechnol*, 54, 751-758.

[92] Pointing, S. B. (2001). Feasibility of bioremediation by white-rot fungi. *Appl Microbiol Biotechnol*, 57: 20-33.

[93] Torres, E., Bustos-Jaimes, I. & Le Borge, S. (2003). Potential use of oxidative enzymes for the detoxification of organic pollutants. *Appl Catal B-Environ*, 46, 12007-05-14.

[94] Rojas-Avelizapa, N. G., Roldán-Carrillo, T., Zegarra-Martínez, H., Muñoz-Colunga, A.M. & Fernández-Linares, L.C. (2007). A field trial for an *ex-situ* bioremediation of a drilling mud-polluted site. *Chemosphere*, 66, 1595-1600.

[95] Jiang, X. Y., Zeng, G. M., Huang, D. L., Chen, Y., Liu, F., Huang, G. H., Li, J. B., Xi, B. D. & Liu, H. L. (2006). Remediation of pentachlorophenol-contaminated soil by composting with immobilized *Phanerochaete chysosporium*. *World J Microb Biot*, 22, 909-913.

[96] Van Gestel, K., Mergaert, J., Swings, J., Coosemans, J. & Ryckeboer, J. Bioremediation of diesel oil-contaminated soil by composting with biowaste. *Environ Pollut*, 125, 361-368.

[97] Puglisi, E., Cappa, F., Fragoulis, G., Trevisan, M. & Del Re, A. A. M. (2007). Bioavailability and degradation of phenanthrene in compost amended soils. *Chemosphere*, 67, 548-556.

[98] Scelza, R., Rao, M. A. & Gianfreda, L. (2007). Effects of compost and of bacterial cells on the decontamination and the chemical and biological properties of an agricultural soil artificially contaminated with phenanthrene. *Soil Biol Biochem*, 39, 1303-1317.

[99] Hatti-Kaul, R., Törnvall, U, Gustafsson, L. & Börjesson, P. (2007). Industrial biotechnology for the production of bio-based chemicals – a cradle-to-grave perspective. *Trends Biotechnol*, 25, 119-124.

[100] Clark, J. H. (2006). Green chemistry: today (and tomorrow). *Green Chem*, 8, 17-21.

[101] Danner, H. & Braun, R. (1999). Biotechnology for the production of commodity chemicals from biomass. *Chem Soc Rev*, 28, 395-405.

[102] Bhat, M.K. (2000). Cellulases and related enzymes in biotechnology. *Biotechnol Adv*, 18, 355-383.

[103] Özdil, N., Özdogan, E. & Öktem, T. (2003). Effects of enzymatic treatment on various spun yarn fabrics. *Fibres Text East Eur*, 11, 58-61.

[104] Olsen, H. S. & Falhot, P. The role of enzymes in modern detergency. *J Surfactants Deterg*, 1, 555-567.

[105] Kamal Kubar, B., Balakrishnan, H. & Rele, M. V. (2004). Compatibility of alkaline xylanases from an alkaliphilic *Bacillus* NCL (87-6-10) with commercial detergents and proteases. *J Ind Microbiol Biotechnol*, 31, 83-87.

[106] Babalola, O. O. (2007). Pectinase and cellulase enhance the control of *Abutilon theophrasti* by *Colletotrichum coccodes*. *Biocontrol Sci Techn*, 17, 53-61.

[107] Inglis, G. D. & Kawchuk, L. M. (2002). Comparative degradation of oomycete, ascomycete, and basidiomycete cell walls by mycoparasitic and biocontrol fungi. *Can J Microbiol*, 48, 60-70.

[108] Shleev, S., Persson, P., Shukamovic, G., Mazhugo, Y., Yaropolov, A., Ruzgas, T. & Gorton, L. (2006). Laccase-based biosensors for monitoring lignin. *Enzyme Microb Tech*, 39, 835-840.

[109] Odaci, D., Timur, S., Pazarlioglu, N., Montereali, A. M., Vastarella, W., Pilloton, R. & Telefoncu, A. (2007). Determination of phenolic acids using *Trametes versicolor* laccase. *Talanta*, 71, 312-317.

[110] Rolle, R. S. (1998). Enzyme applications for agro-processing in developing countries: an inventory of current and potential applications. *World J Microb Biot*, 14, 611-619.

[111] Minussi, R., Pastore, G. M. & Durán, N. (2002). Potential applications of laccase in the food industry. *Trends Food Sci Tech*, 13, 205-216.

[112] Camacho, N. & Aguilar, O. G. (2003). Production, purification and characterization of a low molecular mass xylanase from *Aspergillus* sp. and its application in bakery. *Appl Biochem Biotechnol*, 104, 159-172.

[113] Minussi, R. C., Rossi, M., Bologna, L., Rotilio, D., Pastore, G. M. & Durán, N. (2007). Phenols removals in must: strategy for wine stabilization by laccase. *J Mol Catal B-Enzym*, 45, 102-107.

[114] Celestino, K. R. S., Cunha, R. B. & Felix, C. R. (2006). Characterization of a ß-glucanase produced by *Rhizopus microsporus* var. *microsporus*, and its potential for application in the brewing industry. *BMC Biochem*, 7, art 23.

[115] Alper, N. & Acar, J. (2004). Removal of phenolic compound in pomegranate juices using ultrafiltration and laccase-ultrfiltration combinarions. *Nahrung*, 48, 184-187.

[116] Martínez, A. T., Speranza, M., Ruiz-Dueñas, F. J., Ferreira, P., Camarero, S., Guillén, F., Martínez, M. J., Guitiérrez, A. & del Río, J. C. Biodegradation of lignocellullosics: microbial, chemical, and enzymatic aspects of the fangal attack of lignin. *Int Microbiol*, 8, 195-204.

In: Leading-Edge Environmental Biodegradation
Editor: L. E. Pawley, pp. 249-274

ISBN: 978-1-60021-903-0

Chapter 8

THE PIG HAIR HYDROLYSIS: AN ALTERNATIVE TO LANDFILLING

M. C. Márquez*[*], *M. B. Esteban, J. A. García and P. Ramos

Chemical Engineering Department. Faculty of Chemical Sciences. University of Salamanca. Plaza de los Caídos 1-5, 37008-Salamanca, Spain

ABSTRACT

The European Council Directive on the Landfill of Wastes 1999/31/EEC (LFD), places a requirement on member states to draw up strategies to ensure that the amount of biodegradable municipal waste deposited at landfill progressively reduces over a 15-year period to only 35% of the total amount produced in 1995. An especially problematic organic waste is the pig hair due to the high production generated in the slaughterhouses (34000 tons per year in Spain). In order to reduce this problem, according to the European Council Directive, a study about the transformation of pig hair from slaughterhouses in potential feedstuffs for animals is carried out. Because keratin is the raw material of the hair, proteolysis by commercial protease enzymes is investigated. Trials are conducted in a batch reactor at different operating conditions. Degrees of hydrolysis and amino acid profiles in the hydrolysates are determined. The essential amino acid index, the protein efficiency ratio (PER) and the biological value (BV) in the hydrolysates show the hair waste could be managed as alternative feedstuffs in swine diets, reducing the amount of biodegradable municipal waste going to landfills. From the results a general kinetic equation for the enzymatic hydrolysis of hair proteins is suggested.

[*] Corresponding author. Tel.: +34 923 294 479; fax: +34 923 294 574. E-mail address: mcm@usal.es (M.C. Márquez).

NOMENCLATURE

aa	Amino acid
Ala	Alanine
Arg	Arginine
Asp	Aspartic acid
B	Base consumption
BV	Biological value
CP	Crude protein
Cys	Cysteine
DH	Degree of hydrolysis
E	Enzyme concentration
EAI	Essential amino acid index
E/S	Enzyme – substrate ratio
Glu	Glutamic acid
Gly	Glycine
h	Number of peptide bonds cleaved during the reaction
h_t	Total number of peptide bonds in a given protein
His	Histidine
Ile	Isoleucine
k_1	Coefficient in equation 6
k_2	Coefficient in equation 6
K_1	Kinetic parameter in equation 4
K_2	Kinetic parameter in equation 4
Leu	Leucine
Lys	Lysine
M_p	Mass of protein
Met	Methionine
N_B	Normality of the base
PER	Predicted protein efficiency ratio
Phe	Phenylalanine
Pro	Proline
S	Substrate concentration
Ser	Serine
t	Time
T	Temperature
Thr	Threonine
Trp	Tryptophan
Tyr	Tyrosine
Val	Valine
α	Average degree of dissociation on α-NH_2 groups

1. Introduction

Hundreds of millions of tonnes of swine waste are generated annually worldwide as a by-product of international pork production (Rhyner et al., 1995). During the late 20th Century, a shift occurred in the industry from more traditional limited-confinement pork production techniques to swine confinement facility production in order to meet public demand for leaner pork and to minimize pork production costs. This shift increased swine waste "production concentration", or the amount of waste generated in a limited geographical area available for agricultural land application, referred to as "land-limited conditions". Therefore, swine waste production may exceed the capacity of the local or regional environment to properly assimilate this waste through agricultural land application and/or discharge to natural aquatic systems (Hatfield and Stewart, 1998).

The biggest producer of pigs in the world is China. In 2004 this country produced about 50% of total pig stock in the world. The second producer of pig production is European Union and the next, USA. Pig production in European Union (EU – 15) in 2004 characterized strong regional differentiation by countries (table 1). The important rank in hog livestock production in EU took two countries: Germany and Spain. Pig stock in these countries were bigger than 17300 thousand heads. England, Belgium, France and Italy had the overage level of pig stock (18700 - 17300 thousand heads). The lowest number of pigs was in Luxemburg and Greece, below 88 thousand heads (Witczak, 2006).

Table 1. Pig stock in European Union (EU - 15) in 2004 (Witczak, 2006).

Country	Thousand heads			
	> 17300	17300-8700	8700-88	< 88
Austria			•	
Belgium			•	
Denmark		•		
Finland			•	
France		•		
Germany	•			
Greece				•
Ireland			•	
Italy		•		
Luxemburg				•
Netherlands		•		
Portugal			•	
Spain	•			
Sweden			•	
United Kingdom			•	

The pig hair constitute up to 1% of total pig weight, reaching more than 34000 t/year as a by-product of the pig industry in countries such as Spain. This excessive material is discarded, being a material of difficult degradation that may become an environmental problem because the enormous volumes of waste cannot be assimilated by natural processes, and therefore require special treatment.

At present, the pig hair is landfilled but the European Council Directive on the Landfill of Wastes 1999/31/EEC (LFD) places a requirement on member states to draw up strategies to ensure that the amount of biodegradable municipal waste deposited at landfill progressively reduces over a 15-year period to only 35% of the total amount produced in 1995.

The feeding of biological waste to livestock is an accepted practice and has arisen because of the necessity to reduce costs both in terms of waste disposal and meat production from livestock (El-Boushi and van der Poel, 2000). Rendered feathers, hair, skin, hooves, blood, and intestines can be found in feed, often under catch-all categories like "animal protein products."

Considering that feeds correspond to a significant part of the pig production cost, it is relevant to investigate alternatives to the most expensive dietary ingredients. Since hair has elevated keratin content, the use of this protein source should be considered.

Although pig hair has potential to be an important protein source for animal production, it presents low digestibility. In order to increase the digestibility of the pig hair, a treatment is necessary and the enzymatic hydrolysis of keratin may be a viable alternative.

The application of hydrolytic enzymes seems an attractive means of obtaining better functional properties of hair proteins without deterioration in nutritional value. The degradation of the protein into peptides generally renders the product more soluble. Moreover, other nutritional properties such as amino acid profile, biological value or protein efficiency are also affected and, by choosing the proper set of hydrolysis parameters, specific improvements in these properties can be achieved.

2. Enzymatic Hydrolysis

Proteins are long chains of amino acids linked by peptide bonds. During hydrolysis these peptide bonds are broken. Depending upon the type of enzyme used and processing conditions, smaller chains of amino acids and different amino acid sequences are produced by the process. These hydrolyzed protein chains determine the nutritional and functional characteristics of protein hydrolysates.

In any quantitative work on protein hydrolysis it is necessary to have a measure for the extent of the hydrolytic degradation. It should be evident that the number of peptide bonds cleaved during the reaction, h, is the entity which most closely reflects the catalytic action of proteases.

Then, the degree of hydrolysis, DH, defined as the percentage of peptide bonds cleaved, can be used as measure of the extent of the hydrolysis reaction.

If we denote the total number of peptide bonds in a given protein, h_t (meqv/g), then it follows by the definition of DH and h that:

$$DH = \frac{h}{h_t} 100 \qquad (1)$$

The total number of peptide bonds per weight unit in a protein can be calculated from its amino composition by summing up the mmols of each individual amino acid per gram (Nxf_N). For pig hair, h_t calculated from data shown in table 3 is 7.51 meqv/g.

The number of peptide bonds cleaved during a hydrolysis process h, is called the hydrolysis equivalents and is expressed as equivalents per kg protein (or meqv/g protein). The hydrolysis equivalents can be assayed by determining the increase in the free amino groups (or carboxyl groups). Several methods exist for this, of which the most simple is the pH-stat technique.

The cleavage of peptide bonds generates a pH drop as a result of the liberation of free carboxyl (-COOH) residues (Figure 1). This phenomenon can be used to estimate DH if the enzymatic hydrolysis is pH-controlled using a pH-stat technique.

Briefly, the method consists in hydrolysis at neutral o weakly alkaline pH, using the base consumption for monitoring the process. The quantity of base necessary to maintain pH constant is proportional to the degree of hydrolysis.

Usually, NaOH is recommended but other bases like KOH, $Ca(OH)_2$ or NH_3 may be preferable in special cases. Owing to high buffer capacity of the system, regulation is simple and an inexpensive on-off regulation is sufficient under industrial conditions use.

Figure 2 shows an experimental installation where samples are prepared in the batch reactor by suspending a given amount of substrate in hot water (water temperature = hydrolysis temperature). Suspensions are adjusted to chosen pH before the addition of the enzyme. When the hydrolysis is finished, the enzyme can be inactivated by heating at 100°C for 10 minutes.

1. Opening of the peptide bonds:

$$P_1 - CO - NH - CHP_2 + H_2O \xrightarrow{Enzyme} P_1 - COOH + NH_2 - P_2$$

2. Proton exchange:

$$P_1 - COOH + NH_2 - P_2 \longrightarrow P_1 - COO^- + NH_3^+ - P_2$$

3. Titration of amino group:

$$NH_3^+ - P_2 + OH^- \rightleftharpoons NH_2 - P_2 + H_2O$$

$pK_{25°C} = 7.7$, $pK_{50°C} = 7.1$

Figure 1. The enzyme catalyzed hydrolysis of a peptide bond and the functioning of the pH-stat.

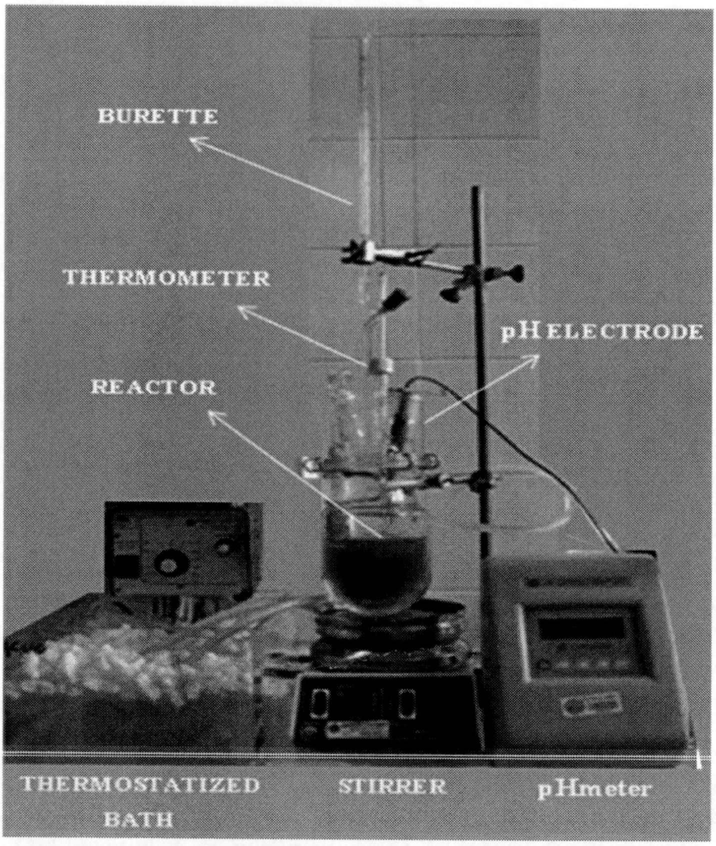

Figure 2. Experimental installation for enzymatic hydrolysis using pH-stat technique.

The pH at which the proteolysis should be carried out depends primarily on the pH-optimum of the enzyme used. However, for a given enzyme, the reaction pH may vary somewhat depending on the type of the protein.

The relationship between DH and base consumption is given by the following equation:

$$DH = \frac{B\, N_B}{M_P} \frac{1}{\alpha} \frac{1}{h_t} 100 \qquad (2)$$

where B is the base consumption, N_B is the normality of the base, M_p is the mass of protein (N x 7.51) and α is the average degree of dissociation on α-NH_2 groups:

$$\alpha = \frac{10^{pH - pK}}{1 + 10^{pH - pK}} \qquad (3)$$

pK is the average pK value of the α-amino groups liberated during the hydrolysis. It depends on the temperature but is fairly independent of the substrate as such.

Table 2. Nutritional composition of the pig hair.

	Moisture	Crude Protein	Fat	Ash
Mean	65.71	32.31	0.84	1.14
Standar deviation	3.35	3.21	0.25	0.26
Number of samples	53	53	17	17

3. HYDROLYSATES

The final hydrolysate depends on:

- The substrate.
- The enzyme.
- The hydrolysis conditions.

3.1. Substrate

Hydrolysis depends upon the nature of the substrate, and rather small structural differences may influence the course of the reaction.

The pig hair, used as substrate, shows two main components: water and protein (table 2). The high content in water is due to the pig treatment before obtaining the hair: pigs are sprayed with boiling water for separating the hair from the skin. In dry basis hair presents a high content in protein (94.21 %), also containing ash (3.32 %) and fat (2.47 %) determined in accordance to specific methods (Madrid et al., 1995). Falla (1994) and Moran et al. (1967) showed similar values for the composition of the hair.

Hair is formed from thin fibres composed mainly of the keratin proteins. Keratins are a family of fibrous structural proteins; they are tough and insoluble (Lehninger, 1988; Rawn, 1989). Hair is covered with overlapping scales, and these scales form a layer called the cuticle. Beneath the cuticle is the cortex, which consists of a rod of five-ply keratin polypeptide chains embedded in a keratin-based protein matrix. Five individual protein molecules about 1 nm across, wind round one another in a helix arrangement to form a protofibril rod of about 2.5 nm in diameter (Figure 3). Five of these rods then twist together to form a microfibril. The microfibrils are bunched into cables, and there is a certain amount of crossing and tangling of microfibrils between cables to fasten the microfibrils together (macrofibrils). It is the twisting of long-chain keratin protein molecules into tightly packed cables that gives hair tensile strength.

The properties which make structural proteins like keratins useful depend on their supermolecular aggregation. These depend on the properties of the individual polypeptide strands, which depend in turn on their amino acid composition and sequence. The α-helix and β-sheet motifs, and disulfide bridges, are crucial to the conformations of globular, functional proteins like enzymes, many of which operate semi-independently, but they take on a completely dominant role in the architecture and aggregation of keratins. Keratins contain a high proportion of the smallest of the 20 amino acids, glycine, whose "side group" is a single hydrogen atom; also the next smallest, alanine, with a small and uncharged methyl group. In the case of β-sheets, this allows sterically-unhindered hydrogen bonding between the amino and carboxyl groups of peptide bonds on adjacent protein chains, facilitating their close alignment and strong binding. Fibrous keratin molecules can twist around each other to form helical intermediate filaments. In addition to intra- and intermolecular hydrogen bonds, keratins have large amounts of the sulphur-containing amino acid cysteine, required for the disulfide bridges that confer additional strength and rigidity by permanent, thermally-stable crosslinking—a role sulphur bridges also play in vulcanized rubber. Extensive disulfide bonding contributes to the insolubility of keratins, except in dissociating or reducing agents such as urea.

The content in amino acids in the pig hair appears in table 3 where glutamic acid, arginine, cysteine, leucine and aspartic acid are prevailing. These values agree with values indicated by Moran et al. (1967). Total amino acids are determined from acid hydrolysis with HCl 6 N (1 % phenol) at 110°C for 65 hours, followed of precolumn derivatization with AccQFluorine reactive (6-aminoquinolil-N-hydroxisuccinimidil carbamate) and separation and quatification with HPLC in reverse phase with a column C18 (WATERS 600 model with UV detector WATERS 2487).

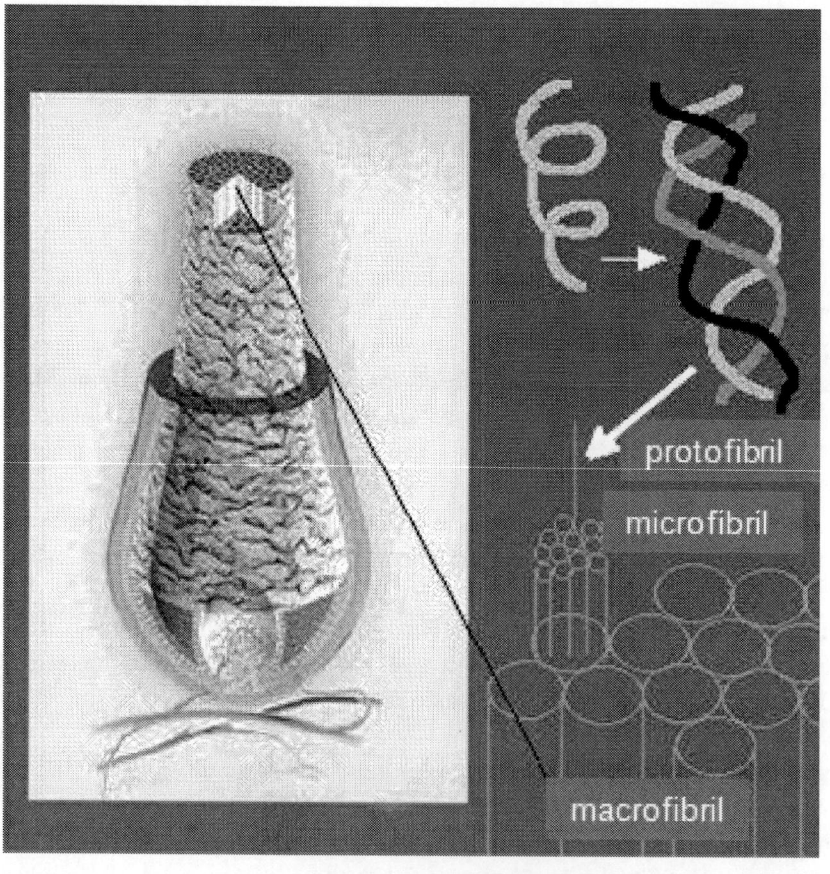

Figure 3. Structure of the hair.

Table 3. Amino acid content of the pig hair.

Amino acid	mg_{aa} / g_{CP}	Amino acid	mg_{aa} / g_{CP}
Ala	42.04	Lys	38.90
Arg	96.22	Met	6.48
Asp	71.86	Phe	31.17
Cys	87.46	Pro	59.32
Glu	155.10	Ser	48.59
Gly	44.66	Thr	48.25
His	13.16	Tyr	34.20
Ile	37.82	Val	58.87
Leu	82.0	Total	956.72

Table 4. Characteristic of some keratinases isolated from different microorganisms.

Microorganism	Source	Type	T °C	pH	M kDa	Reference
Bacillus sp.	Deer skin	Serine protease	40	8-9	40	Suntornsuk & Suntornsuk, 2003
Flavobacterium sp.		Metalloprotease	37			Riffel & Brandelli, 2002
B. licheniformis	Feathers	Serine protease	50	7.5	33	Lin et. al., 1992
Nesternkonia sp.	Feathers	Serine protease	70	7.5-11.5	24	Gessesse et al., 2003
B. pseudofirmus	Feathers	Serine protease	50	11	23	Gessesse et al., 2003
Aspergillus oryzae	Feathers		50	8	60	Farag & Hassan, 2004
Xanthomonas maltophilia	Feathers	Serine protease	60	9	36	Toni et al., 2002
Doratomyces microsporus	Soil	Serine protease	50	8-9	30	Friedrich et al., 2005

The high protein content and the amino acid distribution indicate the nutritional quality of the hair and justifie its use as food substrate.

Keratins present high mechanical stability and resistance to biochemical degradation, qualities dues, to a great extent, to the presence of a considerable amount of disulfide bridges. These connections form a structure that makes difficult the degradation of the most compact areas of the keratinised substrates.

Nevertheless, in the nature an accumulation of keratin does not take place. Several pathogens are known that live in the parts of the men and the animals that contain keratin, such as hair and nails, causing infections. These facts confirm the existence of diverse microorganisms able to degrade the keratin. Several species of fungi and bacteria produce proteolytic enzymes, keratinases that hydrolyze keratin (Onifade et al., 1998). Different keratinases have been isolated from species of fungi both pathogens and no pathogens

(Friedrich et al., 2005; Friedrich and Kern, 2003; Gradišar et al., 2000; Vignardet et al., 1999). Between the bacteria able to produce these enzymes we find Bacillus (Evans et al., 2000; Gessesse, et al., 2003; Lin et al., 1992; Suntornsuk and Suntornsuk, 2003; Varela et al., 1997), Stenotrophomonas (Yamamura et al., 2002 a; Yamamura et al., 2002 b) Xathomonas (Toni et al., 2002), Aspergillus (Farag and Hassan, 2004), Fervidobacerium (Kim et al., 2004).

These microorganisms are extended in the nature although its development is favoured in keratinised substrates. Many of them have been isolated of wastes of the poultry industry, as the feathers, one of the more abundant keratinised wastes coming from animal industry.

Table 4 shows the fundamental characteristics of some keratinases isolated from different microorganisms. These enzymes are not developed commercially.

3.2. Enzymes: Proteases

Since keratinases are not yet developed commercially, the proteases are the enzymes available for the hydrolysis of keratin.

Proteases catalyse the hydrolytic degradation of the proteins generating peptides and/or amino acids:

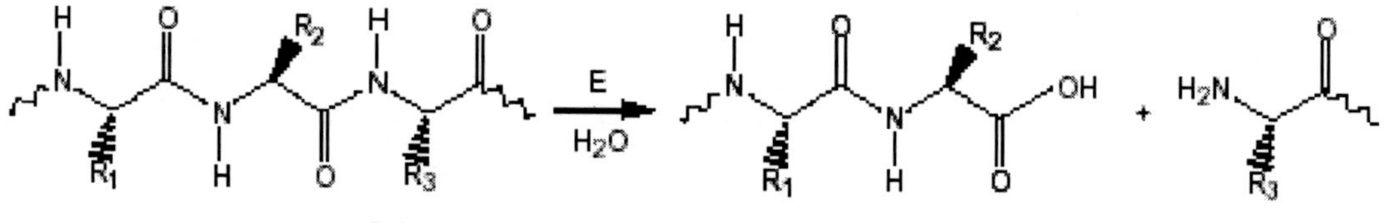

R1, R2, R3 = side (lateral) chains, E = enzyme

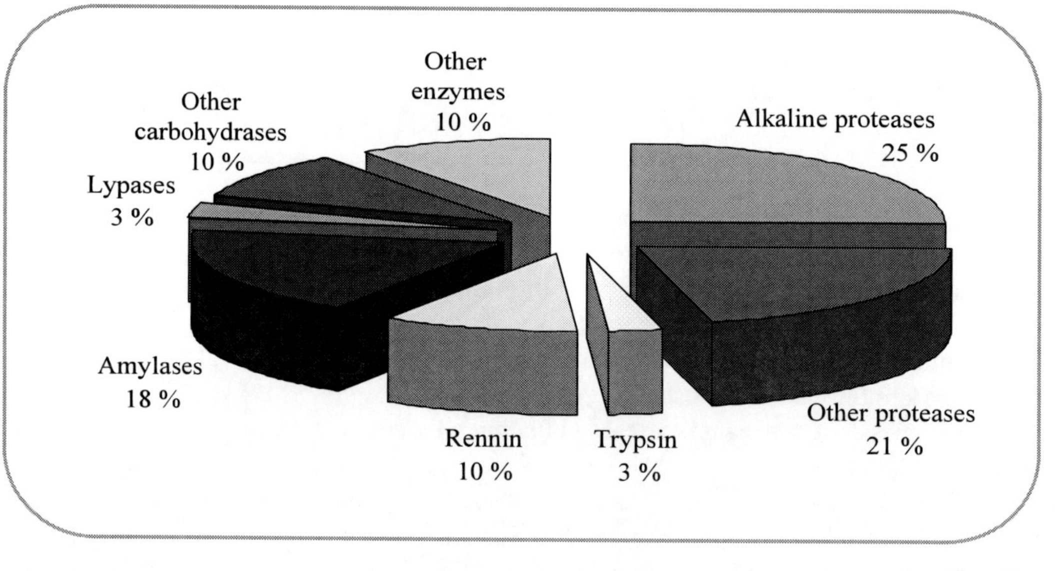

Figure 4. Total enzyme market (Rao et al., 1998).

The number of proteases available for enzymatic hydrolysis is elevated, constituting one of the most important enzyme groups. The 70% of industrial enzymes are hydrolases, an about the 60 % of the world-wide enzyme market are proteases (figure 4) (Rao et al., 1998).

Proteases are classified according to their source (animal, plant, microbial), their catalytic action (endopeptidades or exopeptidases), and the nature of the catalytic side (serine proteases, cysteine proteases, metalloproteases, aspartic proteases

Source

- Proteases from animal source. Trypsin, chymotrypsin, pepsin and rennin are the most known proteases from animal source. They can be obtained with a high degree of purity in great amounts although their yield depends on the number of slaughtered animals.
- Proteases from vegetal source. The papain is the classical example of this type of enzymes. The main disadvantage for availability of proteases from vegetal source is the difficulty for growing: it is necessary a considerable time, availability of land and climatologic conditions adapted.
- Proteases of bacterial origin. The interest by this type of enzymes is increased considerably in the last years representing 40% of the world-wide market (Gupta et al., 2002). In addition, the microorganisms are an excellent enzyme source since they present a fast growth in a small space, ample biochemical diversity and are susceptible of being modified genetically. Different microorganisms produce these enzymes:
  - Bacteria: There are two types of bacterial proteases with commercial significance:
    - Alkaline proteases: They have their maximum activity for pH from 7 to 11.
    - Neutral proteases: hey are active in a range of pH from 5 to 8.
  - Fungi: They produce more variety of enzymes than the bacteria. Proteases from fungi are active in a range of pH from 4 to 11. They are classified in acid, neutral and basic proteases based on pH to which they present an optimal activity.
  - Virus: The importance of this kind of proteases is due to their participation in the synthesis of proteins causing cancer or AIDS. They are endopeptidases (Rao et al., 1998).

The common industrial proteases generally used in the food protein hydrolysis are from bacteria and fungi source (Adler-Nissen, 1986).

Catalytic Action

- Endopeptidases: They degrade the peptide bounds into the chain of protein. The presence of free amino or carboxylic groups has a negative effect on their enzymatic activity. Most of enzymes used industrially contain endopeptidases.
- Exopeptidases: They attack the terminal amino acids. They can attack the terminal C (carboxypeptidases) or N (aminopeptidases).

In food protein hydrolysis, endopeptidases are always used, but occasionally the endopeptidases are combined with exopeptidases in order to achieve a more thorough degradation.

Nature of the Catalytic Side

There are four classes of catalytic sites for endopeptidases of interest in food protein hydrolysis:

- Serine proteases: They have a serine group in the catalytic centre and they present a maximum activity at alkaline pH. The serine proteases are able to degrade substrates of molecular weight from 18 to 35 kDa and iso-electric points between 4 and 6. They are obtained from different fungi and bacteria (S. Cerevisae, Streptomyces, Flavobacterium...), but subtilisins, obtained from Bacillus, are the most known. Two types different from these alkaline proteases have been identified: subtilisin Carlsberg produced by Bacillus licheniformis and subtilisin Novo or bacterial protease Nagase (BPN') obtained from Bacillus amyloliquefaciens.
- Cysteine proteases: They are closely related to serine proteases having –SH instead of –OH in the catalytic centre. Their maximum activity is found at more neutral pH values than serine proteases.
- Metalloproteases: They contain an essential metal atom, usually Zn, and have optimum pH around neutral.
- Aspartic proteases: They have a carboxyl group from aspartic acid in the centre active. They have widespread and have maximum catalytic activity at acid pH.

In order to compare the influence of the type of the enzyme, four commercial food grade enzyme preparations can be assayed:

Enzyme	Source	Catalytic action	Catalytic side	T °C	pH	M kDa	Reference
Flavouryme	*Aspergillus oryzae*	Endo+ exopeptidase	Amino peptidase	50	5.0-7.0	40	Novozymes, 2003a
Neutrase	*Bacillus amyloliquefaciens*	Endo-peptidase	Metallo-protease	45-55	5.5-7.5		Novozymes, 2003b
Protamex	*Bacillus protease*	Endo-peptidase	Metallo-protease +Serine protease	35-60	5.5-7.5	33	Novozymes, 2003c
Alcalase	*Bacillus licheniformis*	Endo-peptidase	Serine protease	40-70	6.5-8.5	24	Novozymes, 2002

3.3. The Hydrolysis Conditions

Given a particular enzyme and a particular substrate, any hydrolysis experiment involves at least five independent variables in the conventional sense. These are time, temperature, pH, substrate concentration and enzyme – substrate ratio.

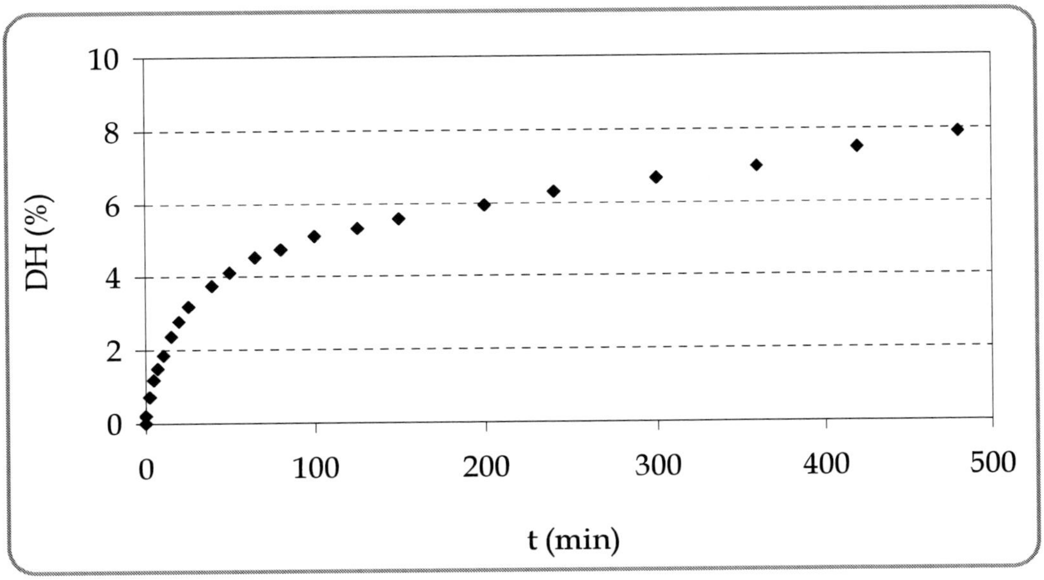

Figure 5. Influence of the hydrolysis time on the degree of hydrolysis (S = 10 g/l, E = 0.5 g/l, T = 50 °C, pH = 8).

The influence of the hydrolysis time, *t*, on the degree of hydrolysis can be shown from a time-course experiment of eight hours long. It is apparent from Figure 5 that the overall rate reaction is decreased by an increase in time tending to a boundary value from 200 minutes. This suggests the use of four hours as the time of reaction rather higher hydrolysis times.

Temperature, *T*, and pH are selected in function of the maxima activity and stability of the enzymes. According to the literature (Novozymes, 2002 and 2003), a temperature suitable for the chosen proteases is 50°C, but pH can be 7 or 8 depending on the enzyme. However, experimental data show pH 8 as the best option in order to obtain the highest degree of hydrolysis for keratin (table 5). This discrepancy is consequence of the substrate used for the assay: haemoglobin in the literature and keratin for the hydrolysis of the pig hair.

Then a common temperature and pH can be fixed for the comparative analysis: 50°C and 8, respectively (figure 6).

Table 5. Hydrolysis degree for keratin: effect of pH for different enzymes.

Enzyme	DH_{240min}	
	pH = 7	pH = 8
Alcalase	5.15	6.44
Flavouryme	4.63	4.80
Neutrase	5.36	5.41
Protamex	5.14	5.41

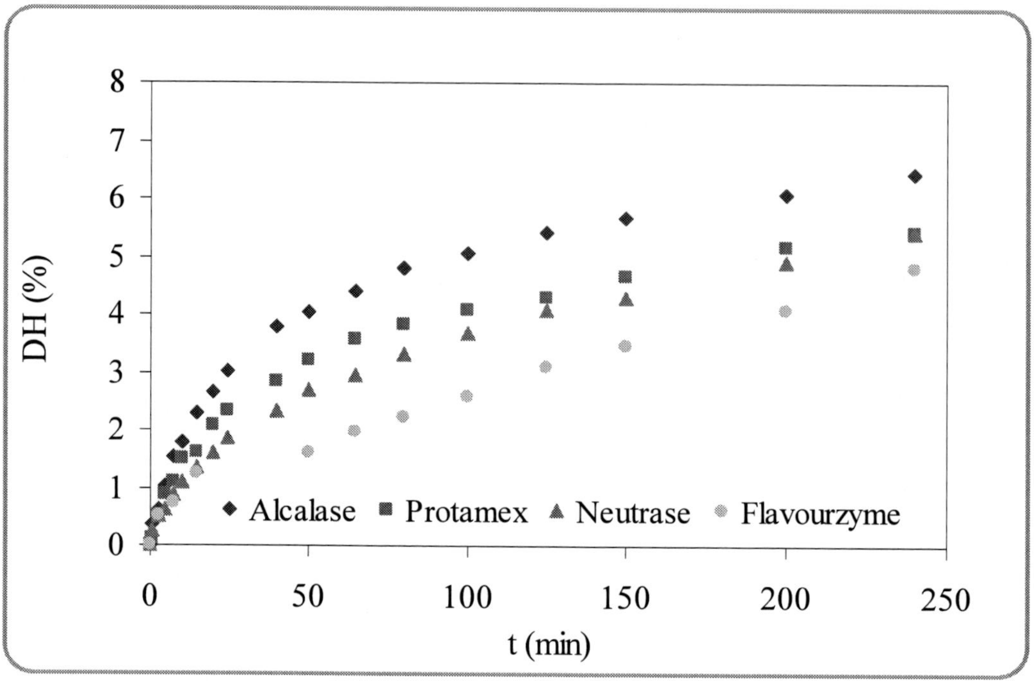

Figure 6. Influence of the type of the enzyme (S = 10 g/l, E = 0.5 g/l, T = 50 °C, pH = 8).

The particular shape of the hydrolysis curves obtained for Flavourzyme suggests the different behaviour of exopeptidases and their lower activity in the hydrolysis of the pig hair.

In the hydrolysis experiments for the three endopeptidases (Alcalase, Neutrase and Protamex), Alcalase shows a somewhat better performance than Neutrase that may be ascribed to a significant and critical role of the serine proteases opposite to the metallo-proteases for the keratine hydrolysis. This is demonstrated on Figure 6 where Protamex (a preparation of serine protease and metallo-protease) is clearly inferior to Alcalase but superior to Neutrase if a high DH is desired.

From the above discussion it seems justified to assume that the serine endopeptidases are the most suitable proteases for the hydrolysis of the keratin coming from the pig hair. Kida et al. (1995) also conclude the same theory when they study the hydrolysis of keratin from hoofs, horns, feathers and wool: serine endopeptidases from Bacillus subtilis (source for Alcalase) allow to obtain the highest hydrolysis degree.

Table 6. Cost of the used enzymes.

Enzyme	Activity (AU/g)	Cost (€/kg)	Cost (€/AU)
Alcalase	2.4	30	0.012
Flavourzyme	0.2	70	0.350
Neutrase	0.8	19	0.024
Protamex	1.5	50	0.033

The fixed amount of enzyme (0.5 g) used for the assays but the different activity of the enzymes (table 6) seems to suggest a different result when the same enzyme activity is used in the hydrolysis. However the higher degree of hydrolysis reached with the serine endopeptidases and the cost of the enzymes confirms Alcalase as the best option for the hydrolysis of the keratin with proteases. Then only Alcalase has been used for the posterior studies.

The substrate concentration, *S*, is usually given as weight of protein (= Nxf_N dry basis) in the total volume of the mixture of reaction at the beginning of the reaction.

For insoluble substrates like hair the relation between the initial hydrolysis reaction rate and the substrate concentration is curvilinear if the enzyme concentration is low, so that the substrate surface is only partially covered by adsorbed enzyme (Figure 7). The reaction rate will level off at high substrate concentrations. For pig hair-Acalase this seems to be achieved at concentrations over 15 g/l.

The range of substrate concentration is limited by viscosity problems: for hair, experiments carried out at S > 15 g/l do not give reproducible curves of hydrolysis.

On the basis that the hair has some content in fat and a high content in water (table 2), it seems reasonable to think that a pretreatment where moisture and fat are removed (figure 8) is favourable for the solubilisation of the keratin. However, the removal of fat by washing of hair (at room temperature for 20 minutes) does not modify the DH reached probably because the low amount of fat does not cover the surface of the hair and does not stop the enzymatic attack. A heat treatment (at 65°C for 24 hours) for removal the water in hair is clearly unfavourable. Although an increase of the DH is usually obtained from a heat treatment of proteins as a result of an unfolding of the original native structure without the disruption of the peptide chain, the behaviour of the myofibrillar proteins like keratin is different: the cross-linking produced from the heat treatment reduces the possibility of an enzymatic attack (Cheftel et al., 1993).

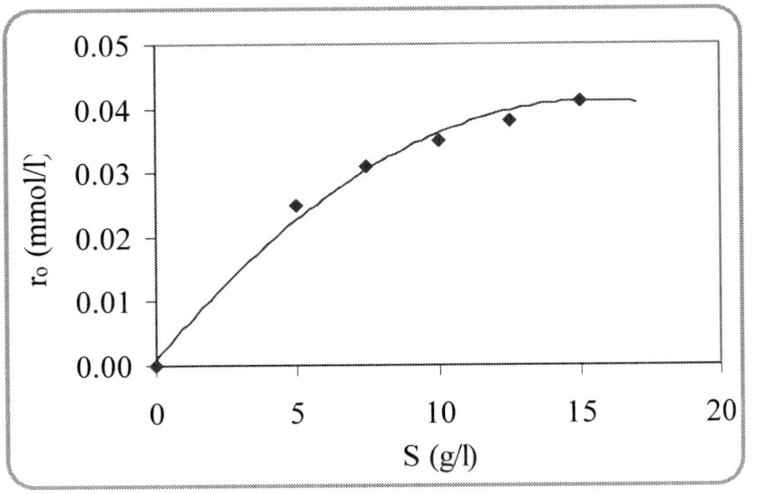

Figure 7. The initial reaction rate as a function of the substrate concentration for an insoluble substrate: hydrolysis of pig hair with Alcalase (E = 0.5 g/l, pH = 8, T = 50°C).

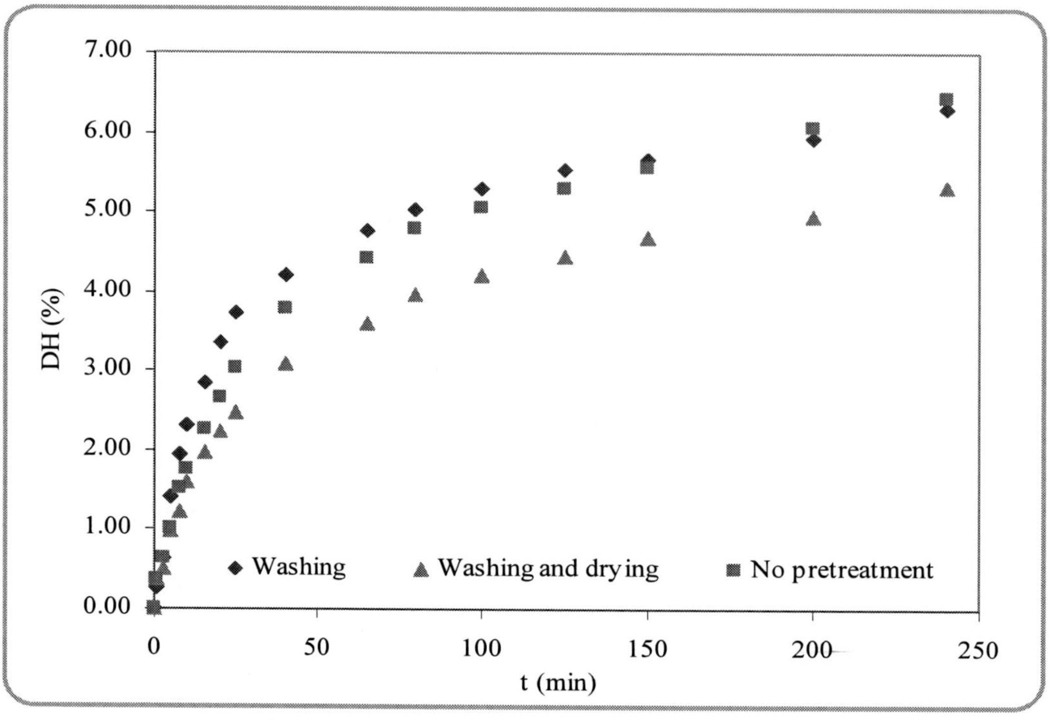

Figure 8. Hydrolysis curves for Alcalase, effect of pretreatment (S = 10 g/l, E = 0.5 g/l, pH = 8, T = 50°C).

The enzyme-substrate ratio (*E/S*), i.e. the concentration of enzyme relative to the concentration of substrate, is usually more characteristic of the reaction rate than the enzyme concentration *per se*. This is because protein hydrolysis processes are often carried out at relatively high substrate concentration where substrate is prevailing (Adler-Nissen, 1986). Thereby the conversion rate, expressed e.g. as the increase in DH per unit of time is proportional to the mass ratio of enzyme to substrate. In addition, this ratio also forms the basis for assessing the contribution to the variable cost from using enzyme. Consequently, the enzyme-substrate ratio is preferred way of expressing the amount of enzyme used in a protein hydrolysis process.

Below 4-6 %, however, the substrate concentration is presumably too low to ensure substrate saturation (AdlerNissen, 1986). The highest substrate concentration possible in the hydrolysis of the pig hair is 1.5 % and, as showed in figure 7, the substrate saturation is not reached. Nevertheless, for a given value of E/S, the hydrolysis time to reach a preset DH-value is fairly independent of S within the range of concentration analysed.

In order to limit the amount of enzyme used, the enzymatic reaction is examined varying the enzyme – substrate ratio (E/S) from 1/5 to 1/100. As shown in figure 9, the hydrolysis time to reach a preset DH-value is inversely proportional to E/S.

When enzyme-substrate concentration ratio versus DH is plotted, a linear relationship is observed (Figure 10). The correlations coefficients obtained for Alcalase at different enzyme concentrations are R^2 = 0.99. From this relationship, the exact ratio of enzyme/substrate necessary to hydrolyse keratin to a required DH, from 60 min to 240 min, could be calculated.

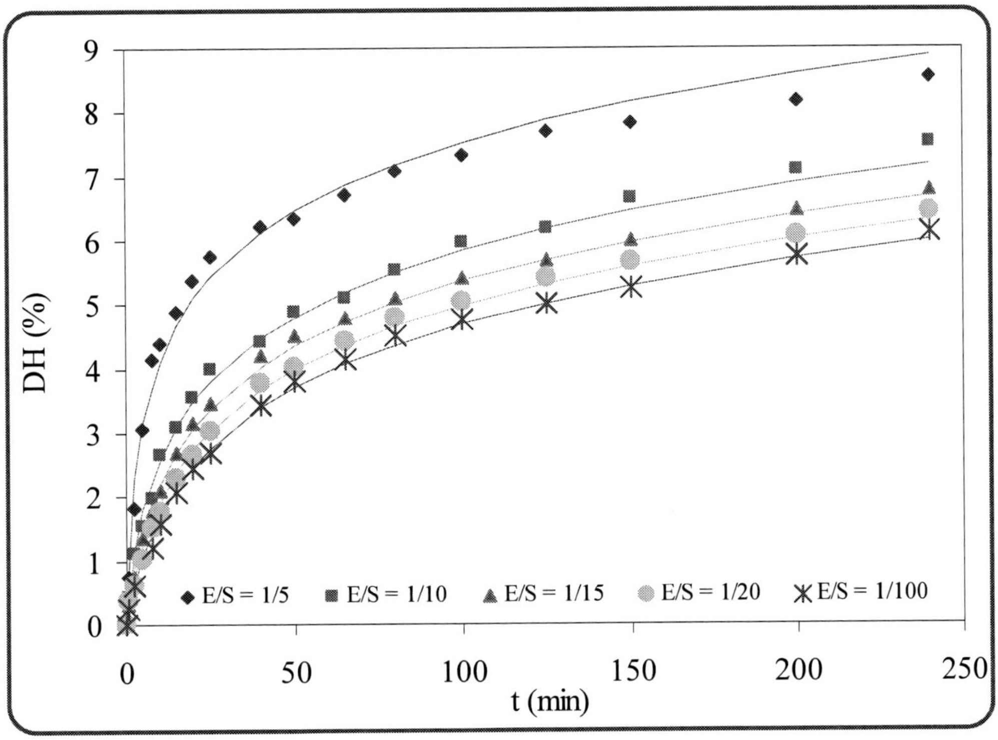

Figure 9. Hydrolysis curves for Alcalase, effect of enzyme/substrate ratio (pH = 8, T = 50°C) (Solid lines are calculated with kinetic model proposed).

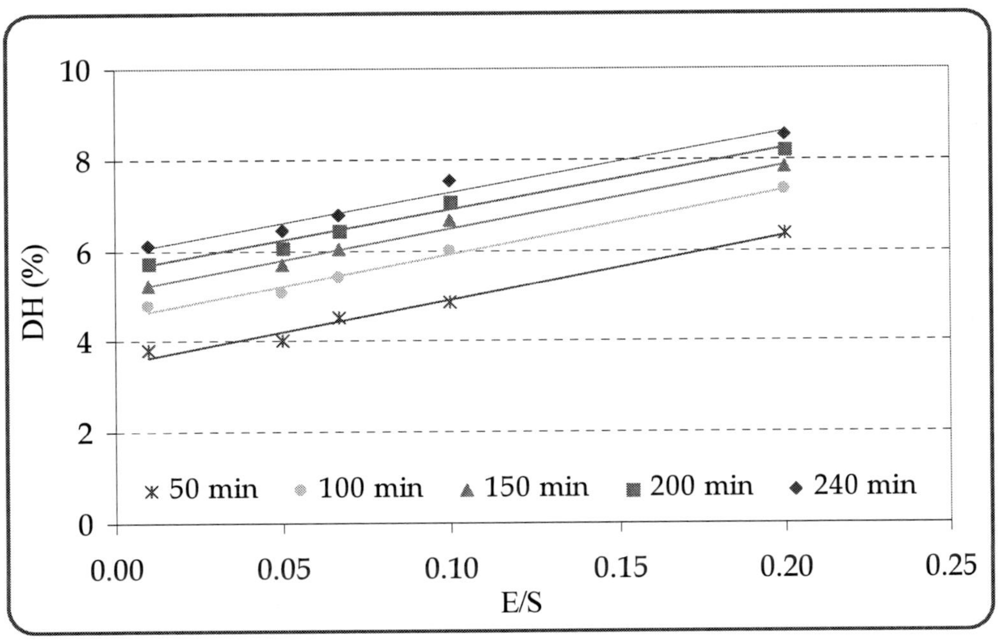

Figure 10. Linearity between E/S and DH.

4. Kinetic Model

To model the hydrolysis curves of keratin with Alcalase, an empirical equation applicable to the enzymatic hydrolysis of proteins can be used (González et al., 1994; Márquez and Fernández, 1993; Márquez and Vázquez, 1999; Wei and Zhimin, 2006):

$$\frac{dDH}{dt} = K_1 \exp(-K_2 DH) \tag{4}$$

where K_1 and K_2 are kinetic parameters with specific values for each operating condition studied. They are determined from nonlinear regression analysis using the "Statgraphics" program. The theoretical curves calculated from the equation (4) are shown as solid lines in Figure 9.

Parameter K_2 does not exhibit any trend when the enzyme and substrate concentrations change and its values lie within a very small range. Thus, K_2 can be considered independent of these variables and constant in the operating conditions with an average value of 0.65±0.04. However, K_1 values calculated from this average value show a clear dependence on the enzyme-substrate ratio (figure 11) according to showed in figure 10.

At a pH of 8 and a temperature of 50°C, the equation (4) can be written as:

$$\frac{dDH}{dt} = 9.36\frac{E}{S}\exp(-0.65\,DH) \tag{5}$$

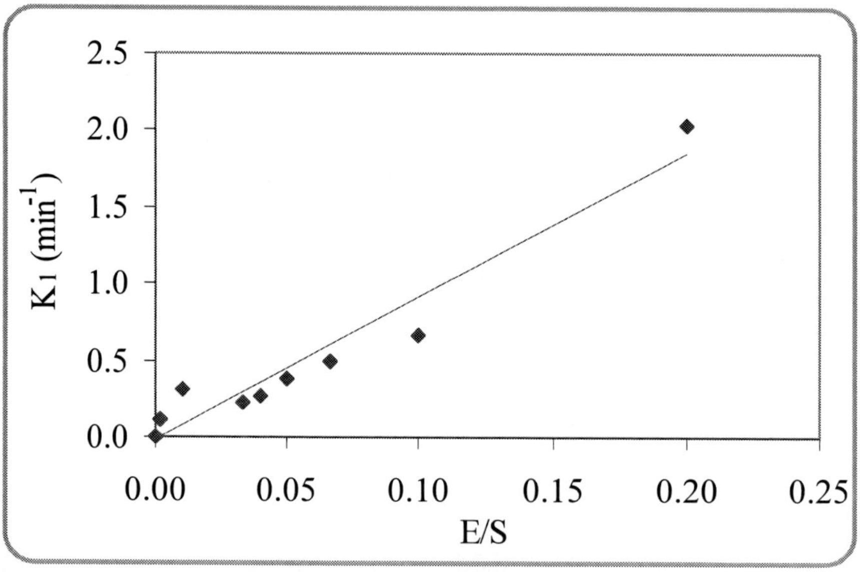

Figure 11. Variation of kinetic parameter K_1 for different E/S values.

5. Aminoacids Distribution

The amino acid composition is the most important factor in defining food protein quality (Sindayikengera and Xia, 2006). To measure the amount of amino acids present in the hydrolysates, they are analysed as shown for the determination of amino acids in the pig hair.

Hydrolysates present a wide distribution of amino acids as showed in figure 12 where distribution is plotted for three E/S values. They show a high content of Lysine, Leucine and Glutamic acid; a high concentration of Methionine, Tyrosine, Valine and Alanine, and a relatively low content of Arginine, Proline and Cysteine. The amino acids amount decreases on decrease E/S.

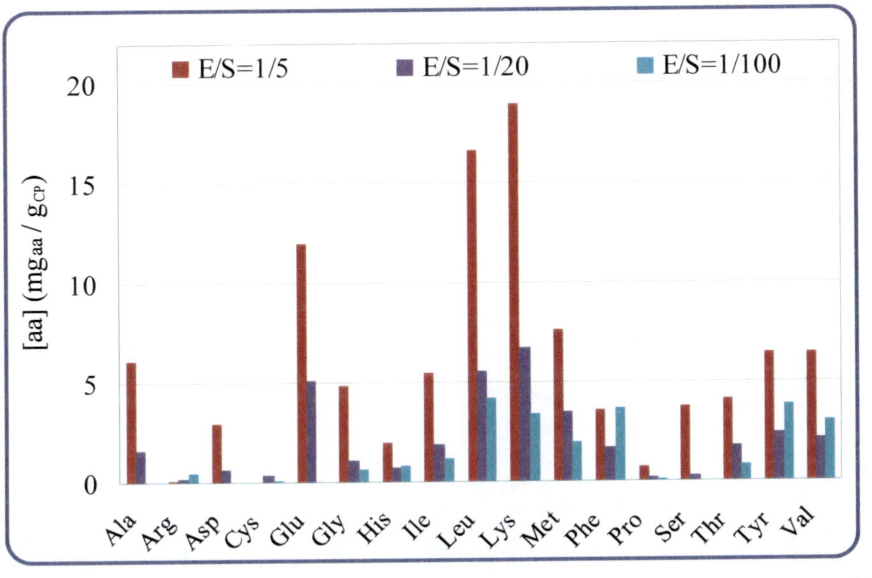

Figure 12. Amino acid profile in hydrolysates of keratin with Alcalase.

Table 7. Correlations between the amino acids concentration and E/S.

Amino acid	$k_1 \cdot 10^2$	$k_2 \cdot 10^1$	R^2
Ala	5.77	-6.10	0.990
Asp	2.65	-1.41	0.999
Gly	3.98	1.91	0.986
His	1.21	5.68	0.914
Ile	3.95	8.75	0.998
Leu	11.77	29.30	0.989
Lys	14.18	26.13	1
Met	5.01	18.84	0.996
Pro	0.68	-0.63	0.913
Ser	4.01	-4.56	0.981
Thr	2.96	7.55	0.992

When enzyme-substrate ratio versus total or individual amino acid concentration in hydrolysate is plotted, a linear relationship is also observed. Figure 13 shows this linearity for several amino acids. The slope (k_1), the intercept (k_2) and the correlation coefficient (R^2) obtained from the linear fitting are summarized in table 7 for the amino acids. From this relationship, it is possible to determine the amount of each amino acid obtained for a determined E/S ratio.

$$[aa] = k_1 \frac{E}{S} + k_2 \tag{6}$$

The total amount of amino acids in the hydrolysates shows a recovery of the 5-10 % of the total amount of amino acids in the pig hair according to the degree of hydrolysis.

Although only a 10 % of the total amino acids in the keratin is recovered in the hydrolysates, it is necessary to emphasize a recovery of more than 70 % of methionine and practically 50 % of lysine. Amino acids such as the phenylalanine, leucine, isoleucine, valine, tyrosine, alanine and histidine are recovered in a percentage from 12 to 25 %, being the recovery of the non essential amino acid lower than 10 %.

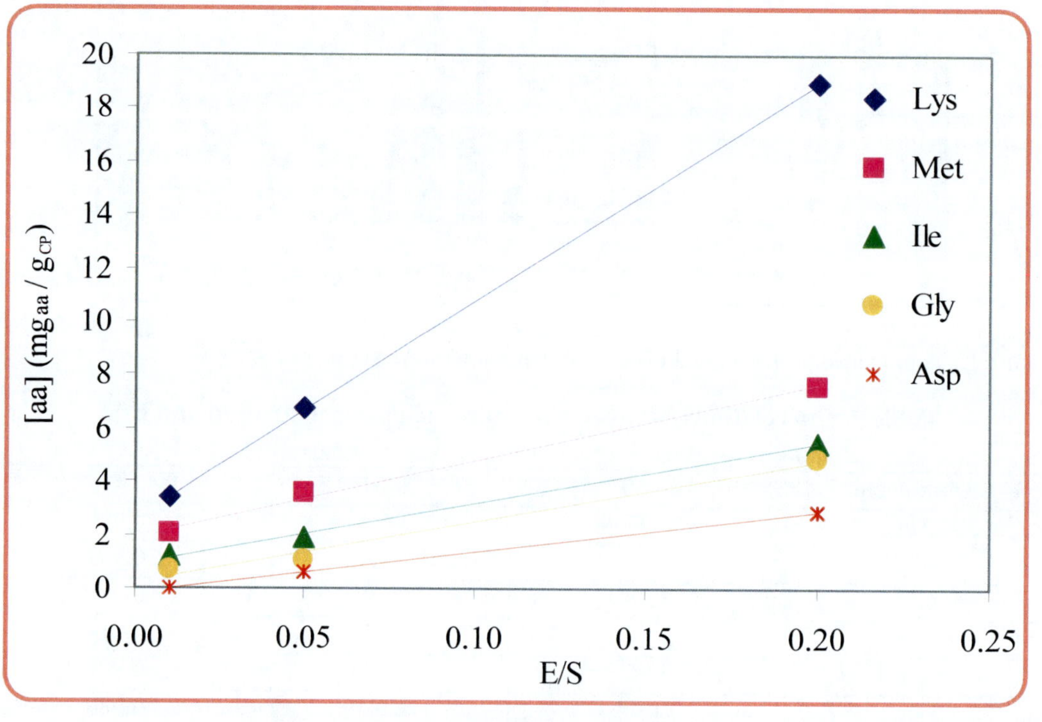

Figure 13. Linearity between E/S and amino acid concentration.

6. Evaluation of Nutritional Parameters of Hair Hydrolysate

The amino acid composition of samples can be used for determination of the nutritional value of hair hydrolysates (Chavan, 2001). Nutritional parameters such as the essential amino acid index, the biological value and the protein efficiency ratio are calculated from the amino acid composition.

An *essential amino acid* is an amino acid that cannot be synthesized *de novo* by the animal organism, and therefore must be supplied in the diet. Foodstuffs that lack essential amino acids are poor sources of protein equivalents.

Essential amino acid index (*EAI*) is a measure of the essential amino acids content in a food. EAI is calculated from the relationship between the essential amino acid concentration (mg/g) and the total amino acid concentration (mg/g):

$$EAI = \frac{[aa]_E}{[aa]_T} = \frac{[Arg]+[His]+[Ile]+[Leu]+[Lys]+[Met]+[Thr]+[Val]}{\begin{array}{c}[Ala]+[Arg]+[Asp]+[Cys]+[Glu]+[Gly]+[His]+[Ile]+[Leu]+[Lys]+\\ [Met]+[Phe]+[\mathrm{Pro}]+[Ser]+[Thr]+]Tr]+[Trp]+[Val]\end{array}} 100 \quad (7)$$

Values obtained for the essential amino acid index in the hidrolysates (Table 9) are similar to the values for other protein sources showed in the literature (Oyarekua and Elenyimi 2004).

The *biological value* is a measure of the proportion of absorbed protein from a food which becomes incorporated into the proteins of the organism's body. It summarizes how readily the broken down protein can be used in protein synthesis in the cells of the organism.

The predicted biological value (*BV*) is calculated from the regression equation developed by Morup and Olesen (1976) and used by several authors (Chavan et al., 2001; Grazziotin et al., 2006):

$$BV = 10^{2.15} q_{Lys}^{0.41} q_{Phe+Tyr}^{0.60} q_{Met+Cs}^{0.77} q_{Thr}^{0.40} q_{Trp}^{0.21} \quad (8)$$

where $q = aa_i$ (sample)/aa_i (reference) for aa_i sample $\leq aa_i$ reference, $q = aa_i$ (reference)/ aa_i (sample) for aa_i sample $\geq aa_i$ reference, aa_i =mg of the amino acid per g of total essential amino acid.

The concentration of the reference amino acid is shown in table 8.

The biological value for the hydrolysates is similar to BV for hydrolysates from substrates with high content in keratin like feathers (Grazziotinet al., 2006).

Protein efficiency ratio is based on the weight gain of a test subject divided by its intake of a particular food protein during the test period. Because PER is a time-consuming bioassay procedure, but is a widely used method for evaluating the quality of protein in food, attempts have been made to predict PER by rapid laboratory techniques.

Table 8. Essential amino acids profile for chickens and pigs.

Essential amino acids	Reference (mg_{aa}/g_{CP})	
	Chickens[1]	Pigs[2]
Arg	54.3	15.5
His	15.2	13.8
Ile	34.8	23.5
Leu	52.2	18.0
Lys	47.8	2.8
Met	21.7	11.5
Met+Cys	39.1	25.0
Phe+Tyr	58.3	40.3
Thr	34.8	27.6
Trp	8.7	7.8
Val	39.1	29.1

[1] NRC, 1994.
[2] NRC, 1998.

The predicted protein efficiency ratio (*PER*) is calculated from the amino acid composition based on the equation developed by Lee et al. (1978):

$$PER = 0.06320 \sum [aa]_E - 0.1539 \tag{9}$$

where $[aa]_E$ = essential amino acid concentration (%):

$$\sum [aa]_E = [Ile] + [Leu] + [Lys] + [Met] + [Phe] + [Thr] + [Val] + [Arg] + [His] + [Trp]$$

PER for enzymatic hydrolisates (table 9) shows values higher than values for other protein sources as consequence of the different amino acids profile.

IAE, BV and PER values do not show a clear dependence on the operating conditions for production of the enzymatic hydrolysates (table 9), showing for all cases high values comparable to other nutritional feed.

Table 9. EAI, BV and PER for the pig hair hydrolysates produced at pH and 50°C.

E/S	EAI	BV		PER
		Chickens	Pigs	
1/5	70.12	15.03	26.8	3.88
1/10	67.6	6.37	10.71	3.59
1/20	75.52	18.69	27.32	4.12
1/30	79.27	6.18	10.11	3.89
1/100	96.92	3.04	5.09	4.94

7. Conclusion

The high nutritional value of the enzymatic hydrolysates of keratin makes them appropriate for feed formulations or as nutritional supplements for animals.

Keratin can be obtained from the hair of the pigs, reducing the amount of this waste to landfill, and making it easy for the observance of the legislation of the European Council regarding the landfill of wastes.

Commercial proteases are successful as enzymes in the keratin hydrolysis, showing serine endopeptidases as the most suitable proteases for the hydrolysis of the keratin coming from the pig hair. When Alcalase (a serine protease from *Bacillus licheniformis*) is used as enzyme, and temperature and pH are fixed according to its maxima activity and stability, the enzyme-substrate ratio (E/S) is the factor that mainly affects the hydrolysis process of pig hair. Although nutritional values of the hydrolysates (essential amino acid index, biological value and protein efficiency ratio) do not seem affected by E/S, the hydrolysis degree and the amino acid profile are directly proportional to the increase of the enzyme-substrate ratio showing linear relationships.

A disadvantage of the use of proteases is the low production of hydrolysate: only 10 % of the hair becomes hydrolysate. The possibility of the future use of other types of enzymes not yet commercialized, like keratinases, could allow an increase in the amount of the hydrolysates maintaining the nutritional value of them.

References

Adler-Nissen, J. (1986). *Enzymic hydrolysis of food proteins*. London, Elsevier Applied Science Publishers.

Chavan, U. D.; Mckenzie, D. B.; Shahidi, F. (2001). Protein classification of beach pea (*Lathyrus maritimus L.*). *Food Chemistry, 75*, 145-153.

Cheftel, J. C.; Cuq. J. L.; Lorient, D. (1993). Aminoácidos, péptidos y proteínas. In: O. R. Semema (Ed.). *Química de los alimentos*. Zaragoza , Editorial Acribia S. A., 374-391.

El-Boushi, A.R. & van der Poel, F.B. (2000). *Handbook of Poultry Feed fromWaste. Processing and Use*. Dordrecht, Kluwer Academic Publishers.

Evans, K.L.; Crowder, J.; Miller, E. S. (2000). Subtilisins of Bacillus spp. hydrolyze keratin allow growth on feathers. *Canadian Journal of Microbiology, 46*, 1004-1011.

Falla, L. H. (1994). Desechos de matadero como alimento animal. Available from: http:// www.fao.org/ag/aga/agap/FRG/APH134/cap7.htm.

Fang, Z.; Smith, R. L.; Inomata, H.; Arai, K. (1999). Phase behavior and reaction of polyethylene terephthalate-water systems at pressures up to 173 Mpa and temperatures up to 490 °C. *Journal of Supercritical Fluids, 15*, 229-243.

Farag, A. M. & Hassan, M. A. (2004). Purification, characterization and immobilization of a keratinase from *Aspergillus oryzae*. *Enzyme and Microbial Technology, 34*, 85-93.

Friedrich, J.; Gradišar, H.; Vreel, M.; Pogačnik, A. (2005). In vitro degradation of porcine skin epidermis by a fungal keratinase of Doratomyces microsporus. *Enzyme and Microbial Technology, 36(4)*, 455-460.

Friedrich, J. & Kern, S. (2003). Hydrolysis of native proteins by keratinolytic protease of Doratomyces microsporus. *Journal of Molecular catalysis B: Enzymatic*, *21*, 35-37.

Gessesse, A.; Hatti-Kaul, R.; Gashe, B. A. M.; Mattiasson, B. (2003). Novel alkaline proteases from alkaphilic bacteria grown on chicken feather. *Enzyme and Microbial Technology, 32*, 519-524.

González-Tello, P., Camacho, F., Jurado, E., Páez, M. P., Guadix, E. M. (1994). Enzymatic hydrolysis of whey protein. I. Kinetic Models. *Biotechnology and Bioengineering 44*, 523-528.

Gradišar, H.; Kern, S.; Friedrich, J. (2000): Keratinase of Doratomyces microsporus. *Applied Microbiology and Biotechnology* , *53*, 196-200.

Grazziotin, A.; Pimentel, F. A.; Jong, E. V.; Brandelli, A. (2006). Nutritional improvement of feather protein by treatment with microbial keratinase. *Animal Feed Science and Technology, 126*, 15-144.

Gupta, R; Beg, Q. K.; Lorenz, P. (2002): Bacterial alkaline proteases: molecular approaches and industrial applications. *Applied Microbiology and Biotechnology*, *59*, 15-32.

Hatfield, J.L. & Stewart, B. (Eds.) (1998). Animal water utilization: Effective use of manure as a soil resource. Chelsea, MI, Ann Arbor Press.

Kida, K.; Morimura, S.; Noda, J.; Nishida, Y.; Imai, T.; Otagiri, M. (1995). Enzymatic hydrolysis of the horn and hoof of cow and buffalo. *Journal of Fermentation and Bioengineering*, *80 (5)*, 478-484.

Kim, J. S.; Kluskens, L. D.; de Vos, W. M. ; Huber, R. ; van der Oost, J. (2004). Cristal structure of Fervidolysin from *Fervidobacterium pennivorans*, a keratinolytic enzyme related to subtilisin. *Journal of Molecular Biology*, *335*, 787-797.

Lee, Y. B.; Elliott, J. G.; Rickansrud, D. A.; Hagberg, E. C. (1978). Predicting protein efficiency ratio by chemical determination of connective tissue content in meat. *Journal of Food Science, 43 (5)*,1359-1362.

Lehninger, A. L. (1988): Bioquímica. Las bases moleculares de la estructura y función celular. Barcelona, Ediciones Omega S. A.

Lin, X.; Lee, C. G.; Casale, E. S.; Shih, J. C. H. (1992): Purification and charactization of a keratinase from a feather-degrading Bacillus licheniformis strain. *Applied and Environmental Microbiology*, *58 (10)*, 3271-3275.

Madrid, A., Madrid, R., Madrid, J.M. (1995). Piensos y alimentos para animales. Madrid, Spain, A. Madrid Vicente, pp. 167–193.

Márquez, M. C. & Fernández, V. (1993), Enzymic hydrolysis of vegetable proteins: Mechanism and kinetics, *Process Biochemistry 28*, 481-490.

Márquez, M. C., Vázquez, M. A. 1999, Modeling of enzymatic protein hydrolysis, *Process Biochemistry 35*, 111-117.

Moran, E. T.; Bayley, H. S.; Summers, J.D. (1967): Keratins as source of protein for the growing chick. 3. The metabolizable energy and amino acid composition of raw and processed hog hair meal with emphasis on cystine destruction with autoclaving. *Poultry Science, 46*, 548-553.

Mørup, I. K., & Olesen, E. S. (1976). New methodfor prediction of protein value from essential amino acidpattern. *Nutritional Report International, 13*, 355–365.

Novozymes A/S (2002): Alcalase, 2001-08281-02. Bagsuaerd, Denmark, Novozymes A/S,.

Novozymes A/S (2003a): Flavourzyme, 2001-08282-05. Bagsuaerd, Denmark, Novozymes A/S.

Novozymes A/S (2003b): Neutrase, 2001-08283-07. Bagsuaerd, Denmark, Novozymes A/S.

Novozymes A/S (2003c): Protamex, 2001-08284-03. Bagsuaerd, Denmark, Novozymes A/S.

NRC (National Research Council) (1994). Nutrient Requirements of Poultry, 9thEdition. Washington, National Academy Press.

NRC (National Research Council) (1998). Nutrient Requirements of Swine, 10thRevised Edition. Washington, National Academy Press.

Onifade, A. A., Al–Sane, N. A., Al-Musallam, A. A., Al-Zarban, S. (1998). A review: Potentials for biotechnological applications of keratin – degrading microorganisms and their enzymes for nutrititonal improvement of feathers and other keratins as livestock feed resources. *Bioresource Technology, 66*, 1-11.

Oyarekua, M. A. & Eleyinmi, A. F. (2004). Comparative evaluation of the nutritional quality of corn, sorghum and Millet ogi prepared by a modified traditional technique. *Food, Agriculture & Environment, 2 (2)*, 94-99.

Rao, M. B.; Tanksale, A. M.; Ghatge, M. S.; Deshpande, V. V. (1998). Molecular and biotechnological aspects of microbial proteases. *Microbiology and Molecular Biology Reviews, 62 (3)*, 597-635.

Rawn, J. D. (1989): Bioquímica. Madrid, McGraw-Hill.

Rhyner, C.R., Schwartz, L., Wenger, R., Kohrell, M. (1995). Waste Management and Resource Recovery. Boca Raton, FL, CRC Lewis Publishers.

Riffel, A.; Brandelli, A. (2002). Isolation and characterization of a featherdegrading bacterium from the poultry processing industry. *Journal of Industrial Microbiology & Biotechnology, 29*, 256-258.

Sindayikengera, S. and Xia, W. (2006). Nutritional evaluation of caseins and whey proteins and their hydrolysates from Protamex. *Journal of Zhejiang University Science B 7(2)*, 90-98.

Suntornsuk, W. & Suntornsuk, L. (2003). Feather degradation by *Bacillus sp.* FK 46 in submerged cultivation. *Bioresource Technology, 86*, 239-243.

Toni de, C.H.; Richter, M. F.; Chagas, J. R.; Henriques, J. A. P.; Termignoni, C. (2002). Purification and characterization of an alkaline serine endopeptidase from a feather-degrading *Xanthomonas maltophilia* strain. *Canadian Journal of Microbiology, 48*, 342-348.

Varela, H.; Ferrari, M. D.; Belobrajdic, L.; Vázquez, A.; Loperena, M. L. (1997). Skin unhairing porteases of Bacillus subtilis: production and partial characterization. *Biotechnology Letters, 19 (8)*, 755-758.

Vignardet, C.; Guillaume, L. M.; Friedrich, J. ; Millet, J. (1999). A first order experimental design to assess soluble proteins released by a new keratinase from Doratomyces microsporus on human substrates. *International Journal of Pharmaceutics, 191*, 95-102.

Wei Q., Zhimin, H. (2006). Enzymatic hydrolysis of protein: mechanism and kinetic model. *Frontiers of Chemistry in China 3*, 308-314.

Witczak, A., (2006). Analysis of regional diferentiation of hog livestock production in Poland in 2004. *Electronic journal of polish agricultural universities 9(2)* Topic Economics.

Yamamura, S.; Morita, Y.; Hasan, Q.; Yokohama, K.; Tamiya, E. (2002a): Keratin degradation: a ccoperative action of two enzymes from *Stenotrophomas* sp. *Bichemical and Biophysical Research Communications, 294*, 1138-1143.

Yamamura, S.; Morita, Y.; Hasan, Q.; Rao, S. R.; Murakami, Y.; Yokohama, K.; Tamiya, E. (2002b). Caracterization of a new kertin-degrading bacterium isolated from deer fur. *Journal of Bioscience and Bioengineering, 93 (6)*. 595-600.

In: Leading-Edge Environmental Biodegradation
Editor: L. E. Pawley, pp. 275-295

ISBN: 978-1-60021-903-0

Chapter 9

BIODEGRADATION MECHANISMS IN FUNGAL DECOLORIZATION OF DYES

***Xueheng Zhao*[1] *and Ian R. Hardin*[2]**

[1]Department of Biology, Jackson State University, Jackson, MS 39232, USA
[2]Department of Textiles, Merchandising, and Interiors, University of Georgia, Athens, GA 30602, USA

ABSTRACT

Synthetic dyes are employed as coloring agents in the textile, food, and pharmaceutical industries, which are also integrated into many industrial products. The release of dyes into the wastewater is highly visible in small quantities due to their brilliance and creates major environmental problems. Tightening government legislation is forcing industries to treat their waste effluent to control water pollution. Dyestuffs in the industrial effluents are highly stable and recalcitrant to conventional treatment. Currently, removal of dyes from effluents is by physical and chemical methods. Such methods, although offering some solutions to the problems, are often very costly and, although the dyes are removed, accumulation of concentrated sludge creates a disposal problem. Biological degradation is recognized as the most effective method for the degrading the dye present in the wastewater from textile industry. Research over the past two decades has provided insight into the various aspects of biological degradation of dyes. Detailed and extensive studies have been made for treatment of dye containing wastewaters by white rot fungi and their enzyme systems. It is observed that white rot fungi have a non-specific enzyme system, which oxidizes the recalcitrant dyes. This chapter reviews the recent progress in the research on understanding the biodegradation mechanism of white rot fungi and their enzymes involved in dye decolorization. Biocatalysis and characteristics of their enzymes and emerging knowledge of the enzymology of these organisms are discussed. Study of fungal degradation products of dyes, espically azo dyes, is reviewed for its importance in the understanding of the nature of bioremediation with fungi. Some of most recent results on biodegradation mechanism of azo dyes and metabolites analysis from authors' group are reported herein.

Perspectives on how to use white rot fungi as an effective, cheaper alternative for dye removal and detoxification are also discussed.

INTRODUCTION

Since the first synthetic dye, mauveine, was developed by William Perkin in 1856, chemical dyes have been employed widely as coloring agents in the textile, pharmaceutical, food, cosmetic industries and are used as biological stains in laboratories and clinics. During application, there are significant amounts of unused dyes left in the wastewater (Hao et al., 2000). It is estimated that about 15 percent of dyes is released into the waste water during textile processing (Wesenberg et al., 2003). Some dyes such as anthraquinone dyes are very stable and remain colored for a long time in the environment (Banat et al., 1996). Because they are highly colored, these dyes in wastewater are readily apparent. Dyes in wastewater not only present aesthetic objections, but they can pose threats to public health. Concern about the toxicity of the polluting dyestuffs sometimes results from the reduction products of some azo dyes such as benzidine derivatives which can induce urinary bladder cancer in humans (Chung, 1983). Furthermore, azo dyes can be metabolized by azoreductases to aromatic amines in the human body or mammalian liver if ingested (Chen, 2006). Some of these aromatic amines are known carcinogens and mutagens.

Dyes are designed to be resistant to light, water and oxidizing agents, so it is difficult to remove them once they are released into the environment. Based on their application, dyes can be classified as direct, disperse, acid, reactive, sulfur, vat, mordant, and basic dyes. Some dyes, i.e. acid, basic, direct, and reactive dyes are water-soluble salts bearing ionic functional groups such as sulfonic, carboxyl, amino, or nitro groups. Disperse dyes, however, are nonionic dyes with extremely low water solubility and are exclusively used for dyeing hydrophobic fibers such as polyester. The trend of increasing use of cotton fabrics in the apparel industry has resulted in wide use of reactive dyes. Reactive dyes are the dye class most found in the wastewater because these dyes undergo hydrolysis during application and the hydrolyzed form has no affinity for fabrics. Based on the structure of the chromophores, most dyes in commercial use fall into two basic and important categories: azo dyes and anthraquinone dyes (Figure 1). Azo dyes consisting of nitrogen to nitrogen double bond (-N=N-) are predominant in the commercial dyes currently in use and accounts for more than half of dyes produced in the world each year. In some cases, transition metals embedded in the dye molecules increase the light and wash fastness but bring difficulties in biodegradation. Other chromophores in commercial dyes include triphenylmethane and phthalocyanine. Several representative dyes with different chromophore structures are listed in Figure 2.

Dye-containing effluents are only slightly decolorized by conventional biological wastewater treatments (Wesenberg et al., 2003; Shaul et al., 1991). The toxicity of textile effluents towards *Daphnia magna* was not changed significantly after treatment by activated sludge (Moran et al., 1997). Typically, textile dyes in wastewater can be physically or chemically decolorized by flotation, adsorption, coagulation, ozonation, ion exchange and ion pair extraction, membrane filtration or wet and Fenton oxidation. Photodegradation is not an efficient option in dye degradation (Porter, 1975). Some decolorization methods such as coagulation and ion pair extraction might be useful but they only accumulate and concentrate dyes and create solid waste, and thus a disposal problem still exists. Chemical techniques, for example, Fenton's reagent and ozonation as well as other advanced oxidation processes have constraints in reaction conditions and limitation in the types of dyes that can be treated. These treatments may produce toxic byproducts and little COD reduction (Lin and Lin, 1993).

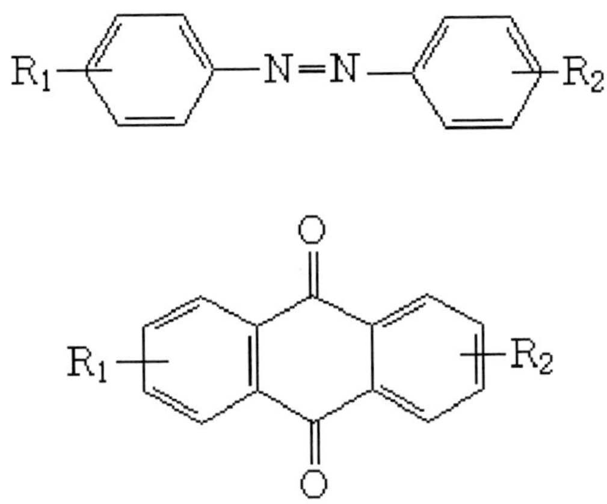

Figure 1. Schematic structures of dyestuffs with a) azo, b) anthraquinone chromophores. R_1, R_2 are auxochromes.

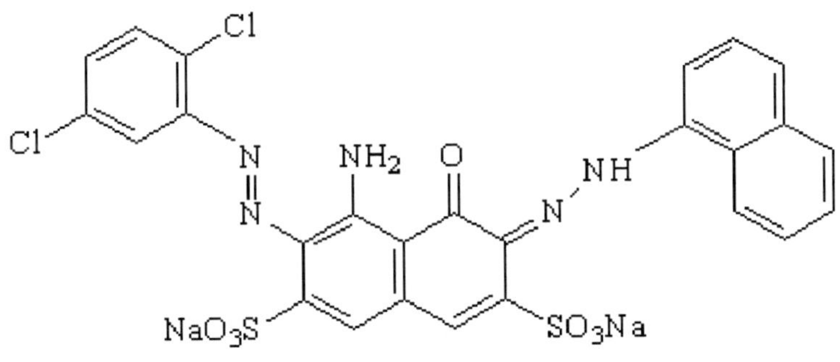

Acid Green 19 (C.I. 20440)

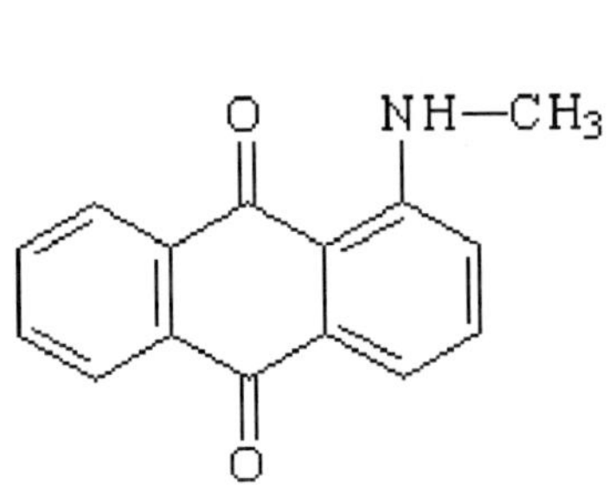

Disperse Red 9 (C.I. 60505)

Figure 2. (Continued)

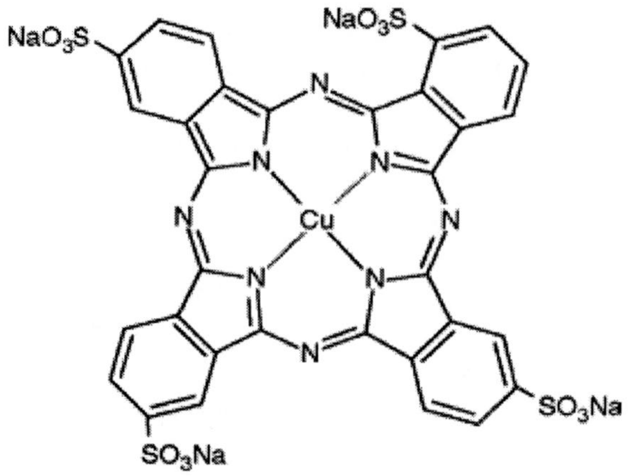

C.I. Reactive Blue 21.

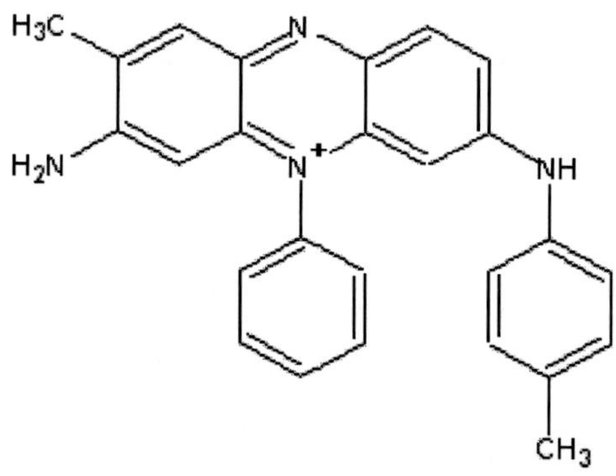

Mauveine A (C.I. 50245).

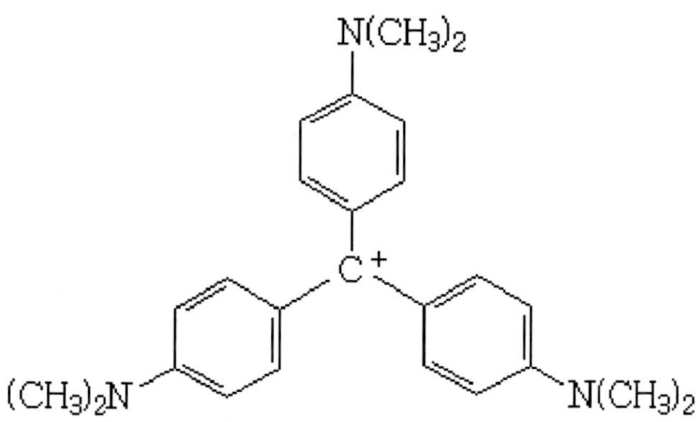

Basic Violet 3 (C.I. 42555).

Figure 2. Representative dyes in different classes with different chromophore structures

Bioremediation of dyes in waste effluents by microorganisms including fungi is an alternative to conventional methods and a very promising area of study because of the relatively low expense involved. It has been shown that microorganisms are able to utilize a wide of variety organic compounds as precursors for the synthesis of their own cell material or as sources of energy, even when such compounds are present at low concentrations in the environment (Egli, 1992).

It is necessary to clarify the concepts of decolorization, degradation, and mineralization of dyes in that these words often occur in the literature and sometimes are misused. Decolorization is simply the disappearance of the color in wastewater with a change in chromophore groups, which does not necessarily mean degradation of the complex dye molecules. Therefore, the decrease in TOC and COD may not be obtained even when complete decolorization is achieved (Hao et al., 2000). Degradation is the conversion of the large dye molecules to smaller components, along with the breakdown of the chromophores. While chromophore groups of dyes may be destroyed in this procedure, in some cases the intermediate produced may be more toxic than the original compounds, e.g. aromatic amines resulting from reduction of azo dyes, and could present significant problems for receiving water bodies. Mineralization means organic compounds are converted to inorganic compounds, i.e., nitrate, carbon dioxide, and water. In this case, a complete detoxification is supposed to be achieved and no secondary pollution will be introduced.

Nevertheless, it is clear that dye decolorization is not equivalent to dye mineralization. There is a gap in current knowledge between decolorization and degradation mechanisms, with a lack of knowledge concerning potentially toxic colorless intermediates (Wesenberg et al., 2003). It is generally agreed that practical application of biodegradation systems using white rot fungi must be preceded by a better understanding of the biodegradation mechanisms involved. Despite much research on the biodecolorization, i.e. by white rot fungi, the biodegradation mechanism, the role of enzymatic systems and cofactors, have not been well understood. Furthermore, the nature of the intermediates of dyestuffs from the biodecolorization and their biodegradability are not clear yet and should be addressed before considering the application of the microorganism. Therefore, the present knowledge on the emzymatic systems involved in fungal degradation and the most recent progress in the metabolite analysis of dyestuff by white rot fungi are reviewed in this chapter to provide some insights to the biodecolorization research.

Fungal Decolorization

A wide variety of microorganisms including bacteria, yeast, and fungi are able to decolorize a diverse array of dyes. By far the single class of microorganisms most efficient in breaking down synthetic dyes is white rot fungi (Wesenberg et al., 2003). The nonspecific nature of the lignin-degrading systems of white rot fungi is a potential advantage for biotreatment of textile effluents, since a mixture of dyes, surfactants and other compounds exist in the wastewater. Another important advantage for degradation of azo dyes using white rot fungi is that ligninolytic enzymes degrade azo dyes by oxidation, in contrast to the reduction pathway in bacterial degradation.

Studies in our laboratory showed white rot fungi could decolorize a range of commercial dyes with different structures (azo, disazo, anthraquinone) (Cao, 2000). Azo dyes are considered recalcitrant xenobiotic compounds due to the presence of a nitrogen double bond (-N=N-) bond and other groups (i.e. sulfonic group) that are not easily biodegraded (Martins et al., 2001). Wastewaters containing these azo dyes are usually resistant to normal biological waste treatment for color removal. Various white rot fungi, including *Phanerochaete chrysosporium, Pycnoporus cinnabarinus, Trametes versicolor, Ceriporiopsis subvermispora, Cyathus stercoreus, Pleurotus ostreatus, Phellinus tremellia, Phellinus oxysponum,* and *Phellinus pini*, were studied for their ability to decolorize commercially used dyes. *Pleurotus ostreatus* has been demonstrated to possess a high potential for decolorizing various azo dyes (Cao, 2000).

Phanerochaete chrysosporium is the most intensively studied dye-decolorizing white rot fungus. It was demonstrated by Cripps et al. (1990) that three azo dyes, Orange II (4-[(2-hydroxy-1-naphthalenyl)azo]-benzenesulfonic acid), Tropaeolin O (4-[(2,4-dihydroxyphenyl)azo]-benzenesulfonic acid), and Congo Red (3,3′- [1,1′- biphenyl]-4,4′-diylbis(azo)]bis[4-amino-1-naphthalenesulfonic acid]) and one heterocyclic dyes, Azure B (3-(dimethylamino)-7-(methylamino)-phenothiazin-5-iumchloride) were decolorized extensively (96-100 percent) during five days of incubation with *Phanerochaete chrysosporium*. Since then, a range of acid, reactive, and mordant dyes with azo, triphenyl methane, heterocyclic, anthraquinonic, indigoid, and polymeric structures have been shown to be decolorized by this ligninolytic fungus or its purified enzymes (Knapp et al., 1995; Chagas and Durrant, 2001; Moreira et al., 2000; Ollikka et al. 1993; Abadulla et al. 2000). Other white rot fungi, such as *Trametes versicolor, Bjerkandera adusta,* and *Thelephora sp*., have also been shown to be efficient in decolorizing different dyes (Selvam et al., 2003; Heinfling et al., 1997). Novotny and co-workers (2001) screened 103 wood-rotting fungal strains for decolorization of azo and other types of textile dyes. *Irpex lacteus* and *Pleurotus ostreatus* degraded dyes from all groups. Application of white rot fungi, particularly *Phanerochaete chrysosporium*, in environmental bioremediation and decolorization has been reviewed (Paszczynski and Crawford, 1995; Wesenberg et al., 2003).

Although *Phanerochaete chrysosporium* and some other strains have been shown to decolorize a wide range of dyestuffs, so far there are only limited white rot fungal strains being studied in detail for decolorization purposes. It is desirable to find more fungi with high capabilities in decolorization of wastewaters containing dyes and other chemicals.

Enzymatic Systems of White Rot Fungi in Decolorization

Ligninolytic fungi are able to degrade numerous aromatic organic pollutants via oxidative mechanisms (Martinset al., 2001; Bogan and Lamar, 1995). A correlation between dye decolorization and the production of ligninolytic enzymes was implied by the work of Wesenberg and co-workers (Wesenberg et al., 2003). The ligninolytic enzymes are extracellularly excreted by the fungi to initiate the oxidation of substrates in the extracellular environment of the fungal cells (Mester and Tien, 2000). Essential extracellular enzymes involved into the degradation of lignin in wood and recalcitrant pollutants (i.e. azo dyes) in the environment are laccase, lignin peroxidase (LiP), and manganese peroxidase (MnP). All of these enzymes are generally believed to form during secondary metabolism of white rot fungi (Hou et al, 2003; Zapanta et al., 1997). LiP is a classical hemeprotein peroxidase containing heme in the active site with a molecular weight between 38 and 47 KDa (Wesenberg et al., 2003). Due to its high redox potential, LiP is able to directly oxidize non-phenolic lignin units (Sarkar et al., 1997). LiP reacts directly with aromatic substrates by abstracting single electrons from their aromatic rings, leading to the formation of a cation radical and subsequent cleavage reactions (Zheng and Obbard, 2002). A characteristic of LiP, which is also shared by non-ligninolytic peroxidases, is its relative unspecificity for substrates such as phenolic compounds and dyes (Martinez, 2002). Whereas phenoxy radicals are the first product of phenolic substrate oxidation by peroxidases, aromatic cations are formed after non-phenolic aromatic ring oxidation by LiP (Martinez, 2002; Kersten et al., 1985).

The catalytic cycle of peroxidases is shown in Figure 3. The ligninolytic peroxidases have a typical enzymatic cycle characteristic of other peroxidases (Mester and Tien, 2000; Tien and Kirk, 1984; Wariishi et al., 1988). The heme group of the native enzyme (ferric form) is oxidized by H_2O_2 with two electrons to compound I. One electron is removed from ferric (Fe(III)) iron to form ferryl (Fe(IV)), while the second electron is withdrawn from the porphyrin ring to form a porphyrin cation radical (Aust, 1995). During this reaction step, H_2O_2 is reduced to water. Compound I oxidizes substrates by one electron and is reduced to compound II. In this step, the porphyrin ring gains one electron. Thus, compound I is able to oxidize substrates with a higher redox potential than compound II (Mester and Tien, 2000). Compound II then reacts with reducing substrates by gaining one electron and changing back to resting enzyme. Substrates (RH) of compound I can also react with compound II and give rise to radical R•. Compound II can also react with H_2O_2 resulting in compound III (Cai and Tien, 1992), which leads to inactivation of the peroxidase (Torres et al., 2003). To avoid inactivation of LiP, veratryl alcohol completes the catalytic cycle of LiP by reducing compound II to resting enzyme. Veratryl alcohol (3,4-dimethoxybenzyl alcohol) is a secondary metabolite synthesized from glucose by several fungi, but mainly the basidiomycetes (Suguimoto et al., 2001; Jong et al., 1994). It is oxidized via two separate one-electron oxidations to give veratraldehyde (Zapanta and Tien, 1997; Tien, 1987). Veratryl alcohol is thought to be produced at the same time as LiP (Fenn and Kirk, 1981). In liquid cultures, the addition of veratryl alcohol causes an increase in LiP activity (Faison and Kirk, 1985). Veratryl alcohol plays a role in the LiP action in three ways. Firstly, it acts as a cation radical redox mediator of remote substrates. Secondly, it acts to protect LiP from inactivation by H_2O_2. Finally, it acts to complete the catalytic cycle of LiP in the presence of substrates that cannot reduce compound II to native enzyme (Zapanta and Tien, 1997; Goodwin et al., 1995; Schoemaker et al., 1994).

Like LiP, MnP is a heme-containing peroxidase (Zapanta and Tien, 1997) and is the most common ligninolytic peroxidase produced by almost all white rot basidiomycetes (Wesenberg et al., 2003). MnP is a glycoprotein with a heme (ferric protoporphyrin) group and a molecular weight of 32 to 62.5 KDa (Hofrichter, 2002; Glenn and Gold, 1985). It shares mechanistic properties with LiP and forms the oxidized intermediates, compound I and compound II (Zapanta and Tien, 1997; Cai and Tien, 1993). It also requires the presence of H_2O_2 to oxidize lignin and lignin-related compounds (Mester and Tien, 2000).

MnP exhibits a dependence on Mn (II) as a substrate for compound II (Wariishi et al., 1988). Mn (II) is also a preferred substrate for compound I. In the degradation of lignin and other substrates, MnP oxidizes Mn (II) to Mn (III) in the presence of H_2O_2, and Mn (III) oxidizes a variety of compounds (Mester and Tien, 2000; Glenn et al., 1986). The chelation of Mn (II) and Mn (III) by organic acids, such as oxalate, is necessary for MnP activity (Zapanta and Tien, 1997). Oxalate is an organic acid chelator produced by white rot fungi at the same time as MnP in the liquid cultures of *Phanerochaete chrysosporium* (Zapanta and Tien, 1997; Wariishi et al., 1992; Kuan and Tien, 1993).

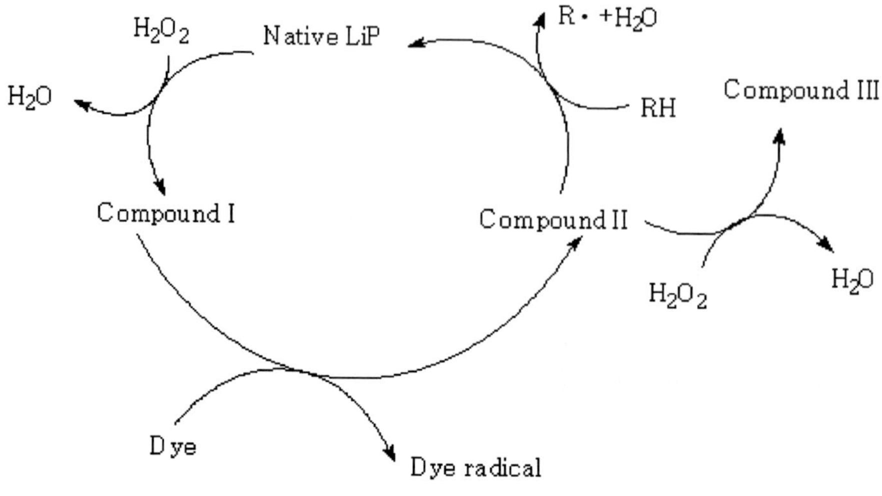

Figure 3. Catalytic cycle of heme-containing peroxidases (e.g. LiP) in the presence of hydrogen peroxide.

Laccase belongs to a group of enzymes called blue copper oxidases with a molecular weight of 60 to 390 KDa (Call and Muke, 1997). Laccase also has broad substrate specificity and is capable of oxidizing phenols and aromatic amines by reducing molecular oxygen (instead of H_2O_2) to water by a multicopper system (Hublik and Schinner, 2000; Hou et al., 2003). In the reaction, the substrates are oxidized by one electron to generate the corresponding phenoxy radicals, which either polymerize to yield a phenolic polymer or are further oxidized by laccase to produce a quinone (Chivukula and Renganathan, 1995). In this procedure, oxygen gets the electron from the substrate and is reduced to water.

Like MnP, laccase acts primarily on phenolic units, but also on non-phenolic ones in the presence of primary co-substrates (Sarkar et al., 1997; Bourbonnais et al., 1996). It was found that a nitrogen limited culture is of benefit to laccase production, and that certain redox mediators, such as 2,2'-azino-bis-(3-ethylbenzothiazole-6-sulfonic acid) (ABTS) and 1-hydroxybenzotriazole (HOBT), can extend this specificity to non-phenolic compounds (Hou et al. 2003, Mester and Tien, 2000). It is still not clear, however, if a laccase mediator system exists in the lignin degradation system (Mester and Tien, 2000).

All three major ligninolytic enzymes are produced in multiple isoforms and are affected by many external factors, such as nutrient level, mediator compounds and metal ions (Wesenberg et al., 2003). It has also been shown that both the composition of culture media and the fermentation time exert a considerable influence on the activity of peroxidases from white rot fungi (Morais et al., 2002).

Other enzymes present in the culture medium or on fungal mycelia may have a role in the decolorization. For example, *Pleurotus ostreatus* also produces other extracellular enzymes, *Pleurotus ostreatus* peroxidase (Kang et al., 1993), Remazol Brilliant Blue R decolorizing peroxidase (Shin et al., 1998), a new kind of enzyme called versatile peroxidase (Martinez, 2002), and glucose oxidase (Shin et al., 1993). *Pleurotus ostreatus* peroxidase and Remazol Brilliant Blue R decolorizing enzyme are H_2O_2 dependent, catalyzing oxidation of phenolic compounds by one electron but having no affinity for nonphenolic compounds. Versatile peroxidase, however, is able to oxidize both LiP and MnP substrates (Martinez, 2002). It was shown that versatile peroxidase could efficiently oxidize some aromatic substrates, including dyes (Martinez, 2002; Heinfling et al., 1998). The available knowledge about fungal azoreductase is still very limited. Its role in decolorization is still unclear yet.

Despite the fact that ligninolytic enzymes have been intensively studied for their role in degradation of lignin and some other organic pollutants, their mechanisms of biodegradation are not well understood. The complex enzymatic systems responsible for dye degradation have not been fully determined and the expression of these enzymes varies with the culture and decolorization conditions.

LiP is considered to be the principal decolorizing enzyme in cultures of *Phanerochaete chrysosporium* and it is clear that ligninolytic enzymes play a significant role in biodegradation of dyes by white rot fungi (Podgornik et al., 1999; McMullan et al., 2001). Whether a compound can serve as a substrate for LiP appears to be determined by the size of the molecule and its redox potential (Mester and Tien, 2000). LiP has a higher redox potential than other peroxidases. The role of purified lignin peroxidase in the decolorization of several azo dyes has been investigated (Rodriguez et al., 1999; Paszczynski et al., 1991). The reaction of the extracelluar ligninolytic enzymes also involves numerous low molecular weight cofactors that may serve as redox mediators (Mester and Tien, 2000).

Robinson et al. (2001) assessed the role of nitrogen in the enzyme production of *Bjerkandera adusta* and *Phlebia tremellosa* in the decolorization of an artificial textile effluent. LiP and laccase appeared to be main ligninolytic enzymes for dye degradation because they were produced in the highest quantities among the three assayed enzymes with or without nitrogen supplementation. They also found that although enzymatic activity is higher in nitrogen rich conditions, no substantial difference could be observed in the amount of degradation for both media. In another study by Wong and Yu (1999) on the laccase decolorization ability, it was demonstrated that anthraquinone dye can be used as direct substrate for this enzyme while azo and indigo dyes were not substrates and the decolorization depended on the presence of small molecular mediators in the solution instead of the specific enzyme activity. Based on the results from Hou et al. (2003), production of laccase from *Pleurotus ostreatus* strain 32 was increased in static cultivation and nitrogen-limited media. They also found that cellobiose and peptone are favorable carbon and nitrogen sources to enhance laccase production. The compound 2,2'-Azino-di-(3-ethylbenzothialozin-6-sulfonic acid (ABTS) was demonstrated as an efficient mediator and Cu^{2+} and tyrosine were useful as inducers for laccase synthesis. In the decolorization of dyestuffs possessing carboxylic or sulphonic groups by *T. versicolor*, activity of laccase, LiP, and glyoxal oxidase were detected while MnP and proteases were never detected (Martins et al 2003).

A better understanding of fungal enzymatic systems will lead to discovery or development of new enzymes with broad substrate specificity and high activity, such as thermostable and pH tolerant characteristics which are especially useful for textile effluent decolorization since these effluents usually have a higher temperature (40-50°C) and a wide pH range.

Fungal Degradation Products and Biodegradation Pathway

Although previous studies show that white rot fungi have a high potential in decolorization, very little is known about the nature of fungal degradation products from organic dyes and the mechanisms by which these products form. For biotreatment applications, it is important to understand the degradation products or intermediates because of concern that these compounds may affect or even increase the aquatic toxicity of the overall effluent. Furthermore, knowledge of the metabolites would help us understand the way that white rot fungi break down the chemically stable and complex dyestuffs. However, the complexity of fungal degradation of dyes makes it difficult to elucidate the mechanism of major enzymatic systems and the structures of metabolites. Studies conducted in our and other research groups have obtained some essential knowledge on the degradation products and intermediates from fungal decolorization.

Fungal degradation includes a complex process of oxidation, reduction, methylation, and hydroxylation. Identification of the degradation products holds the key to understanding these complex biochemical processes and to evaluating the final environmental impact. However, the inherent complexity of dye structures, the different properties of dyes and their metabolites and the not well-understood biotransformation mechanisms, are challenges in studying fungal metabolism of dyes.

Until now, azo dyes have attracted the most attention in the biodegradation studies because of their popularity in the dye industries. Spadaro and Renganathan (1994) described the mechanism for the peroxidase-catalyzed degradation of an azo dye, Disperse Yellow 3. The naphthol analog of Disperse Yellow 3 [1-(4'-acetamidophenylazo)-2-naphthol] was also used as a model compound in their study. Three peroxidases, lignin peroxidase (LiP), manganese peroxidase (MnP), and horseradish peroxidase (HRP), were investigated in the degradation of azo dyes and the products from each system were identified. The peroxidases were indicated to be responsible for the degradation of azo dyes. Asymmetric cleavage of nitrogen-nitrogen double bond (-N=N-) occurred in the degradation. The intermediate compounds generated were suggested to be 4-acetamidophenyldiazene and 4-methyl-1,2-benzoquinone. Phenyldiazene is unstable and is readily oxidized by oxygen. The azo linkage is released as molecular nitrogen and acetanilide, one of the major products from Disperse Yellow 3 and its naphthol analog, is proposed to form through a phenyl radical abstracting a hydrogen radical from the surrounding. Goszczynski and co-workers (1994) studied the mechanism of sulfonated azo dyes degradation by peroxidases of *Phanerochaete chrysosporium* and proposed a cationic radical pathway that was similar to that proposed by Spadaro and Renganathan (1994). However, Goszczynski and co-workers indicated symmetrical splitting of azo linkage as another possibility for initial attack on azo dyes. A nitroso compound and a quinone imine were produced from the symmetric cleavage of the azo linkage.

In our research, ionic and non-ionic azo dyes with different properties have been used in the biodegradation by *Pleurotus ostreatus*. Study of the intermediates of sulfonated azo dyes in the fungal degradation showed that a series of reactions including breakage of azo bond, oxidation, and hydroxylation occurred in this procedure (Zhao et al. 2005; Zhao et al. 2007; Lu and Hardin 2006). With these findings and the accumulated knowledge of the mechanism of fungal degradation, a fungal degradation pathway of sulfonated azo dyes can be proposed for the decolorization mechanism (Figure 4). Acid Orange 7 (AO7) is shown as an example in the proposed pathway. Because the molecule of azo dyes is a fully conjugated system, an initiation with enzymatic excitation could be enough to bring about a stepwise reaction which leads to the breakdown of the whole dye molecule. The reactive intermediates in the degradation undergo a series of redox reactions which result in more stable compounds (Goszczynski et al 1994). In the fungal degradation, enzymes such as LiP extract a proton and an electron from the hydroxyl group of AO7 (I) and form a phenoxy radical, which further transforms to a double bond between the naphthalenic ring and the oxygen (II). The produced carbonium ion (III) is susceptible to nucleophilic attack with H_2O or H_2O_2 to from an unstable tetrahedral intermediate (IV). Azo bond in this unstable intermediate breaks down to produce 4-sulfonate-phenyldiazene (V) and 1,2-naphthoquinone (VI). Phenyldiazene is also not stable and could be easily oxidized to a diazonium salt (VII). Reduction and hydrolysis of this intermediate yield benzenesulfonic acid (VIII) and 4-hydroxybenezene sulfonic acid (IX) that we have observed (Zhao et al. 2005; Zhao et al 2007). The compound 1,2-naphthoquinone (VI) has also been identified in the fungal culture. Several sulfonated azo dyes with similar structures have also been examined in our research (Figure 5). Degradation products identified from these dyes are consistent with the proposed degradation pathway (Zhao et al., 20067; Lu 2006). For example, biodegradation of Acid Orange 8 and Food Yellow 3 produced 3-methyl-4-hydroxy-benzenesulfonic acid, 3-methyl-benzenesulfonic acid, benezenesulfonic acid, 4-hydroxy-benzenesulfonic acid and 6-sulfonate-1,2-naphthoquinone as major degradation products. These evidences from *Pleurotus ostreatus* strongly support the oxidation mechanism of white rot fungi.

Benzenesulfonic acid (VIII) could also result from phenyl radical through obtaining a hydrogen radical from its surroundings. The production of phenyl radical was proposed by Spadaro and Renganathan (1994) and Gavril and Hodson (2007). If this happens, two phenyl radicals will couple to yield disulfonatebiphenyl, but there is no evidence from our or their study for the formation of biphenyl-type products in the degradation. More evidences from intermediate studies are necessary to confirm this proposed pathway. Chivukula and co-workers (1995) reported generation of sulfophenyl hydroperoxides from phenyl radical through reaction with oxygen in lignin peroxidase-catalyzed oxidation. However, such compounds were not found in our studies with the fungal culture *Pleurotus ostreatus*.

Gavril and Hodson (2007) studied the mechanism of the biodecoloration of amaranth by *Trametes versicolor*. To eliminate the possibility of decolorization resulting from existence of hydrogen peroxide, they also carried out experiments with hydrogen peroxide alone in the system. It has been shown that the decolorization could not be achieved by hydrogen peroxide by itself and the enzymatic system of this white rot fungus was responsible for the degradation. They suggested that a phenoxy radical was formed to initiate amaranth's biodegradation, followed by electron redistribution in the dye molecule and opening of aromatic rings, which were catalyzed by peroxidases. It was indicated that the breakage of azo bond released either N_2 or NH_3 in the degradation but was not reduced to –NH2, -N=NH, or =NH. Sulfonic groups could be oxidized to sulfate group and be desulfonated, thus facilitating further degradation.

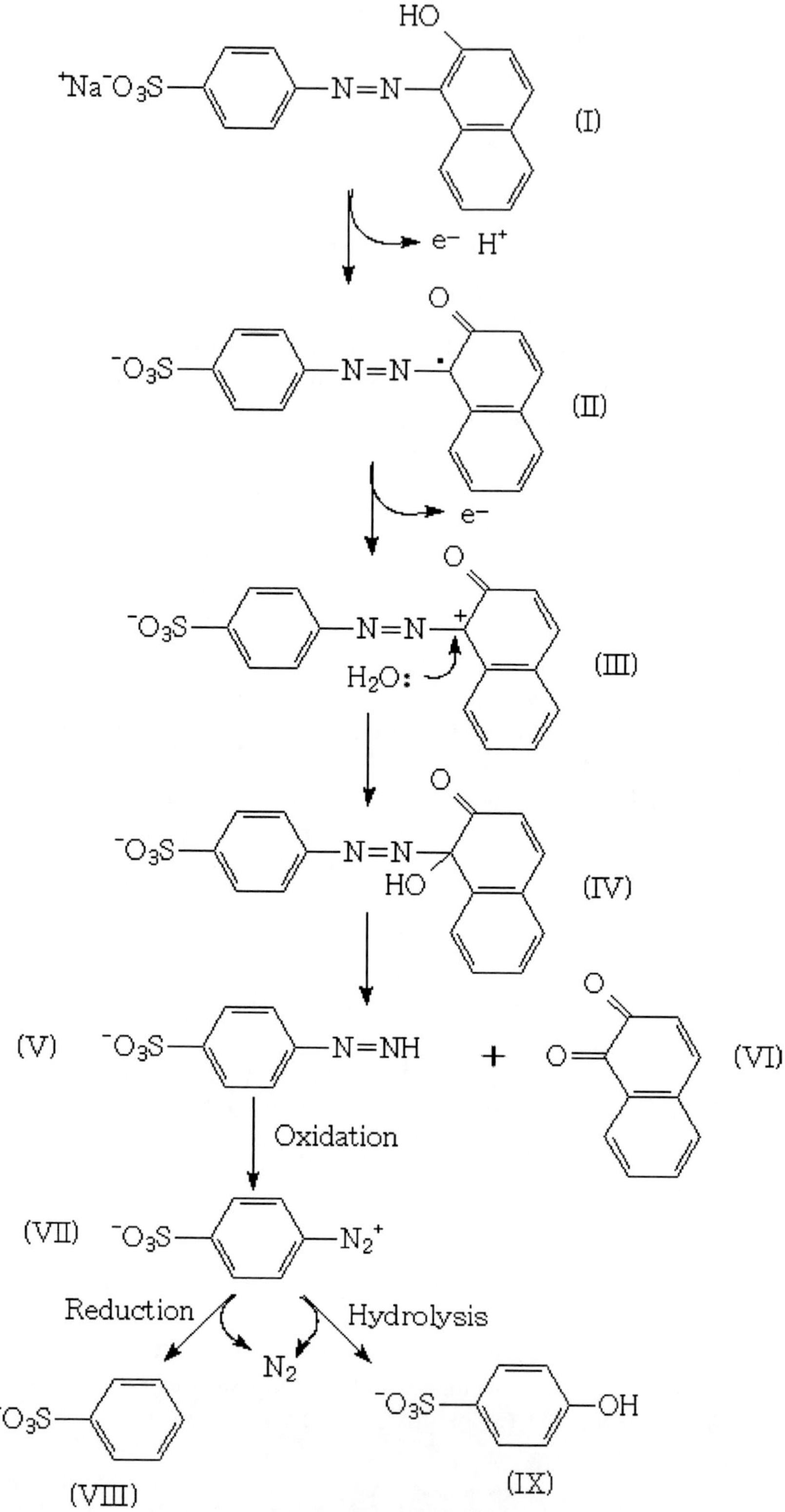

Figure 4. Proposed biodegradation pathway of Acid Orange 7 by *Pleurotus ostreatus.*

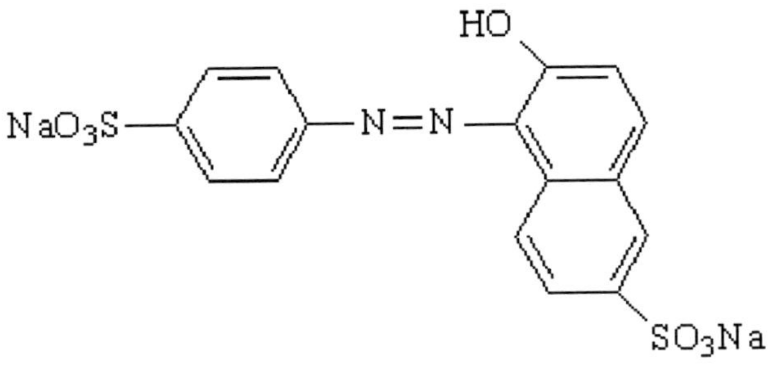

Food Yellow 3.

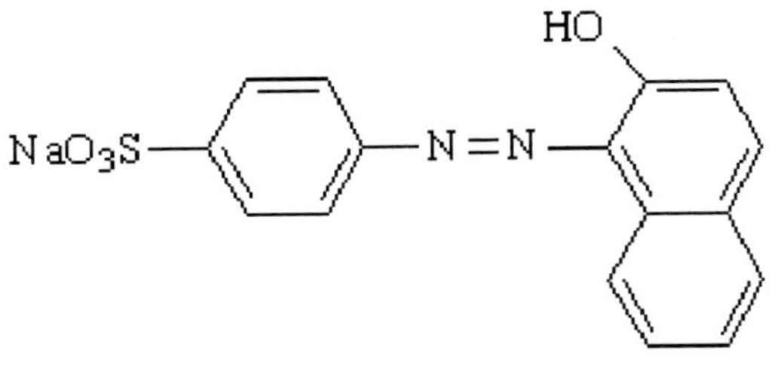

Acid Orange 8.

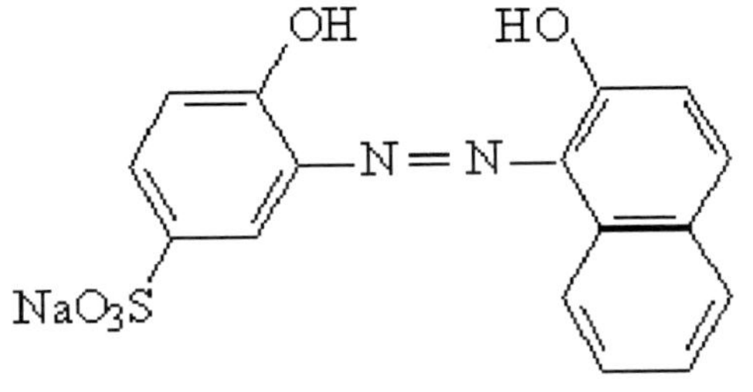

Mordant Violet 5.

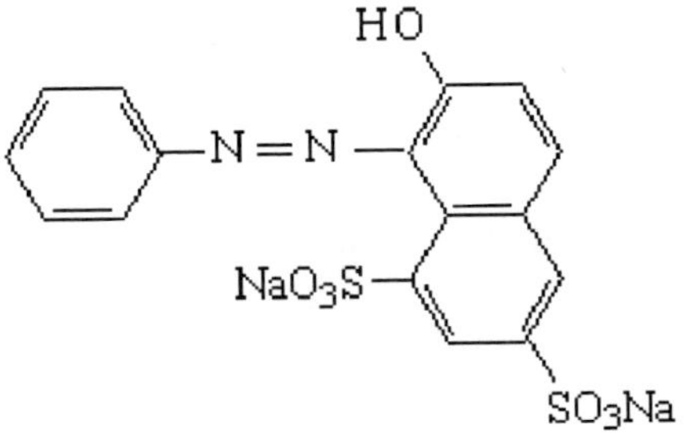

Acid Orange 10.

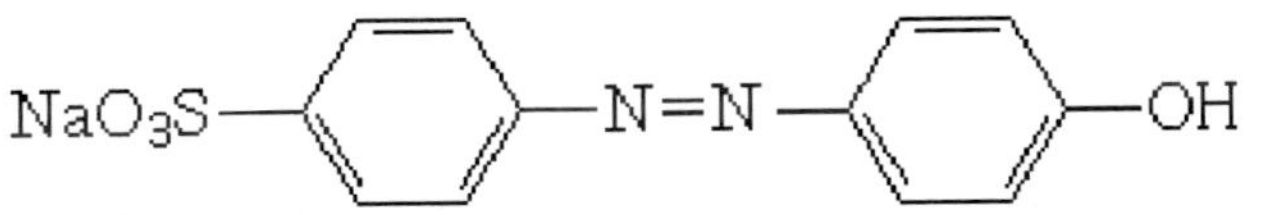

4-[(4-hydroxyphenyl)azo]-benzenesulfonic acid, sodium salt.

Figure 5. Sulfonated azo dyes used in fungal biodegradation in our research.

Examination of the fungal degradation mechanism of non-ionic azo dyes has been conducted in our laboratory (Zhao et al 2006; Lu 2006). Disperse Orange 3 (DO3) and a series of non-ionic dyes with similar structures were degraded with a culture of *Pleurotus ostreatus*. The metabolites of DO3 generated by *Pleurotus ostreatus* were identified as 4-nitroaniline, 4-nitrobenzene, 4-nitrophenol, and 4-nitroanisole. The compound 4-Nitroanisole was the major metabolite when 4-nitrophenol was incubated with *Pleurotus ostreatus*. Methoxylation of aromatic rings facilitates the further opening of these rings and leads to detoxicification of azo dyes and intermediates. Both MnP and laccase have been detected in the culture. Veratryl alcohol, a redox mediator for lignin peroxidase of white-rot fungi, and its oxidant veratraldehyde were also detected in cultures grown in the presence of Disperse Orange 3. The degradation of non-ionic dyes is summarized in Figure 5. Observation of nitrobenzene is in accordance with the degradation pathway proposed by Spadaro and Renganathan (1994) using lignin peroxidase in decolorization. It is suggested that laccase-catalyzed oxidation of azo dyes follows a mechanism similar to that proposed for LiP (Torres et al 2003). It is presumed that peroxidases and laccases were involved in the degradation of Disperse Orange 3 and the methylation of 4-nitrophenol observed in this study. Hydroxylation of the aromatic ring concomitant with cleavage of azo bond is presumed to be responsible for 4-nitrophenol production. The 4-nitroaniline moiety does not seem to be generated by an extracellular lignolytic enzyme. A reduction procedure rather than an oxidation might be responsible for its generation through a symmetric breakage of the azo bond. Rodriguez et al. (1999) suggested that enzymes other than those of the ligninolytic system were engaged in the dye decolorization by *Pleurotus ostreatus*. The formation of 4-nitroaniline may also be related to the reduction ability of this fungus. However, this needs confirmation since no 1,4-phenylenediamine, another likely reduction product, was found in our investigation. Exactly the same degradation products were released from Disperse Red 1, Disperse Red 19, and Disperse Orange 1 by cultures of Pleutotus ostreatus. These are structurally similar to Disperse Orange 3, studied later (Lu and Hardin, 2005). This indicates that this degradation pathway can be used to assess degradation and predict the degradation pathway of other azo dyes with similar structures.

Conneely et al (1999) investigated the metabolism of a copper-phthalocyanine dye Remazol turquoise blue by *Phanerochaete chrysosporium* PC671. Analysis of culture supernatants by HPLC and atomic absorption spectrophotometry showed that approximately half of the copper originally in the dye released into the culture supernatant, which indicated the degradation of the complex dye structure. While they observed a metabolite from Remazol turquoise blue, Conneely and co-workers could not assign an identity to this product. Heinfling-Weidtmann et al. (2001) studied the metabolites from sulfonated phthalocyanine dyes by *Bjerkandera adusta*. The major metabolite identified was sulfophthalimide, which was produced both in fungal degradation and incubation of dyestuffs with purified MnP and LiP.

N-demethylation of triphenylmethane dyes has been found in *Phanerochaete chrysosporium* (Bumpus and Brock 1988). Biodegradation of crystal violet (N,N,N',N',N'',N"-hexamethyl pararosaniline chloride) produced three metabolites by sequential demethylation of the parent compound. Martins et al. (2003) identified hydroxylated metabolites of reactive azo dyes and proposed a metabolic pathway initiated by hydroxylation of the aromatic ring after breakage of the azo bond. It has been suggested that catechol derivatives could be products from these hydroxylated metabolites and opening of aromatic ring could be finally achieved.

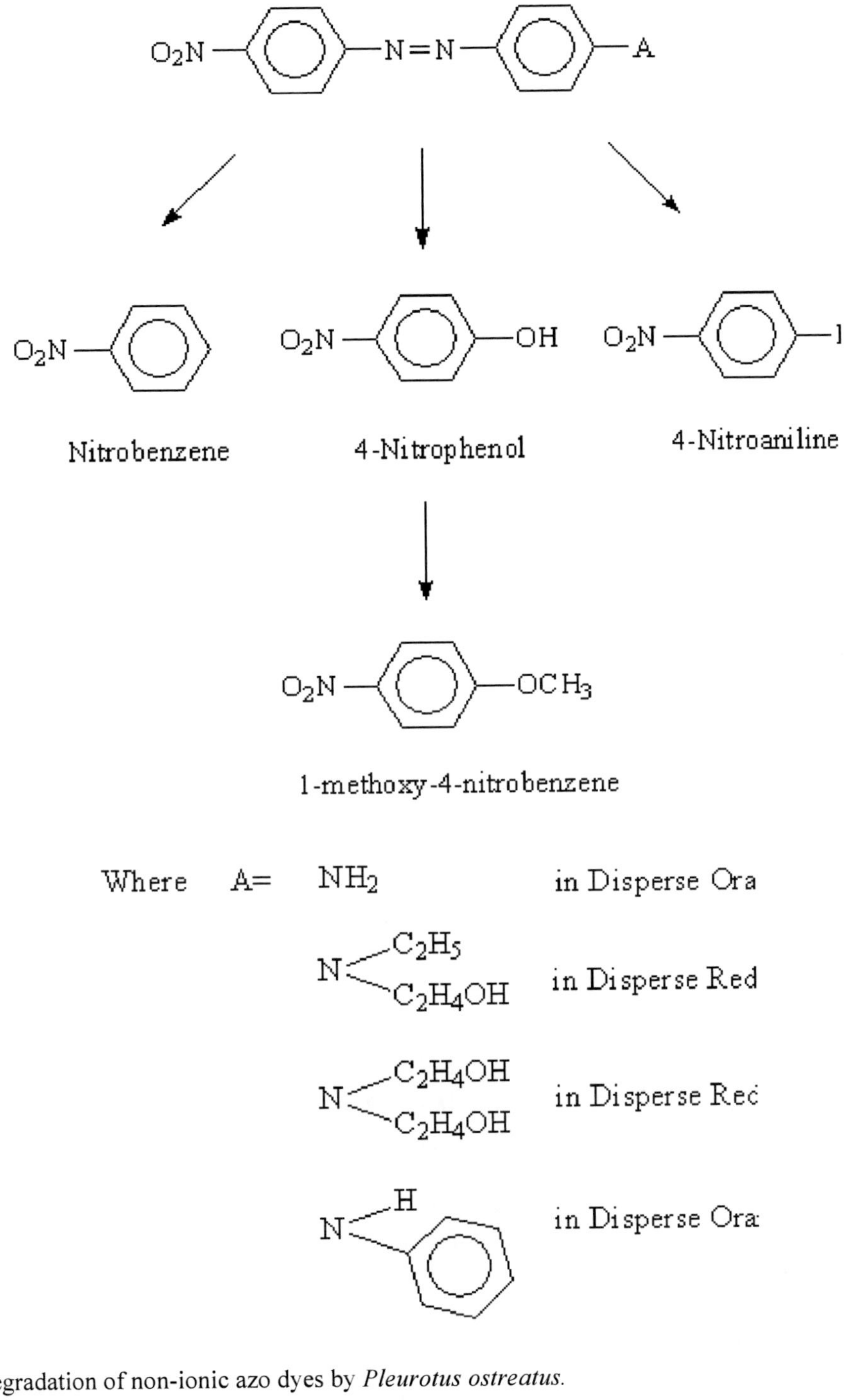

Figure 6. Degradation of non-ionic azo dyes by *Pleurotus ostreatus*.

Due to the complex structure of dyes, few microorganisms can mineralize dyes to CO_2 and H_2O (Hao et al. 2000). Based on our research and others results, it has been shown that the degradation products do not disappear completely. However, it is possible that some dyes can be mineralized all the way to carbon dioxide by white rot fungi (Paszczynski et al., 1992; Spadaro et al., 1992; Ollikka et al. 1993). Spadaro and co-workers (1992) showed that *Phanerochaete chrysosporium* could mineralize the radio-labeled azo compounds 4-phenylazophenol, 4-phenylazo-2-methoxyphenol, Disperse Yellow 3 (N-[4-[(2-hydroxy-5-methylphenyl)azo]phenyl]-acetamide), 4-phenylazoaniline, N,N-dimethyl-4-phenylazoaniline, Disperse Orange 3 (4-[(4-nitrophenyl)azo]-benzeneamine) and Solvent Yellow 14 (1-(phenylazo)-2-naphthalenol) to the extent of 23.1 percent to 48.1 percent. Paszczynski and co-workers (1992) used ^{14}C-labeled azo dyes, such as Orange I (4-[(4-hydroxy-1-naphthalenyl)azo]-benzenesulfonic acid), Orange II (4-[(2-hydroxy-1-naphthalenyl)azo]-benzenesulfonic acid), Acid Yellow 9 (2-amino-5-[(4-sulfophenyl)azo]-benzenesulfonic acid), 4-(3-methoxy-4- hydroxyphenylazo) benzenesulfonic acid, and 4-(2-sulfo-3'-methoxy-4-hydroxyazobenzene-4-azo) benzenesulfonic acid mono sodium salt to determine that the mineralization of these dyes by *Phanerochaete chrysosporium* was in the range of 17.2 percent to 34.8 percent. It is obvious that some of the aromatic rings can be opened and be further degraded to smaller compounds while others remain intact.

It has also been demonstrated that substitute groups with electron giving or withdrawal characteristics significantly affect the degree of biodegradation by white rot fungi. Pasti-Grigsby and co-workers (1992) found that various sulfonated azo dyes were not equally susceptible to microbial attack. Twenty-two azo dyes were studied to determine the influence of aromatic substitution on azo dye degradation by *Phanerochaete chrysosporium*. They found that dyes with hydroxyl and azo groups in the 1,2-positions were degraded fastest among test dyes. Spadaro and Renganathan (1992) indicated that aromatic rings with substituents such as hydroxyl, amino, acetamido, or nitro groups were degraded to a greater extent than unsubstituted rings. Studies in our laboratory showed that different electronic characters of substituents and solubility parameters of moieties of dye molecules could affect the biodegradation (Lu 2006). It is seems that the configuration of dyes has an effect on the biodegradation pathway and that different degradation products are generated from trans- and cis- Mordant Violet 5 (Lu 2006).

White rot fungi have a high potential for decolorization of recalcitrant dyestuffs in the environment. However, full understanding of the biodegradation mechanism of these microorganisms is still to be realized. It has been demonstrated that some products accumulate after decolorization of azo dyes by white rot fungi. The biotransformation products that resist further degradation may be toxic or genotoxic. Thus, the toxicity of dyes after fungal degradation should be assayed. The relationship between the structures of parent pollutants and toxicity after treatment should be studied, and methods to detoxify the toxic products are needed. Furthermore, characterization of unknown enzymes involved in the decolorization is desirable to understand the biodegradation procedure. Examining other factors that may influence degradation, both enzymatic and non-enzymatic, such as mediators and the presence of small molecules in the enzymatic mechanism by white rot fungi, would be helpful as well as the study of molecular biology to improve the activity and stability of these enzymes. Identifying the intermediates and degradation products is difficult and tedious work, but critical in mechanism studies. More sensitive and advanced analytical methodologies need to be developed to facilitate the detection of the metabolites and intermediates from fungal decolorization.

REFERENCES

Abadulla, E., Robra, K., Gubitz, G.M., Silva, L.M., Cavaco-Paulo, A. 2000) Enzymatic decolorization of textile dyeing effluents. *Textile Research Joural* 70, 409-414.

Aust, S.D. (1995) Mechanisms of degradation by white rot fungi. *Environmental Health Perspectives*, 103, 59-61.

Banat, I.M., Nigam, P., Singh, D., and Marchant, R., (1996) Microbial decolorization of textile-dye-containing effluents: a review. *Bioresource Technology,* 58, 217-227.

Bogan, B.W. and Lamar, R.T. (1995) One-electron oxidation in the degradation of creosote polycyclic aromatic hydrocarbons by Phanerochaete chrysosporium. *Applied and Environmental Microbiology,* 61, 2631-2635.

Bourbonnais, R., Paice, M.G., Freiermuth, B., Bodie, E., Bornemann, S. (1996) Reactivities of various mediators and laccases with kraft pulp and lignin model compounds. *Applied and Environmental Microbiology,* 63, 4627-4632.

Cai, D. and Tien, M. (1993) Lignin-degrading peroxidases of Phanerochaete chrysosporium. *Journal of Biotechnology,* 30, 79-90.

Cai, D. and Tien, M. (1992) Kinetic studies on the formation and decomposition of compound II and III. Reactions of lignin peroxidase with H_2O_2. *Journal of Biological Chemistry,* 267, 11149-11155.

Call, H. P. and Mucke, I. (1997) Minireview: history, overview and applications of mediated ligninolytic systems, especially laccase-mediator-systems (Lignozym-Process). *Journal of Biotechnology,* 53, 163-202.

Cao, H. (2000) *Decolorization of textile dyes by white rot fungi.* Ph.D. dissertation, University of Georgia, Athens, GA.

Chagas, E.P. and Durrant, L. R. (2001) Decolorization of azo dyes by Phanerochaete chrysosporium and Pleurotus sajorcaju. *Enzyme and Microbial Technology,* 29, 474-477.

Chen, H. (2006) Recent advances in azo dye degrading enzyme research. *Current Protein and Peptide Science,* 7, 101-111.

Chivukula, M. and Renganathan, V. (1995) Phenolic azo dye oxidation by laccase from Pyricularia oryzae. *Applied and Environmental Microbiology,* 61, 4374-4377.

Chivukula, M.; Spadaro, J.T.; Renganathan, V. (1995) Lignin peroxidase-catalyzed oxidation of sulfonated azo dyes generates novel sulfophenyl hydroperoxides. *Biochemistry,* 34, 7765-7772.

Chung, K.T. (1983) The significance of azo-reduction in the mutagenesis and carcinogenesis of azo dyes. *Mutation Research/Reviews in Genetic Toxicology* 114, 269-281.

Conneely, A.; Smyth, W. F.; McMullan, G. (1999) Metabolism of the phthalocyanine textile dye Remazol Turquoise Blue by Phanerochaete chrysosporium. *FEMS Microbiology Letters,* 179, 333-337.

Cripps, C.; Bumpus, J. and Aust, S. (1990) Biodegradation of azo and heterocyclic dyes by Phanerochaete chrysosporium. *Applied and Environmental Microbiology,* 56, 1114-1118.

Egli, T. W. (1992) General strategies in the biodegradation of pollutants. In *Metal ions in biological systems.* H. Sigel and A. Sigel, Eds., pp. 1-41, Marcel Dekker, New York, NY.

Faison, B.D. and Kirk, T.K. (1985) Factors involved in the regulation of ligninase activity in Phanerochaete chrysosporium. *Applied and Environmental Microbiology,* 49, 299-304.

Fenn, P. and Kirk, T.K. (1981) Relationship of nitrogen to the onset and suppression of ligninolytic activity and secondary metabolism in Phanerochaete chrysosporium. *Archives of Microbiology,* 130, 59-65.

Glenn, J. K. and Gold, M. H. (1985) Purification and characterization of an extracellular Mn (II)-dependent peroxidase from the lignin-degrading basidiomycete, Phanerochaete chrysosporium. *Archives of Biochemistry and Biophysics,* 242, 329-341.

Glenn, J. K.; Alileswaran, L. and Gold, M. H. (1986) Mn(II) oxidation is the principal function of the extracellular Mn-peroxidase from Phanerochaete chrysosporium. *Archives of Biochemistry and Biophysics,* 251, 688-696.

Goodwin, D.C.; Aust, S.D. and Grover, T.A. (1995) Evidence for veratryl alcohol as a redox mediator in lignin peroxidase-catalysed oxidation. *Biochemistry,* 34, 5060-5065.

Goszczynski, S.; Paszczynski, A.; Pasti-Grigsby, M.B.; Crawford, R.L. (1994) New pathway for degradation of sulfonated azo dyes by microbial peroxidases of Phanerochaete chrysosporium and Streptomyces chromofuscus. *Journal of Bacteriology,* 176, 1339-1347.

Gavril, M. and Hodson, P.V. (2007) Chemical evidence for the mechanism of the biodecolorization of Amaranth by Trametes versicolor. *World Journal of Microbiology and Biotechnology,* 23, 103-124.

Hao, O. J., Kim, H., and Chiang, P. (2000) Decolorization of wastewater. *Critical Reviews in Environmental Science and Technology,* 30, 449-505.

Heinfling, A.; Bergbauer, M. and Szewzyk, U. (1997) Biodegradation of azo and phthalocyanine dyes by Trametes versicolor and Bjerkandera adusta. *Applied Microbiology and Biotechnology,* 48, 261-266.

Heinfling, A.; Ruiz-Duenas, F.J.; Martinez, F.J.; Bergbauer, M.; Szewzyk, U. and Martinez, A.T. (1998) A study on reducing substrates of manganese-oxidizing peroxidases from Pleurotus eryngii, and Bjerkandera adusta. *FEBS Letters,* 428, 141-146.

Hofrichter, M. (2002) Review: Lignin conversion by manganese peroxidase (MnP). *Enzyme and Microbial Technology,* 30, 454-466.

Hou, H.; Zhou, J.; Wang, J.; Du, C.; and Yan, B. (2003) Enhancement of laccase production by Pleurotus ostreatus and its use for the decolorization of anthraquinone dye. *Process Biochemistry,* 39, 1415-1419.

Hublik, G. and Schinner, F. (2000) Characterization and immobilization of the laccase from Pleurotus ostreatus and its use for the continuous elimination of phenolic pollutants. *Enzyme and Microbial Technology,* 27, 330-336.

Jong, E. D.; Field, J.M. and De Bont, J.A.M. (1994) Aryl alcohols in the physiology of ligninolytic fungi. *FEMS Microbiology Review,* 13, 153-188.

Kang, S.O.; Shin, K.S.; Han, Y.H.; Youn, H.D. and Hah, Y.C. (1993) Purification and characterization of an extracellular peroxidase from white rot fungus Pleurotus ostreatus. *Biochimica et Biophysica Acta,* 1163, 158-164.

Kersten, P.J.; Tien, M.; Kalyanaraman, B. and Kirk, T.K. (1985) The ligninase of Phanerochaete chrysosporium generates cation radicals from methoxybenzenes. *Journal of Biological Chemistry,* 260, 2609-2612.

Knapp, J.S.; Newby, P.S. and Reece, L.P. (1995) Decolorization of dyes by wood-rotting basidiomycete fungi. *Enzyme and Microbiological Technology,* 17, 664-668.

Kuan, I.C. and Tien, M. (1993) Stimulation of Mn peroxidase activity: A possible role for oxalate in lignin biodegradation. *Proceeding of the National Academy of Sciences of the United States of America,* 90, 1242-1246.

Lu, W. (2006) Studies of the Degradation of Nitrophenylazopheneylamine Dyes with White Rot Fungus Pleurotus Ostreatus. Ph.D. dissertation, University of Georgia, Athens, GA.

Lu, Y. (2006) The effect of molecular structure on the biodegradation of selected sulphonated phenylazonaphthol dyes by white rot fungus Pleurotus ostreatus. Ph.D. Dissertation, University of Georgia, Athens, GA.

Lu, Y.; Hardin, I.R (2006) 31Analysis of sulfonated azo dyes degraded by White rot fungus Pleurotus ostreatus. *AATCC Review,* 6, 31-36.

Lin, S. and Lin, C. (1993) Treatment of textile waste effluent by ozonation and chemical coagulation. *Water Research,* 27, 1743-1748.

Martinez, A.T. (2002) Molecular biology and structure-function of lignin-degrading heme peroxidases. *Enzyme and Microbial Technology,* 30, 425-444.

Martins, M.A.M.; Ferreira, I.C.; Santos, I.M.; Wueiroz, M.J. and Lima, N. (2001) Biodegradation of bioaccessible textile azo dyes by Phanerochaete chrysosporium. *Journal of Biotechnology,* 89, 91-98.

Martins, M.A.M.; Lima, N.; Silvestre, A.; Queiroz, M. (2003) Comparative studies of fungal degradation of single or mixed bioaccessible reactive azo dyes. *Chemosphere,* 52, 967-973.

McMullan, G.; Meehan, C.; Conneely, A.; Kirby, N.; Robinson, T. and Nigam, P.(2001) Mini-review: microbial decolourisation and degradation of textile dyes. *Applied Microbiology and Biotechnology,* 56, 81-87.

Mester, T. and Tien, M. (2000) Oxidation mechanism of ligninolytic enzymes involved in the degradation of environmental pollutants. *International Biodeterioration & Biodegradation,* 46, 51-59.

Morais, H.; Ramos, C., Forgacs, E.; Cserhati, T. and Oliviera, J. (2002) Using spectrophotometry and spectral mapping technique for the study of the production of manganese-dependent and manganese-independent peroxidases by Pleurotus ostreatus. *Journal of Biochemical and Biophysical Methods,* 50, 99-109.

Moran, C., Hall, M., Howell, R. (1997) Effects of sewage treatment on textile effluent. *Journal of Society of Dyers & Colourists* 113, 272-274.

Moreira, M.T.; Mielgo, I.; Feijoo, G. and Lema, J.M. (2000) Evaluation of different fungal strains in the decolorisation of synthetic dyes. *Biotechnology letters,* 22, 1499-1503.

Novotny, C.; Rawal, B.; Bhatt, M.; Patel, M.; Sasek, V. and Molitoris, H.P. (2001) Capacity of Irpex lacteus and Pleurotus ostreatus for decolorization of chemically different dyes. *Journal of Biotechnology,* 89,113-122.

Ollikka, P., Alhonmakl, K., Leppanen, V.M., Glumoff, T., Rajjola, T., Suominen, I. (1993) Decolorization of azo, triphenyl methane, heterocyclic, and polymeric dyes by lignin peroxidase isoenzymes from Phanerochaete chrysosporium. *Applied Environmental Microbiology,* 59, 4010-4016.

Pasti-Grigsbty, M. B., Paszczynski, A., Gosczynski, S., Crawford, D. L., Crawford, R. L. (1992) Influence of aromatic substitution patterns on azo dye degradability by Streptomyces chromofuscus and Phanerochaete chrysosporium. *Applied Environmental Microbiology,* 58, 3605-3613.

Paszczynski, A. and Crawford, R.L. (1991) Degradation of azo compounds by ligninase from Phanerochaete chrysosporium: involvement of veratryl alcohol. *Biochemical and Biophysical Research Communications,* 178, 1056-1063.

Paszczynski, A. and Crawford, R.L. (1995) Potential for bioremediation of xenobiotic compounds by the white-rot fungus Phanerochaete chrysosporium. *Biotechnology Progress,* 11, 368-379.

Paszczynski, A; Pasti-Grigsby, M.B.; Goszczynski, S.; Crawford, R. L. and Crawford, D.L. (1992) Mineralization of sulfonated azo dyes and sulfanilic acid by Phanerochaete chrysosporium and Streptomyces chromofuscus. *Applied and Environmental Microbiology,* 58, 3598-3604.

Podgornik, H.; Grgic, I. and Perdih, A. (1999) Decolorization rate of dyes using lignin peroxidases of Phanerochaete chrysosporium. *Chemosphere,* 38, 1353-1359.

Porter, J. (1975) Stability of vat, disperse and sulfur dyes to light and water. *Textile Chemist and Colorist,* 7, 130-135.

Robinson, T., Chandran, B., and Nigam, P. (2001) Studies on the production of enzymes by white-rot fungi for the decolourisation of textile dyes. *Enzyme and Microbial Technology,* 29, 575-579.

Rodriguez, E.; Pickard, M.A. and Vazquez-Duhalt, R. (1999) Industrial dye decolorization by laccases from ligninolytic fungi. *Current Microbiology,* 38, 27-32.

Sarkar, S.; Martinez, A. T. and Martinez, M. J. (1997) Biochemical and molecular characterization of a manganese peroxidase isoenzyme from Pleurotus ostreatus. *Biochimica et Biophysica Acta,* 1339, 23-30.

Schoemaker, H.E.; Lundell, T.K.; Hatakka, A. and Piontek, K. (1994) The oxidation of veratryl alcohol, dimeric lignin model compounds and lignin peroxidase: the redox cycle revisited. *FEMS Microbiology Review,* 13, 321-332.

Selvam, K.; Swaminathan, K.; and Chae, K. S. (2003) Decolourization of azo dyes and a dye industry effluent by a white rot fungus Thelephora sp. *Bioresource Technology,* 88, 115-119.

Shaul, G.M.; Holdsworth, T.J.; Dempsey, C.R.; Dostall, K.A. (1991) Fate of water soluble azo dyes in the activated sludge process. *Chemosphere,* 22, 107-119.

Shin, K.S. and Kim, C.J. (1998) Decolorisation of artificial dyes by peroxidase from the white-rot fungus, Pleurotus ostreatus. *Biotechnology letters,* 20, 569-572.

Shin, K.S.; Youn, H.D.; Han, Y.H.; Kang, S.O. and Hah, Y.C. (1993) Purification and characterization of D-glucose oxidase from white rot fungus Pleurotus ostreatus. *European Journal of Biochemistry,* 215, 747-752.

Spadaro, J. T.; Gold, M. H. and Renganathan, V. (1992) Degradation of azo dyes by the lignin-degrading fungus Phanerochaete chrysosporium. *Applied Environmental Microbiology,* 58, 2397-2401.

Spadaro, J.T. and Renganathan, V. (1994) Peroxidase-catalyzed oxidation of azo dyes: mechanism of Disperse Yellow 3 degradation. *Archives of Biochemistry and Biophysics.* 312, 301-307.

Suguimoto, H. H.; Barbosa, A.M.; Dekker, R.F. and Castro-Gomez, R.J. (2001) Veratryl alcohol stimulates fruiting body formation in the oyster mushroom, Pleurotus ostreatus. *FEMS Microbiology Letters* 194, 235-238.

Tien M. and Kirk, T.K. (1984) Lignin-degrading enzyme from Phanerochaete chrysosporium: purification, characterization, and catalytic properties of a unique H_2O_2 –requiring oxygenase. *Proceeding of the National Academy of Sciences of the United States of America,* 81, 2280-2284.

Tien, M. (1987) Properties of ligninase from Phanerochaete chrysosporium and their possible applications. *CRC Critical Reviews in Microbiology,* 15, 141-168.

Torres, E., Bustos-jaimes, I., and Borgne, S. L. (2003) Potential use of oxidative enzymes for the detoxification of organic pollutants. *Applied Catalysis B: Environmental,* 46, 1-15.

Wariishi, H.; Akileswaran, L. and Gold, M.H. (1998) Manganese peroxidase from the basidiomycete Phanerochaete chrysosporium: spectral characterization of the oxidized states and the catalytic cycle. *Biochemistry,* 27, 5365-5370.

Wariishi, H.; Valli, K. and Gold, M.H. (1992) Manganese(II) oxidation by manganese peroxidase from the basidiomycete Phanerochaete chrysosporium. *Journal of Biological Chemistry,* 267, 23688-23695.

Wesenberg, D.; Kyriakides, I.; and Agathos, S.N. (2003) White-rot fungi and their enzymes for the treatment of industrial dye effluents. *Biotechnology Advances*, 22, 161-187.

Wong, Y. and Yu, J. (1999) Laccase-catalyzed decolorization of synthetic dyes. *Water Research,* 33, 3512-3520.

Zapanta, L.S. and Tien, M. (1997) The roles of veratryl alcohol and oxalate in fungal lignin degradation. *Journal of Biotechnology,* 53, 93-102.

Zhao, X.; Hardin, I. (2007) HPLC and spectrophotometric analysis of biodegradation of azo dyes by Pleurotus ostreatus. *Dyes and Pigments,* 73: 322-325.

Zhao, X.; Lu, Y.; Phillips, D.; Hwang, H.; Hardin, I. (2007) Study of biodegradation products from azo dyes during fungal degradation with capillary electrophoresis/ electrospray mass spectrometry. *Journal of Chromatography,* A. 1159 217-224 .

Zhao, X.; Hardin, I.; Hwang, H. (2006) Biodegradation of a model disperse azo dye by the white rot fungus Pleurotus ostreatus. *International Biodegradation and Biodeterioration,* 57: 1-6.

Zhao, X.; Lu, Y.; Hardin, I. (2005) Determination of biodegradation products from sulfonated azo dyes by Pleurotus ostreatus using capillary electrophoresis coupled with mass spectrometry. *Biotechnology Letters,* 27: 69-72.

Zheng, Z and Obbard, J.P. (2002) Oxidation of polycyclic aromatic hydrocarbons (PAH) by the white rot fungus, Phanerochaete chrysosporium. *Enzyme and Microbial Technology,* 31, 3-9.

In: Leading-Edge Environmental Biodegradation
Editor: L. E. Pawley, pp. 297-315

ISBN: 978-1-60021-903-0

Chapter 10

NEW TRENDS IN ASSESSMENT OF COMPOSTABILITY OF BIODEGRADABLE POLYMERIC AND PACKAGES

Gaurav Kale, Thitisilp Kijchavengkul and Rafael Auras*

School of Packaging, Michigan State University, East Lansing, MI-48824-1223, USA

ABSTRACT

There is growing potential for the adoption of biodegradable polymers in packaging applications due to municipal solid waste concerns and packaging sustainability initiatives. Since fossil fuel based polymers are difficult to recycle when soiled with contaminants, the use of biodegradable polymers opens the opportunity to successful recycling by composting them along with other compostable materials. However, to better understand this alternative, it is necessary to investigate the compostability performance of biodegradable polymers and packages in real composting conditions as a function of time. This information would provide a time-frame for composting these packages and in general provide a better representation of biodegradation trend. This chapter discusses the current ASTM international and European standards for evaluating biodegradation of plastics in simulated composting conditions, where the biodegradation is correlated to the conversion of carbon content in the polymer into carbon dioxide. There are different methods of recording carbon dioxide such as cumulative measurement (which is recommended by ASTM/ISO), direct measurement (using a computerized control), and gravimetric measurement (weight change due to CO_2 absorption). All three methods and associated results are discussed in this article. Actual testing of biodegradable packages in real composting conditions is also described.

Keywords: compostability, biopolymers, packaging, biodegradation

* Phone: (517) 432-3254; F: (517) 353-8999; E-mail: aurasraf@msu.edu

1. INTRODUCTION

Composting is a natural process in which organic material such as leaves, wood waste, manure, and food scraps are decomposed into a soil like substance called humus and yields CO_2. This process occurs because of a variety of mesophilic and thermophilic microorganisms which consume carbon present in organic material as a food source and nitrogen for building cells. A 30:1 carbon nitrogen ratio results in efficient composting. Apart from carbon and nitrogen sources, heat, moisture, and oxygen are also essential for reproduction of microorganisms which help in the decomposition of organic matter. The composting process occurs in two phases: active composting and curing period stage. During active composting, the temperature raises as long as organic matter is properly mixed and sufficient supply of moisture and oxygen is achieved. The temperature is usually stable and stays around 55°C depending on the ways the compost pile is set up such as windrow composting, turned windrow, static windrow, or even in-vessel composting system. Sometimes turning is done frequently to promote better aeration and mixing of organic matter which enhances the decomposition. At the end of active composting, when microorganisms activity decreases, the curing stage starts. In the curing stage, microorganisms consume the remaining carbon until all the carbon is converted into carbon dioxide, although the activity at this stage is at much slower rate and at lower temperatures as no more turning is involved [1].

Organic materials which are commonly composted are yard waste (leaves, grass clippings, garden wastes, tree trunks, etc), food waste (residential which are mostly food waste generated from individual or homes, or commercial), and manure waste (organic material excreted by animals, feces or bodily waste); out of which yard waste is the most commonly composted material in the USA. In the United States, source separated composting has gained much attention; since by offering flexibility in classification of waste streams, it can make composting more successful. In countries such as Germany, Austria, and France source separated composting is very common, and home collecting schemes are in place to facilitate this process. Currently in both the USA and Europe, there is no facilitating infrastructure to deploy polymeric packaging materials in the composting process as most of them are not biodegradable, and they are considered as contaminants. In fact, plastics' separation in compost is an important process for compost feedstock and end product to eliminate any contamination caused due to plastics [2].

In 2005, 13.7 million tons (5.5%) of plastics packaging waste was generated in the municipal solid waste (MSW) stream (245.7 million tons) in the United States out of which only 9.4% was recovered through recycling [3]. Plastics in packaging are normally used for packaging of food, medical, and consumer goods applications, and since they are soiled with contaminants at the end of their life cycle; this makes recycling difficult and they are usually landfilled. Hence, there has been a recent growth in interest and use of biodegradable plastics in packaging applications, which opens a new area of waste disposal, composting. As with other composting amendments, microorganisms consume carbon present in plastics and evolve CO_2. Standards have been developed by ASTM International (originally known as ASTM or the American Society for Testing and Materials) and ISO (International Organization for Standardization) to evaluate the biodegradability of biodegradable plastics based on CO_2 evolution or oxygen demand [2].

In the past, there has been confusion about what is "biodegradable" and how to measure the biodegradability of a material. Biodegradability, for instance, was measured by the reduction of mechanical properties and/or the reduction in molecular weight. Therefore, ASTM and ISO introduced standards related to compostable plastics to clarify the confusion whether certain packages/coatings are compostable and to be used as standard specification, guideline, and testing methods for assessing the compostability of those materials [2]. These standards describe terminology and definitions, testing guidelines, procedures, testing conditions, significance, limits, and results interpretation to assess biodegradation of materials.

This chapter will briefly explain the details of both series of standards, including advantages and limitations, and it will introduce new trends and modifications to these standards, especially the use of vermiculite in the compost mixtures or as inoculums sources. After that, three types of equipments used for testing biodegradation will be presented, including (1) a standard cumulative measurement respirometric system (CMR), (2) a new in-house built direct measurement respirometric system (DMR), and (3) a novel gravimetric measurement respirometric system (GMR). The results obtained from those systems will also be discussed. Finally, an innovative system to test biodegradation of polymeric packages under real composting conditions will be presented.

a) ASTM International Standards

The heart ("holy grail") of ASTM standards related to compostable plastics is ASTM D 6400-04 [4] "Standard Specification for Compostable plastics" for plastic product/packages and ASTM D6868-03 [5] "Standard Specification for Biodegradable Plastics Used as Coatings on Paper and Other Compostable Substrates" for coating substrates. In these standards, in order to be identified as compostable, the polymeric material must meet three requirements, including satisfactory disintegration, biodegradation, and terrestrial and aquatic safety (i.e., no adverse impact from the compost) [2]. According to ASTM D 6400 and D 6868, disintegration and biodegradation of the test materials have to be tested based on ASTM D 5338-98 [6] (reapproved 2003) test method, in which the test materials are incubated with the compost mixture in the close vessels called "bioreactors" for 45 days or until reaching plateau stage, but not more than 180 days at 58°C. The ratio of compost mixture to test materials is 6:1. At least 12 bioreactors are required, 3 blanks, 3 negative controls such as polyethylene, 3 positive controls such as corn starch or cellulose, and 3 for the test material, to test one material, or 15 bioreactors for testing 2 materials and so forth. To pass the disintegration requirement there should be no more than 10% of the original material's dry weight remaining, after sieving the final compost. To pass the biodegradation requirement at the end of the test, the % mineralization (i.e., the cumulative percentage of organic carbon converted into carbon dioxide gas) of the test material must be greater than 60% for a material containing a single polymer (homopolymer and random copolymer) or 90% for a material containing block/graft copolymer, blend, or low molecular weight additives. The standards do not define properly what low molecular weight additives are, and how low the molecular weight of the additives should be at the end of the test. Finally, to pass the safety requirement, the concentration of heavy metals in the plastic materials must be less than 50% of those in 40CFR§503.13 (Table 1) [7], and the germination rate and the plant biomass of the sample compost conducted based on section 6.5 om ASTM D 6002 [7] and OECD guideline 208 "Terrestrial plant Growth Test" must not be less than 90% of that of the control compost [2, 4]. Tomato plants are normally used for the test.

Table 1. Pollutant concentration from table 3 of 40CFR§503.13 [8]

Pollutant	Monthly average concentration (mg/Kg)
Arsenic	41
Cadmium	39
Copper	1500
Lead	300
Mercury	17
Nickel	420
Selenium	100
Zinc	2800

Note: The concentrations of heavy metals in plastic materials should be less than half of those in the table 1.

b) European and ISO Standards

Similarly to ASTM standards, the core of the ISO standards is EN 13432:2000 [9] which describes the specification, test scheme, and guidelines for the testing. To be identified as compostable, the materials must pass four criteria, including material characterization, disintegration, biodegradation, and compost quality or ecotoxicity. These four criteria are much similar to the ASTM's three requirements. The details of biodegradation test are described in ISO 14855-1:2005 [10] and ISO 14855-2:2006 [11]. The ISO 14855-1 is similar to ASTM D5338 except that: (1) a negative control material is not required in the test, (2) the acceptance criterion for % mineralization is 90% of the maximum of that of the positive control at the plateau stage, and (3) activated vermiculite can be used instead of the mature compost. In the ISO standard, the disintegration test must be conducted in the full-scale composting and not under the laboratory scale conditions, but the acceptance criteria are still the same at 10% remnants of the dry weight. ISO 14855-2 is a standard under development and is different than ISO 14855-1 in terms of measurement of CO_2 and amount of compost and sample used for testing in laboratory conditions. This will be explained later in this chapter.

There are several advantages to both series of the standards. First, they provide the basic knowledge and understanding of the key terminology and definitions, especially the meaning of "biodegradable" and "compostable", since these words are different but sometimes are used interchangeably. According to ASTM D 6400 [4], the term "biodegradable plastic" is defined as "*a plastic whose degradation results from the action of natural microorganisms such as bacteria, fungi, and algae.*" On the other hand, "compostable plastic" is *"a plastic that biodegrade(s) during the composting to yield CO_2, water, inorganic compound, and biomass at a rate consistent with other known compostable materials and leave(s) no visible, distinguishable, or toxic residue"*[4]. It can be said that for a polymer to be certified as compostable it must be biodegradable, but the opposite is not true. Compostable plastic is a subset of the biodegradable plastic. Second, they provide the scientific and measureable criteria to determine whether such material is compostable or not, and, furthermore, these standards are used by the certified bodies, such as the Biodegradable Products Institute (BPI), DinCertco, and Vinçotte, to certify/label the products as compostable. Therefore, composters can be ensured that any products with this label are compostable. An example of the certification labels obtained after the certification process is shown in the Figure 1. Testing according to standards, such as ASTM D6400 and EN 13432, is a time-consuming process. The biodegradation test may take from 2 to 6 months to complete, plus 3 months for the plant growth test; therefore, the total testing time may take up to 9 months or a year. In addition, the equipment for biodegradation or respirometer test is very costly. Availability of the approved testing laboratories is another drawback of the standards. According to BPI in January 2005, there were only 13 approved testing laboratories globally. Only two of them are located in the U.S. (Advanced Materials Center, Inc. in Illinois and Organic Waste Systems in Ohio), while the majority of them are located in Europe, with 6 labs in Germany and one in Belgium, and the rest are in Japan and Korea [12].

Both ASTM and ISO standards are intended for plastics designed to be composted in municipal or industrial composting facilities, and not for backyard composting. Also, both of the biodegradation tests (ASTM D 5338 and ISO 14855) are not intended to (1) simulate the environment in any composting system specifically and (2) infer the time required for biodegradation in a real composting facility.

c) Modifications and New Trends in Existing Standards

Standards by ASTM and ISO are currently being modified, under development or in phase of development to establish new methods for successful representation of biodegradation of plastics. ISO 14855-1 [10] allows the use of vermiculite as a composting medium instead of compost when the biodegradation is affected by a priming effect from the test materials and when the final carbon balance and retrieval of test material residues need to be performed [10]. Vermiculite is a clay material known to be a microbial carrier and the heated and expanded vermiculite has high water holding capacity. It can be categorized by size into three types: concrete, medium, and fine. The concrete type, which has density of 80 ± 16 kg/m^3 and 80% of its particles size are between 4 and 12 mm, is recommended for the biodegradation test [10, 11].

Prior to using the vermiculite instead of the compost mixture in a biodegradation test, the vermiculite has to be activated by inoculating it with a solution containing organic, inorganic nutrients, and mature compost extract for 3-4 days at 50 ± 2°C [10]. The mixture solution used for preparing the activated vermiculite is shown in Table 2-4 [10]. When using activated vermiculite the ratio of vermiculite to the dry weight of the sample is reduced to 4:1, instead of the 6:1 required when using a compost mixture.

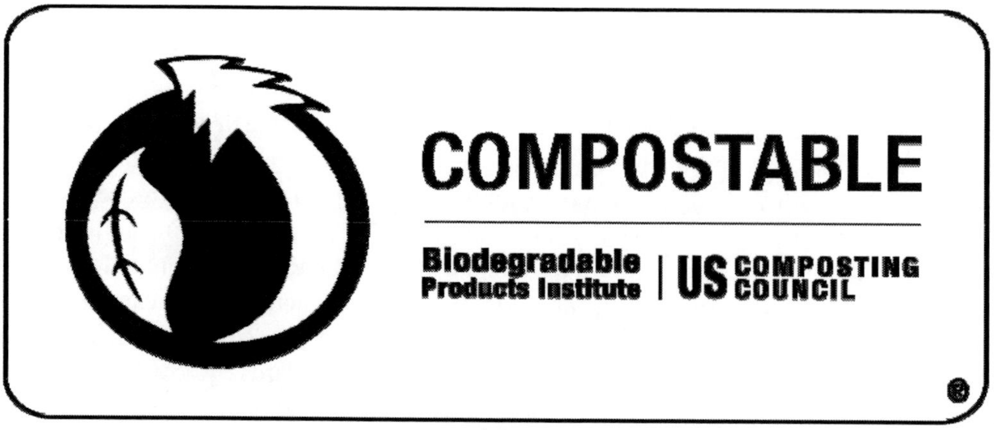

(a)

(b)

(c)

Figure 1. (a) BPI compostable logo, (b) DIN CERTCO compostable logo, (c) Vinçotte OK compost logo. Reprinted with permission of BPI (New York, US), DIN CERTCO (Berlin, Germany) and Vinçotte (Vilvoorde, Belgium).

Table 2. Composition of 1 L inoculum solution [10]

Constituent	Mineral Solution (from table 3)	Suitable nutrient broth	Urea	Corn Starch	Cellulose	Compost extract
Amount	500 ml	13 g	5.8 g	20 g	20 g	500 ml

Table 3. Composition of 1 L mineral solution [10]

Constituent	KH_2PO_4	$MgSO_4$	$CaCl_2$ (10% solution)	NaCl (10% solution)	Trace element solution (from table 4)
Amount	1 g	0.5 g	1 ml	1 ml	1 ml

Table 4. Composition of 1 L trace element solution [10]

Constituent	H_3BO_3	KI	$FeCl_3$	$MnSO_4$	$(NH_4)_6Mo_7O_{24}$	$FeSO_4$
Amount	500 mg	100 mg	200 mg	400 mg	200 mg	400 g

The method of measuring CO_2 can be the same as the one proposed in ISO 14855-1 for samples in compost. ISO 14855-2 (gravimetric method) is the second part of ISO 14855-1 (general method) and also correlates biodegradation by the amount of evolved CO_2 from compost and sample. The gravimetric method differs from the general method in the following aspects: (a) the way the CO_2 is measured, (b) the amount of compost and sample used which is 1/10 of the one used in ISO 14855-1, (c) the amount of replicates of the sample, two for controls and two for blanks, (d) the use of vermiculite or inert material such as sea sand for better aeration in compost, and (e) the taking out and turning of the compost mixture to homogenize humidity and improve aeration[11]. The final validity criteria for ISO 14855-2 test is: (a) the mineralization of the positive control should be more than 70% after 45 days and (b) the difference between replicates of both positive controls should be less than 20%. More on methods and results on ISO 14855-2 will be discussed later. No standard except ISO 14855-2 advices to mix vermiculite or any other inert material with compost; however, mixing vermiculate provides added advantages to the simulated composting process such as; (a) very good water holding capacity as compared to compost, (b) less CO_2 than mature compost thereby reducing background noise, and (c) better aeration through mixing and less channeling of air flow through it thereby providing better air supply to the whole cross section [11].

2. Biodegradation under Simulated Conditions

The key advantages of testing biodegradation under simulated conditions in laboratories are that: (1) the operator can have greater control over the temperature and moisture content of the compost mixture and (2) testing under real conditions using a compost pile may not be practical in some areas, where an institutional composting facility is not available. ASTM and

ISO provide the guidelines for the biodegradation test under simulated conditions, including the apparatus and testing conditions. About the testing apparatus, one may recall one of the experiments in biology class where a grass hopper, cricket, or chick is kept in a closed jar connected with a plastic tubing filled with water, and the critter or animal's respiration rate is measured based on how far the water is moved in the tube. That setup is called a respirometer or a respirometric system. To measure the biodegradation of plastic, this very same instrument principle is used, but in a much larger scale.

Respirometric System

When polymers aerobically biodegrade, carbon in the polymer molecules is converted by microorganisms into biomass or humus, water, carbon residues, and carbon dioxide gas as defined by equation 1 [11].

$$C_{polymer} + O_2 \quad \rightarrow \quad CO_2 + H_2O + C_{residue} + C_{biomass} \tag{1}$$

Therefore, the degree of aerobic biodegradation can be determined by measuring the total amount of carbon dioxide (CO_2) gas evolved from the polymer and can be calculated as a percentage of mineralization as defined earlier in this chapter and described by the equations 2 and 3. The % carbon content of the test material can be measured by elemental analysis or theoretically calculated from the chemical formula of the material, where the former method is preferred.

The respirometric system or respirometer can be used to measure the amount of CO_2 gas evolving from biodegradable polymers during the aerobic biodegradation. The respirometric system is a device used for monitoring and/or measuring the respiration activity of living organisms by either measuring oxygen (O_2) consumption, CO_2 evolution or both. It consists of 3 major components: (1) an air supply, (2) an air-tight closed vessel called the "bioreactor", which contains the living test organisms, and (3) a measuring device, to quantify the amount of O_2 uptake or CO_2 release. The basic concept or mechanism of using the respirometric system to measure the biodegradability of the polymers is that the microorganisms living in the bioreactor will use the oxygen from the air supply system and consume carbon from the biodegradable polymer as food source and generate CO_2, which can be measured by different techniques and is explained later [13].

Percent mineralization is calculated on the basis of CO_2 evolved from blank and test materials compared to the theoretical amount of CO_2 that should have evolved. The theoretical value of CO_2 for a test material is determined by:

$$ThCO_2 = m \text{ x } w_c \text{ x } \frac{44}{12} \tag{2}$$

where;

m = Mass of test material, introduced in test system

w_c *= Carbon content of test material*

Percent mineralization (%B) from amount of CO_2 evolved is given by:

$$\%B = \frac{\sum(CO_2)_T - \sum(CO_2)_B}{ThCO2} \text{ x } 100 \quad (3)$$

where;

$\sum (CO_2)_T$ = *cumulative amount of* CO_2 *in grams, evolved from test material between the start and the end of the test*

$\sum (CO_2)_B$ = *mean cumulative amount of* CO_2 *in grams, evolved from blank between the start and the end of the test.*

According to ASTM D5338 [6] and ISO 14855 [14], respirometric systems can be categorized by their measuring technique as: (1) cumulative measurement respirometric system (CMR), (2) direct measurement respirometric system (DMR), and (3) gravimetric measurement respirometric system (GMR).

a) Cumulative Measurement Respirometric System (CMR)

In the cumulative technique or CMR (see Figure 2), the CO_2 present in ambient air has to be eliminated from the reactor supplied air by using a scrubber system such as NaOH solution, NaOH pellets, soda lime, or calcium hydroxide ($Ca(OH)_2$) pellets. The acid-base titration method is normally used to determine the amount of evolved CO_2. The evolved CO_2 gas is trapped in a sodium hydroxide (NaOH) or barium hydroxide ($Ba(OH)_2$) solution. Then, the trap solution is titrated using a known concentration of hydrochloric acid (HCl). The advantages of the CMR system are: (1) easy to build, (2) simple calculation method, and (3) economically favorable. On the down side, this system requires a lot of time and labor to properly operate due to the number of titrations. For instance, during the first week of the test, the compost mixture releases large amounts of CO_2 and an operator has to titrate the traps quite often, probably every other day, for all the three replicates. This number is multiplied if there are couples of samples running at the same time, or the operator may need to titrate more than once for each replicate to reduce the titration errors.

A biodegradation test on polylactide (PLA) bottles was conducted in a CMR system as per ASTM D5338 at Michigan State University (East Lansing, MI). The PLA bottle (25 grams) was cut into pieces (0.01m x 0.01m) and mixed with the compost mixture. Mixture of manure compost (400 grams dry weight) and vermiculite (100 grams) was used with a moisture level of 90% of its water holding capacity. The test was carried out in controlled conditions of 58°C and 60±13% RH. The CMR results for the PLA bottle and cellulose for 40 days are shown in Figure 3. Cellulose reached 60% mineralization value on the 30th day, whereas PLA reached 60% between the 38th and 40th days. The compost sample ratio was very big (16:1) as compared to 6:1 as recommended by ASTM/ISO. One advantage of a higher compost sample ratio is that the more is the surface area or quantity of the compost, the more is the reproduction of microorganisms. Similar biodegradation studies were also conducted on the same PLA bottles using different methods of CO_2 measurement and are discussed later in the chapter.

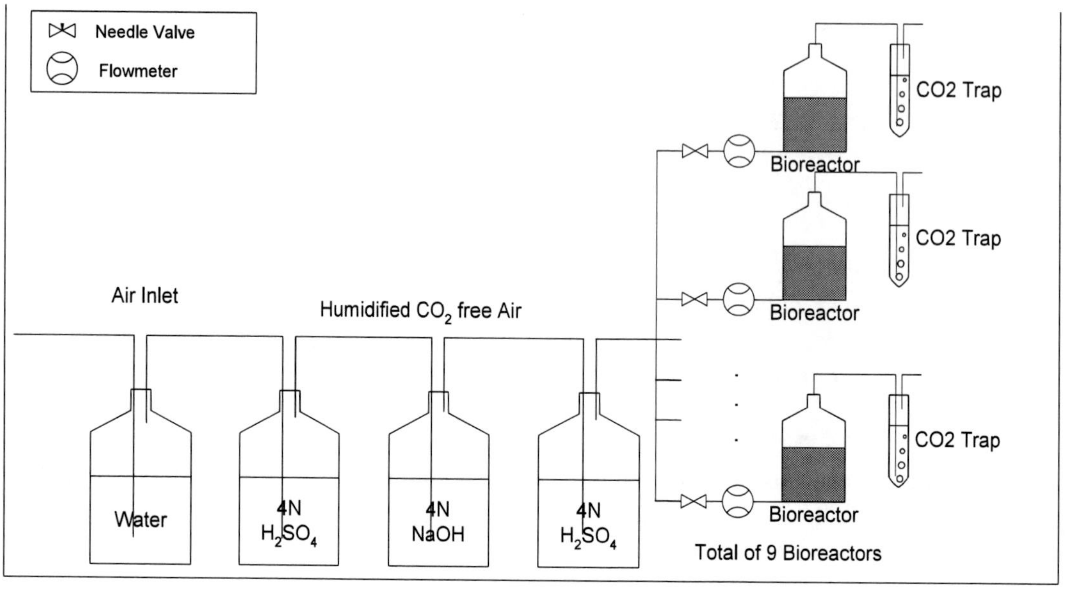

Figure 2. Schematic drawing of Cumulative Measurement Respirometric (CMR) system. Reprinted from Polymer Testing, 25(2006), T. Kijchavengkul et al., Development of an Automatic Laboratory-Scale Respirometric System to Measure Polymer Biodegradability, pp 1006-1016, Copyright 2006, with permission from Elsevier.

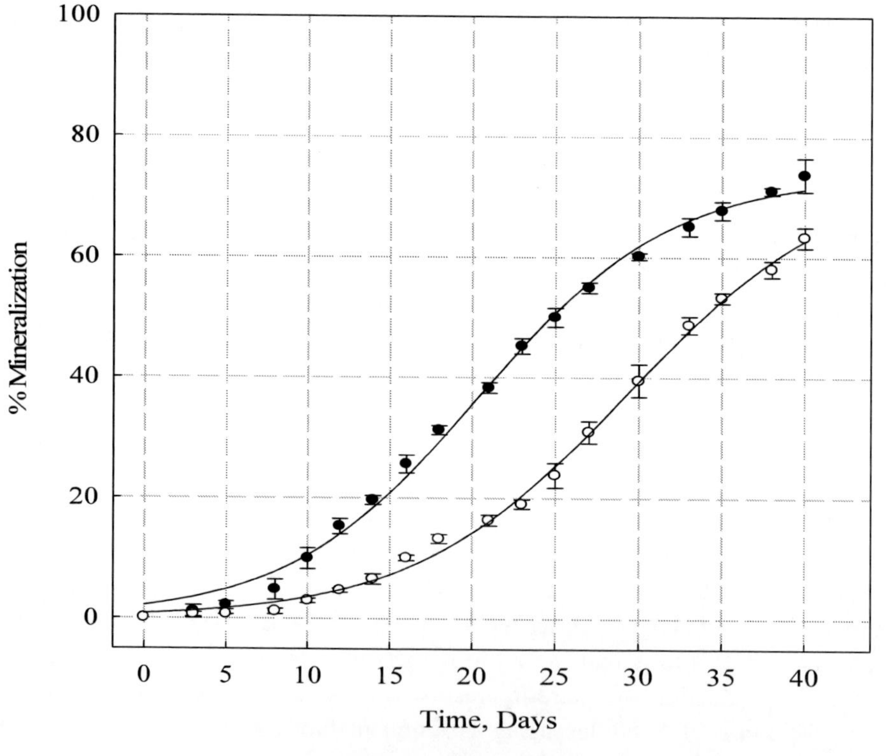

Figure 3. Percentage of mineralization of PLA bottles (○) and cellulose (●) in the CMR system.

b) Direct Measurement Respirometric System (DMR)

In the DMR, the concentration of CO_2 is directly measured from the exhaust of the bioreactor using gas chromatography (GC), or an in-line infrared gas analyzer, such as a non-dispersive infrared (NDIR) gas analyzer, as shown in Figure 4. For this method, the CO_2 concentration of the exhaust gas of each bioreactor is measured at least once every six hours [6]. This allows the use of a computerized switching system, which can simultaneously facilitate the biodegradation testing of many samples. In addition according to ASTM D5338, the ambient CO_2 does not need to be eliminated, since the detector can correctly measure the concentration of CO2. The advantages of DMR system over CMR system are: (1) ease of use since all the operations are computerized and the data is already converted from concentration to mass of carbon dioxide evolution and mineralization making it easy to interpret and analyze and (2) eliminates human intervention and errors especially from titration. On the other hand, this system is much costlier than the CMR due to the detection system and the switching system. Also, programming all the DMR operation sequences is time-consuming and requires special expertise to do this task. Commercially DMR systems are available.

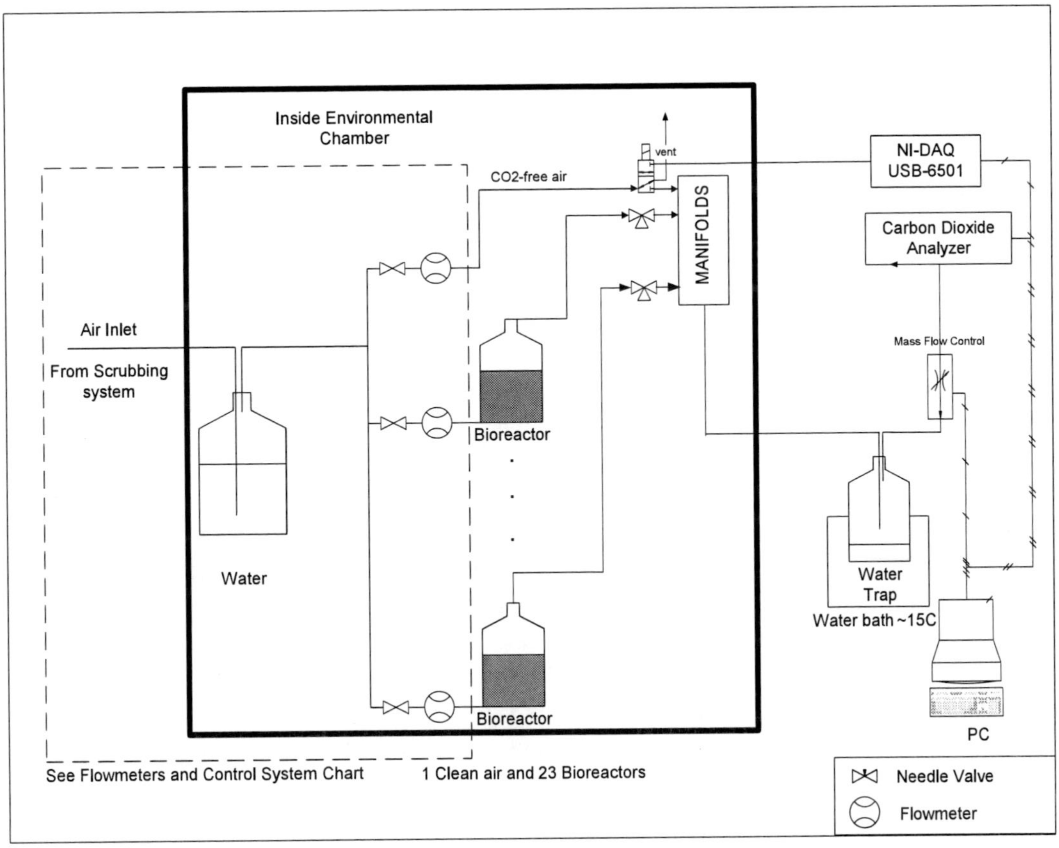

Figure 4. Schematic drawing of direct measurement respirometric (DMR) system. Reprinted from Polymer Testing, 25(2006), T. Kijchavengkul et al., Development of an Automatic Laboratory-Scale Respirometric System to Measure Polymer Biodegradability, pp 1006-1016, Copyright 2006, with permission from Elsevier.

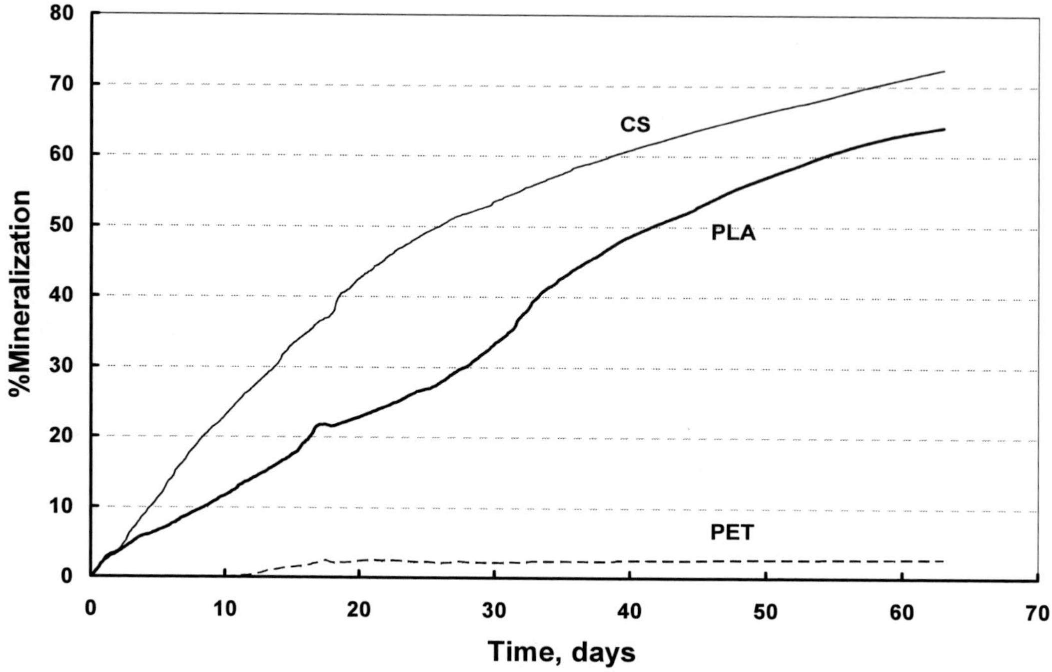

Figure 5. Percentage of mineralization of PLA, PET, and corn starch (CS) at 58±2°C and approximately 55±5% RH in yard waste compost. Reprinted from Polymer Testing, 25(2006), T. Kijchavengkul et al., Development of an Automatic Laboratory-Scale Respirometric System to Measure Polymer Biodegradability, pp 1006-1016, Copyright 2006, with permission from Elsevier.

Results by some of the authors of biodegradation test of PLA bottles compared to corn starch (positive control) and poly(ethylene terephthalate) or PET bottles conducted using an in-house built DMR system at the School of Packaging, Michigan State University (MSU) are shown in Figure 5 [13]. The testing conditions were 58±2°C and 50-60%RH, and yard waste compost was used as medium [13]. The PLA bottles were biodegraded within 63 days, since at that time the % mineralization was greater than 60% and that of the positive control (corn starch) was greater than 70%. This result complied with the disintegration requirement of ASTM and the biodegradation requirement of both ASTM and ISO standards, but the test could not confirm the compostability of the PLA bottles, since no compost quality and safety testing was conducted after the biodegradation test.

c) Gravimetric Measurement Respirometric System (GMR)

The GMR technique is a standard method (under development) for determining the biodegradability of a plastic material in simulated controlled conditions (i.e., temperature, humidity, and aeration ratio). In this method a small scale reactor (6×10^{-3} m^3) is used to determine the biodegradability of plastic material correlating its biodegradation to the CO_2 evolution from the reactor. The test material is mixed with a mixture of mature compost and inert material such as vermiculite or sea sand. The amount of CO_2 evolved is captured by CO_2 absorption columns and is based on weight increase in the columns. Two blanks, two known reference material, and two test material reaction columns are recommended in this standard. The test is considered as valid if the percent biodegradation of known reference material reaches more than 70% after 45 days, and the difference between the percent biodegradation of both reference material columns is less than 20% at the end of the test [11]. A system built under this specification has been manufactured by Hissan Trading Company Ltd, Japan known as "Microbial Oxidative Degradation Analyzer" (MODA) [15]. MODA consists of four reaction columns; one for blank, one for positive control and two for test material. The schematic of the MODA system is shown in Figure 6.

In this system, pressurized air is passed through a decarbonation column containing soda lime to remove CO_2 in air, later it is bubbled through deionized water to generate humidity and sustain moisture levels in the reaction chamber. The humidified air is passed through a reaction column containing a mixture of mature compost, inert material and test/reference material. In the reaction column, oxygen from air is consumed by microorganisms and evolves into CO_2. This evolved CO_2 is passed through 1M H_2SO_4 to remove ammonia; later through columns of silica gel and calcium chloride ($CaCl_2$) to remove moisture, and finally through CO_2 ad/absorption columns. CO_2 is mainly absorbed by the column containing soda lime and sodium hydroxide. While CO_2 is absorbed, water is released in the reaction which is later captured by the column containing calcium chloride. The weight of both columns is considered for correlating the biodegradation rate [11].

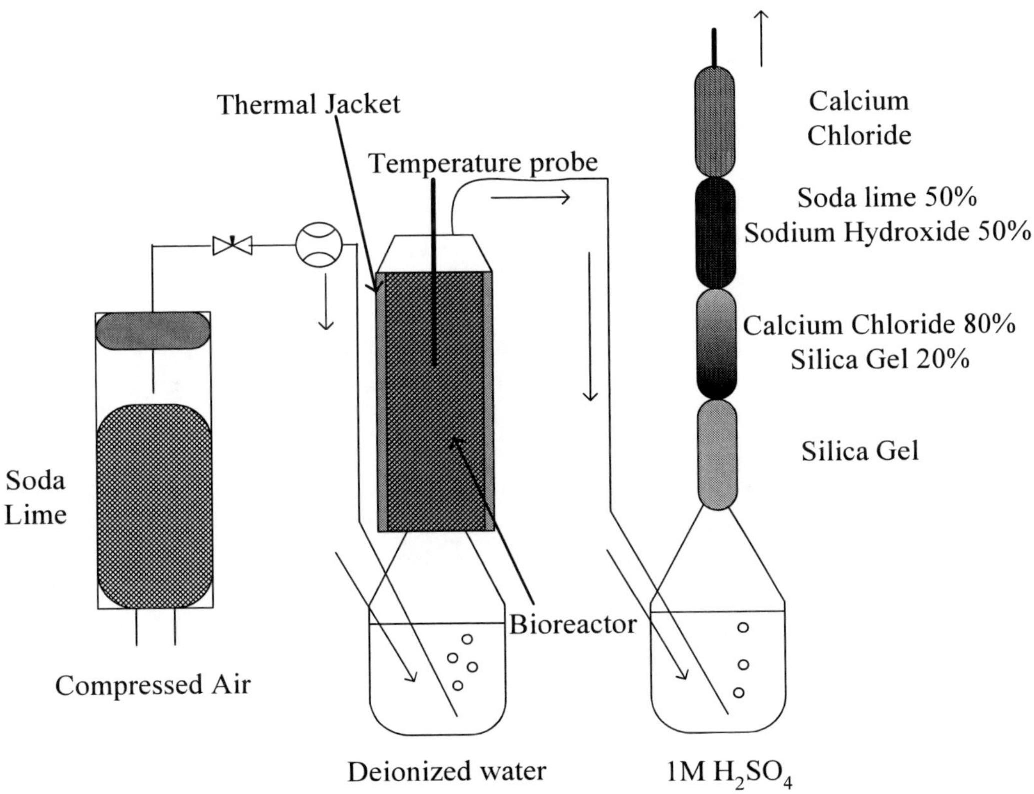

Figure 6. Schematic of GMR system based on ISO 14855-2.

(a) Preparation of Inoculum and Test Material for MODA

Mature compost 3-4 months aged is recommended for use in this test. The compost should be free of glass, stones and metals; and before use it should be sieved through a 3 mm screen, which is different from the 5-10 mm one used in ISO 14855-1/ASTM D5338. The pH of the compost should be between 7 and 9. Inert material such as sea sand or vermiculite is used for mixing with compost to improve aeration and moisture [11].

(b) Measurement of Carbon Dioxide

As mentioned before, CO_2 from the bioreactor is absorbed by the CO_2 absorption columns and is then correlated to the level of biodegradation. The absorption column consists of soda lime (NaOH immobilized to slaked lime) and soda talc (NaOH immobilized to talc) in a 1:1 ratio and the produced CO_2 is completely absorbed by the following reaction [11, 15]:

$$CO_2 + 2NaOH \rightarrow Na_2CO_3 + H_2O \tag{4}$$

The separated H_2O obtained in this reaction is further trapped by the $CaCl_2$ column and the weight of those two columns is used to calculate the weight of produced CO_2. The weight of both columns is measured frequently until the whole organic carbon content in the polymer is consumed by the microorganisms. This consumption should ideally take the biodegradation up to 100%. CO_2 measurement in this case is quite easy as compared to the CMR system's which is a laborious task due to titration. However, it is very important to maintain the accuracy and precision in the measurement of weight. Apart from that, it is also important that the chemicals are properly secured in the column and that there are no losses in the middle of the experiment. Even losses of 1 gram of NaOH or $CaCl_2$ pellet from the reaction column would result in the need to repeat the experiment from the beginning.

A study with the GMR system was carried out at Michigan State University on PLA bottles (10 grams divided into 0.01m x 0.01m pieces) and the results are shown in Figure 7.

In this case, cellulose shows again a faster initial biodegradation as compared to PLA and reached 80% mineralization on the 67th day. PLA initially during its hydrolysis action shows very slow mineralization rates as seen before in the CMR and DMR systems. However, after the 20th day there was a rapid increase in the biodegradation and 80% mineralization was reached on the 62nd day. Two PLA bottles samples were used in this study; hence the results shown in Figure 7 are an average of both. The error difference at the end of the 67th day was ±10%, the same since the beginning of the experiment. Hence, concluding biodegradation and its time representation on the basis of two samples with large error becomes difficult.. Kunioka et al. [16] also conducted a biodegradation test on PLA and cellulose in a similar GMR system and they found similar and close results for cellulose. PLA reached 80% biodegradation on the 40th day itself and went up to 95% on the 60th day. However, PLA used in their case was in the form of powder (average particle size of 92.5μm) whereas in the above case pieces from PLA bottle (0.01m x 0.01m) were used; which affects the degradation rate. In fact, Kunioka et al. [16] showed that the size of PLA particles affects the biodegradation rate, the smaller the particles the faster is the biodegradation. For example, they used PLA powders in the ranges of 0-125μm, 0-250μm, 125-250μm, and 250-500μm and found that 0-125μm degraded much rapidly (100% mineralization in 60 days) as compared to the other particle sizes [16].

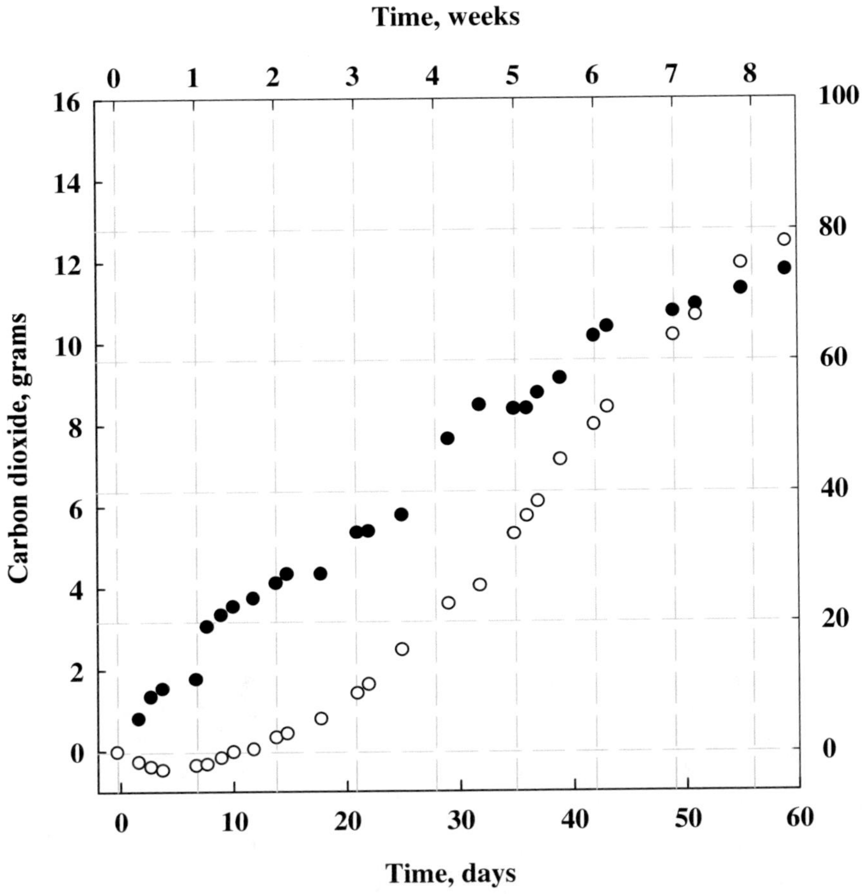

Figure 7. Carbon dioxide in grams and % mineralization as a function of time in weeks for PLA bottles (○) and cellulose (●) in the GMR system.

A similar study was conducted on polycaprolactone (PCL) powder by Funabashi et al. [17] on an GMR system for its biodegradation with cellulose as a standard. Cellulose powder showed similar biodegradation trends as mentioned before in both studies and reached an 80% biodegradation value in around 55-60 days. PCL was very fast as compared to cellulose and reached an 80% biodegradation value in a mere 10 days. Those results were also compared with the CMR system (using the 14855-1 standard) and different results for PCL (80% biodegradation in 20 days) were found. Variability in particle size and its effect on biodegradation was also studied by Funabashi et.al on PCL powder (246.3μm) and pellets (3mm) and found that there was a negligible effect of the particle size on its biodegradation value which is the opposite of the case of PLA as mentioned before [16, 17].

3. Biodegradation in Real Conditions

The standards and results discussed above are all in simulated composting conditions where parameters such as temperature and humidity are controlled. The report based on those results gives us an understanding of mineralization of plastic material and also the time it takes to be mineralized completely in microbial environments such as composting. However, it does not completely depict the performance of plastics in real composting conditions because real conditions are governed by atmospheric conditions, compost raw materials, surface area of composting, and microbial reproduction.

A study on biodegradability of biodegradable packages such as PLA bottle, trays and deli container was carried out at MSU by some of the authors [18, 19]. A composting pile made up of wood shavings, wood waste and cow manure waste was built at MSU's composting facility. The size of the compost pile was 6 x 24 x 3 meters and initial temperature, relative humidity and pH were 65 ± 5°C, 63 ± 5%, and 8.5 ± 0.5, respectively [18, 19]. Due to the huge size of the compost pile, it was difficult to track and trace the packages in the pile. Hence, an open wooden box with mesh on its bottom was used to place packages and monitor their biodegradability performance through visual inspection and variations in their physical properties. The PLA bottles were placed in the compost pile in duplicate sets of seven. Those sets were taken out on days 1, 2, 4, 6, 9, 15, and 30, and every time bottles were visually inspected for changes in shape, color, and texture. Later the samples were collected for analyzing changes in physical properties such as molecular weight, glass transition, melting, and decomposition temperatures. For more information on methodology, readers can refer to references [18, 19].

Figure 8. Biodegradation of PLA bottles in real composting conditions. Reprinted from Macromolecular Bioscience, 7 (3; 2007, G. Kale et al., Compostability of bioplastic packaging materials: An Overview. pp 255-277, Copyright 2007, with permission from John Wiley & Sons, Inc.

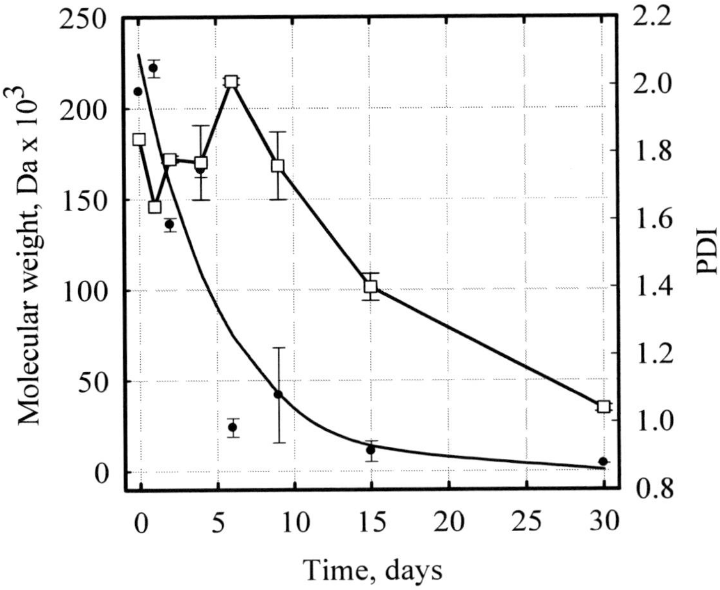

Figure 9. Variation in molecular weight (●) and PDI (□) for PLA bottles in real composting conditions.

Biodegradability performance of PLA bottles is shown in Figure 8. Degradation in those bottles can be seen already in the first day as reflected by changes in the bottles' shapes. Initially, PLA undergoes hydrolysis and random non-enzymatic chain scission of ester groups occurs. As a result, there is a reduction of molecular weight. This process is called fragmentation. These low molecular weight oligomers are consumed by microorganisms to yield CO_2, water and soil-like substance called humus [18, 19]. These reactions can be explained in more detail through looking at variation in physical properties such as molecular weight.

Figure 9 shows the variation in molecular weight and polydispersity index (PDI) for the PLA bottles in real composting conditions. Fragmentation of PLA bottles can be observed on the 15th day, when the molecular weight decreased to 11,000 Daltons from 210,600 Daltons [18, 19]. This could be correlated to the hydrolysis reaction as mentioned before. Later these low molecular weight bottle residuals were consumed by microorganisms and CO_2 evolved during this process. This type of trend was also seen in the case of PLA bottles' biodegradability analysis in CMR, DMR, and GMR techniques, where initially CO_2 evolution was very little, and it increased after the initial fragmentation. However, in all the simulated techniques, the temperature and humidity were controlled which give a more even degradation trend. Also, on the other hand, in real conditions the whole package (PLA bottle) was tested, whereas in simulated conditions (i.e., CMR, DMR, and GMR) small pieces (0.01m x 0.01m) of bottles were tested. If the PLA bottles are placed in the composting pile in the form of small pieces then the fragmentation and hence, biodegradation occurs much faster. Variation in PDI is also shown on the right axis in Figure 9. PDI was reduced to 1.04 on the 30th day from 1.84 on day zero, mainly because of the polymer fragmentation which narrows the molecular weight distribution.

With regards to the variation in molecular weight in each day of the study, we can see that there was an initial rise in molecular weight of 11,400 Daltons in the first days which could be due to recombination reactions in PLA [18-21]. Also on the 4th, 6th, and 9th days major variation in the molecular weight is also observed which could be due to different microbial reproduction, temperature, and humidity at different places in the compost pile. Hence, optimum conditions are essential for successful and even composting of those packages.

REFERENCES

[1] R. Rynk, *On-farm composting handbook*, ed. R. Rynk. 1992: Natural Resource, Agriculture and Engineering Service (NRAES) Cooperative Extension.

[2] G. Kale, et al., *Compostability of bioplastic packaging materials: an overview*. Macromolecular Bioscience, 2007. *7*: p. 255-277.

[3] EPA. *Municipal Solid Wastes (MSW) in United States: 2005 facts and figures*. 2006 [cited 2006 11-20-2006]; Available from: http://www.epa.gov/garbage/pubs/ex-sum05.pdf

[4] *ASTM D 6400-04 Standard Specification for Compostable Plastics*. 2004, ASTM International: West Conshohocken, PA.

[5] *ASTM D 6868-03 Standard Specification for Biodegradable Plastics Used as Coatings on Paper and Other Compostable Substrates*. 2003, ASTM International: West Conshohocken, PA.

[6] *Standard Test Method for Determining Aerobic Biodegradation of Plastic Materials Under Controlled Composting Conditions*. 1998. *ASTM D 5338-98*.

[7] *ASTM D 6002-96 Standard Guide for Assessing the Compostability of Environmentally Degradable Plastics*. 2002, ASTM International: West Conshohocken, PA.

[8] EPA, *Standards for the Use or Disposal of Sewage Sludge*.

[9] *EN 13432 Packaging-Requirements for packaging recoverable through composting and biodegradation-Test scheme and evaluation criteria for the final acceptance of packaging*. 2000, European Committee for Standardization: Brussels, Belgium.

[10] *ISO 14855-1 Determination of the ultimate aerobic biodegradabilty and disintergration of plastic materials under controlled composting conditions-Method by analysis of evolved carbon dioxide*. 2005, International Organization for Standardization: Geneva, Switzerland.

[11] *ISO/DIS 14855-2 Determination of the ultimate aerobic biodegradability of plastic materials under controlled composting conditions - method by analysis of evolved carbon dioxide. Part 2: Gravimetric measurement of carbon dioxide evolved in a laboratory scale test. (under development)*, ISO/TC61/SC5, Editor. 2006, International Organization for Standardization: Geneva, Switzerland.

[12] *BPI Logo Program*. [cited March 14, 2007]; Available from: http://www.bpiworld.org/BPI-Public/Program.html.

[13] T. Kijchavengkul, et al., *Development of an automatic laboratory-scale respirometric system to measure polymer biodegradability*. Polymer Testing, 2006. *25*(8): p. 1006-1016.

[14] ISO, *ISO 14855-1 Determination of the ultimate aerobic biodegradabilty and disintergration of plastic materials under controlled composting conditions-Method by*

analysis of evolved carbon dioxide. Vol. ISO 14855-1:2005. 2005: International Standards Organization.

[15] *Microbial Oxidative Degradation Analyzer*. [cited 04-15-07]; Available from: http://www.hissan.co.jp/business/moda/e_main.html.

[16] M. Kunioka, F. Nimomiya, and M. Funabashi, *Biodegradation of poly (lactic acid) powders proposed as the reference test material for the international standard of biodegradation evaluation methods.* Polymer Degradation and Stability, 2006. *91*: p. 1919-1928.

[17] M. Funabashi, F. Nimomiya, and M. Kunioka, *Biodegradation of Polycaprolactone powders proposed as reference test materials for international standard of biodegradation evaluation method.* Journal of Polymers and the Environment, 2007. *15*: p. 7-17.

[18] G. Kale, R. Auras, and S.P. Singh, *Degradation of commercial biodegradable packages under real composting and ambient exposure conditions.* Journal of polymers and environment, 2006. *14*: p. 317-334.

[19] G. Kale, R. Auras, and S.P. Singh, *Comparison of the degradability of poly (lactide) packages in composting and ambient exposure conditions.* Packaging Technology and science, 2007. *20*: p. 49-70.

[20] G.H. Kai-Lai, et al., *Field Exposure Study of Polylactid Acid (PLA) Plastic Films in the Banana Fields of Costa Rica 1.* Journal of Environmental Polymer Degradation, 1999. *7*(4): p. 167 - 172.

[21] X. Zhang, et al., *An Investigation of Poly(lactic acid) Degradation.* Journal of Bioactive and Compatible Polymers, 1994. *9*: p. 80 - 100.

In: Leading-Edge Environmental Biodegradation
Editor: L. E. Pawley, pp. 317-331
ISBN: 978-1-60021-903-0

Chapter 11

CHLORINE IN THE FOREST ECOSYSTEM (A RADIOTRACER STUDY)

Miroslav Matucha
Institute of Experimental Botany of Academy of Sciences of the Czech Republic, Videnska 1083, 14220 Prague, Czech Republic

ABSTRACT

In nature, chlorine does not occur only as chloride or bound in substances of anthropogenic origin in the polluted environment but also in many hundreds of compounds of natural origin. Chloride deposited in the forest ecosystem from the atmosphere reacts with soil organic matter (SOM) under the mediation by enzymes and/or microorganisms (to a smaller degree also abiotically), forming chloroacetic acids (CAA), chloromethane, chloroform, other aliphatic and aromatic compounds and chlorinated humic substances. Chloride, to which Norway spruce is sensitive, is thus mostly blocked in organochlorines, which even prevail over chloride in soil. Chlorine is thus involved in SOM degradation, leading to smaller SOM decay products and finally to their mineralization. Closely related microbial processes were met in methodically similar soil bioremediation studies conducted before. The aim of the investigations was first (i) to elucidate the uptake, effects and fate of phytotoxic trichloroacetic acid (TCA, previously considered to be only a secondary atmospheric pollutant) in the Norway spruce/soil-system, (ii) to address the role of CAA, and finally (iii) to consider the global role of chlorine in coniferous forest ecosystems. Radiotracer methods were used with advantage.

Using carbon 14, the fastest uptake of TCA was found to occur in the youngest spruce shoots due to a higher transpiration stream; TCA is then rapidly biodegraded in the forest soil. This microbial degradation depends on TCA concentration, soil humidity and temperature as well as on the soil character and microorganisms. The biodegradation of TCA took place also in the phyllosphere of the needles; TCA elimination in needles may be caused by slow decarboxylation. On the other hand, TCA may be formed by biooxidation of absorbed atmospheric tetrachloroethylene in needle chloroplasts.

Using chlorine 36 it was shown for the first time that chlorination of SOM yields TCA and also dichloroacetic acid (DCA), which is biodegraded even faster than TCA. The chlorination process was shown to proceed more microbially than abiotically, basidiomycetes being suggested as the responsible microorganisms. It is not clear whether the process is mediated by extracellular enzymes or intracellularly, to what degree the abiotic chlorination proceeds, whether dehalogenation of chlorinated substances takes place, and what further influences affect the ecosystem under study.

A connection between carbon and chlorine cycle was clearly shown: products of chlorination of SOM are further mineralized and contribute to the litter decay and a part of chlorine is released from the forest ecosystem again. The previous chlorine level is thus restored. Moreover, the monitoring of chloride, TCA, AOX and other substances in forest soil, needles and soil solution in forest stands, especially near salted roads, showed an adverse effect of these substances on coniferous trees, especially spruce.

INTRODUCTION

Chlorine is one of the most abundant elements on the surface of the Earth and some authors (e.g. Winterton 2000 and Öberg 2003) compare its biological role to that of phosphorus. In forest ecosystems, it is mainly present as chloride ion (Cl_{in}) in the soil and in the biosphere or bound to organic matter (Cl_{org}). As will be shown, it contributes to the degradation of soil organic matter (SOM), which represents the major sink of carbon in the forest ecosystem. Forest SOM is formed from litter, which is decomposed by basidiomycetes (de Jong and Field 1997). An abiotic chlorination of SOM has also been suggested (Keppler et al. 2000, Fahimi et al. 2003). Soils contaminated by chlorinated substances like PCBs may be remediated by degrading microorganisms (Šašek et al. 1993, Vyas et al. 1994a).

The main source of chloride ions in the forest ecosystem is the oceans (in spite of the fact that more chlorine is contained in the Earth's crust and mantle – Winterton 2000), whereas Cl_{org} originates from various sources. The content of chloride in the environment depends on the geographic situation and Cl_{org} is connected with the biogeochemical cycling (Öberg 1998, 2002, 2003). Chlorinated compounds were formerly considered to be of anthropogenic origin only; however, the occurrence of more than two thousand natural chlorinated compounds has now been reported (Winterton 2000, Gribble 2003). Well known are toxic xenobiotics such as PCBs – we have previously studied their degradation by white-rot fungi using PCBs labelled with carbon 14 and suitable analytical methods (Šašek et al. 1993, Vyas et al. 1994a, Zachař et al. 1996, Kubátová et al. 1996a,b, 1998) - or pesticides, e.g. DDT and lindane, or solvents, e.g. tetrachloroethene and methylchloroform. Less known is the natural origin of chlorinated dibenzodioxins and dibenzofurans (Silk et al. 1997), of the antibiotic chloramphenicol and of the methylating agent chloromethane (Wintertone 2000). Several research fields have addressed this issue. They include studies of water disinfection by chlorination showing, among other things, the main chlorination products of humic substances (e.g. Christman et al. 1983, de Leer et al. 1985): Investigations have also been made concerning the effect of trichloroacetic acid (TCA) and chlorinated solvents (Frank and Frank 1989, Frank 1991, Frank et al. 1994, Weissflog et al. 2003, 2004 and 2005) on coniferous forest, the occurence of AOX or Cl_{org} in the forest soil (Öberg 1998, Öberg et al. 2005), chlorination and the litter degradation ability of basidiomycetes (de Jong and Field 1997), the fate of chlorine 36 as a uranium fission product in soils (Lee et al. 2001) and the bleaching of kraft pulps (Solomon 1996 and Juuti 1997).

The role of chlorine in the forest ecosystem drew increased attention after apparent problems with coniferous forest decline in central and northern Europe in the eighties and the nineties. Injuries to conifers such as chlorosis and needle loss, without evident damage caused by coal burning, were observed in regions marked by a higher Cl content in needles and were ascribed to chlorinated solvents (Figge 1989 and 1990) produced industrially and used in large amounts at that time, especially to TCA, their atmospheric photooxidation product (Sidebottom and Franklin 1996, Mc Culloch 2002, Hoekstra 2002). It was presumed that TCA, a compound with phytotoxic properties, arises by atmospheric photooxidation of chlorinated C_2-hydrocarbons - methylchloroform, tri- and tetrachloroethylene - used on a broad scale in large amounts (about 1 million tons/yr at that time) as cleaning and degreasing solvents in the electronic, metallurgical and textile industries, affecting the health state of coniferous forest (McCulloch 2002, Matucha and Uhlířová 2002). The effect of TCA on conifers was studied in Germany (Figge 1989 and 1990, Frank 1984, 1988, 1989 and 1991, Frank and Frank 1985 and 1986, Frank et al. 1989, Frank et al. 1990b, 1991, 1992a,b, and 1994, Plümacher et al. 1993, Plümacher and Schröder 1994, Schröder and Plümacher 1998, Schröder et al. 2003), in Finland (Juuti et al. 1994, 1995 and 1996, Juuti 1997, Norokorpi and Frank 1993 and 1995, Sutinen et al 1995 and 1997), in the Czech Republic (Uhlirova et al. 1995, 1996 and 2002, Matucha et al. 2001, Forczek et al. 2001) and later in Scotland (Cape et al. 2003, Dickey et al. 2004, Heal et al. 2003, Stidson et al. 2004a,b). Natural occurrence of Cl_{org} was indicated many times (e.g.Hoekstra and de Leer, E. W. B.1995, Hoekstra et al. 1995, Gribble 2003). It was found that TCA is also of natural origin and occurs in the forest soil among other places (Hoekstra et al. 1999, Matucha et al. 2007a), It was predicted by Schöler et al. (2003).. This later led to investigations of chlorination of SOM, formation of AOX and associated problems (Öberg et al. 2002, Bastviken et al. 2007, Gryndler et al. 2007, Rohlenová et al. 2007).

Related studies by Weissflog et al. (2001) on the influence of tetrachloroethene on conifers in highly polluted regions of Germany and Russia showed natural formation of PER by microbiological processes in salt lakes (Weissflog et al. 2005), its transformation to TCA in needles and chloroplasts and its effect on conifers (Weissflog et al. 2007, Forczek et al. 2007).

Of similar importance are the studies of Cl_{org} (chlorinated compounds or SOM) in the aquatic and forest ecosystems carried out by the group at Linköping University (Asplund and Grimvall 1991, Öberg et al. 2002, Öberg 2003, Johansson et al. 2001, Öberg and Sandén 2005), showing an occurence of amounts higher than 200 mg Cl_{org}/kg forest soil. They showed that relatively low chloride contents in precipitation or forest soil may lead to a higher Cl_{org} content in forest soil and run-off water (estimated as AOX; an aromatic character of some part of AOX was also indicated).

The approach of the group at Wageningen Agricultural University, whose attention was drawn to chlorination of litter or SOM by basidiomycetes, suggested an ecologically significant role of these microorganisms in decomposition of forest organic matter connected with chlorination activity of the fungi (Field et al. 1995, de Jong and Field 1997, Verhagen et al. 1998). It was shown that the chlorination products are smaller molecules of aromatic and aliphatic character, such as methyl chloride, chloroform, TCA and chlorinated anisyl derivatives (CAMs), as well as chlorinated humic acids of smaller size (Lee et al 2001). They may behave like intermediates, i.e. they may be at the same time microbially degraded and mineralized (Verhagen et al. 1998). A possibility of intracellular microbial chlorination was also mentioned (Steinberg et al. 2006). An abiotic chlorination of SOM (similar to microbial chlorination) was suggested by the group from the University of Heidelberg (Keppler et al. 2000, Fahimi et al. 2003).

METHODOLOGY

Forest is a very complex ecosystem and it is thus very difficult to sample its compartments, e.g. soil. It is much easier to study processes in vitro than to investigate them in the field or by lysimeter experiments. –An additional problem with field studies is that it is almost impossible to find a representative, 'mean' part of the ecosystem because of its diversity and inhomogeneity, as indicated by Rodstedth et al. (2003), Öberg et al. and Öberg and Sandén (2005). They attempted to solve this problem by analysing many samples or investigating many soil cores, which is laborious, expensive and perhaps not representative enough. Previous experience and the instrumentation available led us to an application of radiotracer techniques, which have several advantages.

Carbon 14 and chlorine 36, radioisotopes with properties suitable for long-term radioanalytical studies of the role of chlorine in the complex forest ecosystem, i.e. nuclides with long half-lives (5760 and 3.01×10^5 y, resp.) and sufficiently hard β-emitters (E_{max} = 156 and 710 keV) were used. Use of carbon 14 has a long tradition since its discovery and use in the studies of photosynthesis in the fortieth by M. Calvin et al., chlorine 36 was used as a conservative tracer for translocation studies in plants by Gage and Aronoff (1960) and in hydrology (Nyberg et al. 1999).

Syntheses of [1,2-^{14}C]TCA and [1,2-^{14}C] DCA (Bubner et al. 1996 and 2001) of high specific radioactivity (100 mCi/mmol) and radiochemical (>98%) and chemical purity and two analytical methods (Matucha et al. 2004 and 2006) as prerequisities of the investigations were developed for the intended research. Another method developed by Bastviken at al. (2007) is still being optimized. The most important advantage of using radioisotopes in environmental research is the possibility of tracing the element or labelled compound examined (following its transport, translocation and distribution), its reactions and metabolism, and also checking the radioactivity budget in the system under study. Importantly, the isotopic exchange of chloride with Cl_{org} was also proved for typical compounds (Matucha et al. 2007b).

Common liquid scintillation was used for measurements of the β-radioactivity of both radioisotopes (Beckman LS 6500). The determination of carbon 14 in soil was conducted by combustion of the soil sample at 900°C in a stream of oxygen and absorption of the developed $^{14}CO_2$ in a suitable liquid scintillator cocktail for radioactivity counting (Zinsser). A similar method was used for determining mineralized $^{14}CO_2$ in the atmosphere after biodegradation of [1,2-^{14}C] chloroacetic acids (CAAs); the flow-through method using absorption in 1.5 M KOH (Stork et al. 1997) was adapted (Matucha et al. 2003).

Extraction of polar compounds from soil and needle samples was conducted by repeated extraction with water or potassium nitrate solution; apolar or less polar substances were extracted by *tert*-butyl methyl ether. Humic substances were extracted by 1M KOH

(Stevenson 1981) or 0.1 M NaOH (Bastviken et al. 2007). The alkaline extraction of humic substances with their subsequent acidic precipitation led, however, to the loss of bound chlorine (Rohlenová et al. 2007). Before extraction of chloride from soil microorganisms their cell walls were destroyed by freezing to -20°C or below and slow thawing (Matucha et al. 2004 and 2007, Bastviken et al. 2007).

Volatile substances such as chloroform (Hoekstra et al. 1998a,b, Svensson et al. 2007) and chlorocarbons were determined by capillary GC, using cryogenic or SPME techniques: detection was conducted mainly by ECD (Varian 3400) and MS. For $CHCl_3$ separation from soil samples before determination a lyophilization was used (Rohlenová and Fuksová, non-published results). Identification was carried out by GC-MS (Ion trap Finnigan 800). TCA in environmental samples was determined after extraction and esterification also by capillary GC combined with ECD or MS detection (Frank et al. 1990 and 1995). AOX and TOX were determined by coulometric titration of chloride after soil sample combustion at 1000°C in an oxygen stream (Euroglas). Thin-layer chromatography after preconcentration by evaporation was used for separation of CAAs. Distribution of radioactivity on TLC chromatograms was measured by a radioluminescence imaging technique (Bio-Imager BAS 5000 Fuji). The methods used are described in detail in the corresponding articles.

Results and Discussion

The applications of radioactively labelled compounds for environmental research, which were methodically prepared several years (Matucha et al. 1998, Bubner et al. 2001, Kubátová et al. 1998 and Matucha and Uhlířová 2002) were first connected with the microbial remediation of PCB contaminated soil by white-rot fungi. In these investigations we verified our capabilities to synthesize and analyze ^{14}C-labelled substances as well as complex PCB mixtures (Šašek et al. 1993, Vyas et al. 1994a, Zachař et al. 1996, Kubátová et al. 1996a,b, 1998).

Our studies using radiotracer techniques and [1,2-^{14}C]TCA in the field of forest research started with the investigation of TCA uptake by Norway spruce shoots and TCA distribution in needles (Uhlířová et al. 1996), and were followed by studies of the behavior of TCA in the spruce/soil-system and then of chlorination of SOM with ^{36}Cl-chloride.

The study of the role of chlorine in the forest ecosystems included investigations of uptake, effects and fate of TCA (first used as a herbicide, see Ashton and Crafts 1973, Lignell et al. 1984, McGrath 1976) and then considered as a secondary atmospheric polutant (Frank 1991, Frank et al. 1994, McCulloch 2002), in Norway spruce/soil system (Matucha et al. 2001). ^{14}C-labelled TCA and DCA were synthesized for this purpose (Bubner et al. 1996 and 2001).

It was shown that the phytotoxic TCA, a polar and well water soluble compound, is readily taken up by roots from soil, then translocated by transpiration stream and accumulated in spruce needles (Uhlířová et al. 1996, Matucha et al. 2000 and 2001, Forczek et al. 2001 and 2004, Schröder et al. 2003). Its higher concentrations cause chlorosis, necrosis and needle loss and lower shoot increments (Uhlířová et al. 2002). Microbial degradation of [1,2-^{14}C] TCA in soil leads to a release of $^{14}CO_2$ into the atmosphere, influencing thus substantially the experiments with spruce plants in closed growth boxes by assimilation of the released $^{14}CO_2$

(Matucha et al. 2001). Field experiments with the plant/soil-system (with plants in a pot) were therefore preferred. It was also found that the „soil“ TCA translocated through xylem by transpiration stream and accumulated in needles (most quickly into current needles) is eliminated relatively slowly (Matucha et al. 2001, Forczek et al. 2001, Forczek et al. 2004, Matucha et al. 2006), whilst in the forest soil it is rapidly degraded by microorganisms; for this reason, TCA biodegradation in soil was investigated thoroughly (Matucha et al. 2001 and 2003, Forczek et al. 2001, Schröder et al. 2003).

An investigation of TCA biodegradation in soil has shown that the biodegradation depends on TCA concentration (previously unknown in herbicide applications) (Forczek et al. 2001), humidity, temperature and origin of the forest soil, the fastest (aerobic) microbial degradation was found in the fermentation horizon of mildly humid forest soils (Matucha et al. 2003, Schröder et al. 2003, Matucha et al. 2007b), the microbial activity in inorganic horizon being lower (Matucha et al. 2007b). In sterilized soils TCA degradation was negligible – a clear evidence of microbial character of the degradation (Matucha et al. 2007b). In needles we determined TCA elimination by microbial degradation (Forczek et al. 2004), and in addition also by slow thermal decarboxylation of TCA (Matucha et al. 2006).

As shown recently, there is still another possibility of TCA formation and accumulation in needles: uptake of tetrachloroethylene – a primary air pollutant of both anthropogenic and natural origin and a precursor of TCA – and its penetration through the needle cuticula and all the cell membranes into chloroplasts, followed by its biooxidation there to TCA with its adverse effect on the photosynthetic apparatus as a result of its decay to toxic HCl and CO_2. Compared with the „soil“ TCA transported through xylem, the TCA in chloroplasts has a much stronger adverse impact on spruce health due to the effects of protons from toxic HCl formed in the most sensitive needle subcellular structures (Weissflog et al. 2007, Forczek et al. 2007).

Finally, it was found - and proved by our application of chlorine 36 (Matucha et al. 2007a) - that TCA was formed in forest soil together with DCA and chloroform by microbial chlorination of organic matter, i.e. of aromatic resorcinolic structures as suggested earlier (Hoekstra et al. 1999b). It is caused by hypochlorous acid (i.e. by chlorine radicals) arisen by enzymatic oxidation of chloride at forest soil humidity; probably only a minor part of SOM is chlorinated abiotically by Fenton reaction (Matucha et al. 2007a,b) as shown by chlorination of sterile and non-sterilizied soils. This chlorination mechanism resembles the mechanism of chlorination of dissolved humic substances in water affording the same products (e.g. during water disinfection), however, in the forest soil under natural conditions it was proved for the first time. The combination of both processes – of formation and degradation - means that DCA and TCA represent intermediates in the degradation of SOM (Matucha et al. 2007a) and that their level in the forest soil may corespond to an equilibrium or show fast degradation when absent (Peters 2003). A preliminary examination of behavior of chloride after its uptake in lower forest plants showed also a possibility of formation of volatile organochlorines there (Laturnus and Matucha, 2007). Chloride absorption by microorganisms and formation of chlorinated humic acids in forest soil was reported recently (Bastviken et al. 2007). The author, often cooperating on an international scale, participated in three review articles from this field (Laturnus et al. 2005, Cape et al. 2006, Clarke et al. 2007).

The processes studied - concerning among others chlorine and carbon cycling in the coniferous forest ecosystem - showed that chloride and chloroacetic acids are not inert at all and that they are rapidly translocated into plants, that they are there microbially converted and function as stressors and their reactions in soil lead (concerning chloride content in soil and litter) in the forest ecosystem always to an equilibrium, the main ecological principle of the nature. These studies bring among others also useful practical results showing the formation in soil of chloroform that may be translocated into ground water (and is evaporated to some degree into the atmosphere), and documenting that road salting causes a direct damage to conifers (Kayama et al. 2003).

At the end, the results of the study explain the presence of Cl_{org} including AOX, in the forest soil and show that chlorine (and especially TCA and chloride) in the temperate and boreal forest ecosystems may also be represented as an anthropogenic stressor. It was further shown that chlorine also takes part in the natural SOM decay (and carbon cycle) and some of its metabolic and microbiological mechanisms of this degradation process were elucidated (Matucha et al. 2003, Matucha et al. 2004, Matucha et al. 2007a,b, Rohlenová et al. 2007, Gryndler et al. 2007). Radiotracer methods facilitated the (otherwise not always easy) investigations of the above problems and pointed out the main processes in the forest ecosystems under chlorine participation that represent a substantial part of the chlorine and carbon biogeochemical cycling there. The study also indicated some practical questions connected with analysis (thermal decarboxylation of TCA, TCA reaction with acetone - see Cape et al. 2006, TCA instability during drying of soil samples), with road salting and handling chlorinated solvents.

It is evident that this research field is a developing one without such efforts and responses to the issue as e.g. investigations on toxic PAHs (polycyclic aromatic hydrocarbons) (Vyas et al., 1994, 52 citations), nevertheless, its environmental importance is doubtless.

CONCLUSIONS

The research carried out revealed that

1. The results achieved show the main ecological principle of the nature – the priciple of equilibrium conservation. Chloride deposited in the coniferous forest ecosystems is not conservative; however, it is reactive and contributes to the degradation of SOM, i.e. of litter, which was mediated mostly through microbial processes.
2. Application of radioactively labelled compounds makes possible to follow the processes taking place in the ecological systems studied (TCA in the plant/soil-system, SOM chlorination in the forest ecosystem as well as mycoremediation of PCB contaminated soils) and to check their completeness by means of a radioactivity budget.
3. Several analytical methods were elaborated (elimination of TCA biodegradation in forest soil samples by their freezing, determination of TCA in environmental studies using 14C and 36Cl, extraction of chlorine and Clorg from forest soil, determination of chloroform in soil samples) and some reactions in soil and needles (TCA decarboxylation, reaction of TCA with acetone, substitution of OH-group in veratryl alcohol by chlorine) were explained.

4. TCA and DCA as phytotoxic substances are formed in coniferous forest soil mainly by enzymatic or microbial chlorination of soil organic matter; in addition, TCA is produced in needle chloroplasts by biooxidation of penetrating tetrachloroethylene.
5. TCA uptake from soil by roots is caused by the transpiration stream and is highest in the current year needles and slower in older needles. It can be slowly retransported from needles through phloem back into the soil.
6. TCA in soil is microbially degraded to carbon dioxide and chloride and used as the source of carbon and energy. The biodegradation rate depends on TCA concentration, soil character and humidity, temperature and microbial activity. The degradation at low environmental concentrations proceeds much more rapidly than was shown by investigations of TCA used as herbicide in the past.
7. The chlorination process explains the formation of chlorinated compounds, including AOX, and - because of their biodegradation – it may be considered as one of the processes in SOM (litter) decomposition.
8. Chlorination of soil organic matter is a slow process mediated by microorganisms, above all basidiomycetes, which take up chloride from soil first. Chlorination (proceeding also abiotically) leads to chlorinated aliphatic and aromatic substances (including chloroacetic acids and chlorinated phenols) as well as to smaller chlorinated molecules of humic substances. Some of these are volatile (e.g. chloromethane, chloroform).
9. It is obvious that some environmental processes may be very slow and take place also at low concentrations (e.g. TCA decarboxylation in needles). The results achieved show again the main ecological principle of nature – the priciple of equilibrium conservation.
10. A practical result was also achieved: knowledge of the behavior of chloride in soil is important for road salting and its impact on afflicted vegetation.

REFERENCES

[1] Ashton, F.M., Crafts, A.L. (1973) „*Mode of Action of Herbicides*" Wiley, N.Y., London, Sydney, Toronto.

[2] Asplund, G., Grimvall, A. (1991) "Organohalogens in nature, more widespread than previously assumed" *Environ. Sci. Technol. 25*, 1346-1350.

[3] Asplund, G.; Christiansen, J. V.; Grimvall, A. B. (1993) A chloroperoxidase-like catalyst in soil. Detection and characterization of some properties. *Soil Biology & Biochemistry 25*, 41-46.

[4] Bastviken, D., Thompsen, F., Karlsson, S., Svensson, T., Sandén, P., Shaw, G., Matucha, M., Öberg, G. (2007) "Chloride Retention in Forest Soil by Microbial Uptake and by Natural Formation of Organochlorines", *Geochim. Cosmochim. Acta 71*, 3182-3192. doi:10.1016/j.gca.2007.04.028.

[5] Bubner, M., Meyer, M., Fuksová, K., Matucha, M., Heise, K.H., Nitsche, H. (1998) „Synthesis of Uniformly ^{13}C- and ^{14}C-Labelled PCB-Congeners", *Ecological Chem. (St. Petersburg), 7*, 65-68

[6] Bubner, M., Matucha, M., Šašek, V. Heise, K. H., Nitsche, H., Velemínský, J., Spížek, J. (1997) „Investigation of Microbial Degradation of PCB Congeners Using Isotopically Tracers“, *Ecological Chem. (St. Petersburg) 6* , 204-211 (1997)

[7] Bubner, M., Fuksová, K., Matucha, M., Heise, K. H., Bernhard, G. (2001) "Synthesis of [1,2-^{14}C] Trichloroacetic Acid“, *J. Labelled Cpd. Radiopharm. 44*, 811-814.

[8] Cape, J.N., Forczek, S.T., Gullner, G., Mena-Benitez, G., Schröder,P., Matucha, M. (2006) „Progress in understanding the sources, deposition and above-ground fate of trichloroacetic acid.“ *Environ. Sci. Pollut. Res. 13*, 276-286.

[9] Christman, R.F., Nordwood, D.L., Millington, D.S., Johnson, J.D. (1983) „Identity and yields of major halogenated products of aquatic fulvic acid chlorination“ *Environ. Sci. Technol. 17*, 625-628.

[10] Clarke, N., Fuksová, K., Gryndler, M., Liste, H.-H., Matucha, M., Rohlenová, J., Schroll, R. (2007) „The formation and fate of chlorinated organic substances in temperate forest soils“ *Environ. Sci. Pollut. Res.*, submitted.

[11] de Jong E., Field, J.A. (1997) „Sulfur Tuft and Turkey Tail: Biosynthesis and Biodegradation of Organohalogenes by Basidiomycetes“ *Annu. Rev. Microbiol. 51*: 375-414.

[12] de Leer, E.W.B., Damste, J.S., Erkelens, C., de Galan, L. (1985) „Identification of intermediates leadingto chloroform and C-4 diacids of the chlorination of humic acids“ *Environ. Sci. Technol. 19*, 512-522.

[13] Dickey, C.A., Heal, K.V., Stidson, R.T., Koren, R., Cape, J.N., Schröder, P., Heal, M.R. (2004) „Trichloroacetic acid cycling in Sitka spruce (Picea sitchensis) saplings and the effects on tree health following long term exposure“, *Environ. Pollution 130*, 165-176.

[14] Fahimi, I.J., Keppler, F., Schöler, H.F. (2003) „Formation of chloroacetic acids from soil, humic acids and phenolic moities“ *Chemosphere 52*, 513-520.

[15] Field, J.E., Verhagen, F.J.M., de Jong, E. (1995) "Natural organohalogen production by basidiomycetes" *TIBTECH 13*, 451-456.

[16] Figge, K. (1989) „Wirkungen halogenierter Aliphaten auf Pflanzen“ *VDI Berichte* Nr. 745, 527-561.

[17] Figge, K. (1990) „Luftgetragene, organische Stoffe in Blattorganen - Vorgang der ad/absorptiven Anreicherung“ *UWSF-Z. Umweltchem. Ökotox. 2*, 200-207.

[18] Forczek, S.T., Matucha, M., Uhlířová, H., Albrechtová, J., Fuksová, K., Schröder P. (2001) "Biodegradation of Trichloroacetic Acid in Spruce/Soil-System", *Biologia Plantarum 44*, 317-320.

[19] Forczek, S.T., Uhlířová, H., Gryndler, M., Albrechtová, J., Fuksová, K., Vágner, M., Schröder, P., Matucha, M. (2004) "Trichloroacetic Acid In Norway Spruce/Soil-System. II. Degradation in the Plant" *Chemosphere 56*, 327-333.

[20] Forczek, S.T., Schröder, P., Weissflog, L., Rohlenová, J., Uhlířová, H., Fuksová, K., Krüger, G., Matucha, M. (2007) "Trichloroacetic acid of different origin in needles of Norway spruce (Picea abies /L./ Karst.): content in chloroplasts and effects on the plant“ *Biologia Plantarum*, accepted.

[21] Frank, H. (1984) „Waldschäden durch Photooxidantien?“ *Nachr. Chem. Tech. Lab. 32*, 298-305.

[22] Frank, H.; Frank, W. (1985) "Chlorophyll-Bleaching by Atmospheric Pollutants and Sunlight - A Probable Cause of Tree Damages Observed in Recent Years" *Naturwissenschaften 72*, 139-141.

[23] Frank, H., Frank, W. (1986) „Photochemical activation of chloroethenes leading to destruction of photosynthetic pigments" *Experientia 42*,1267-1269.

[24] Frank, H. (1988) „Trichloressigsäure im Boden: eine Ursache neuartiger Waldschäden" *Nachr. Chem. Tech. Lab.* 36, 889.

[25] Frank, H.; Frank, W. (1989) „Uptake of Airborne Tetrachloroethene by Spruce Needles" *Environ. Sci. Technol. 23*, 365-367.

[26] Frank, H.; Vital, J.; Frank, W. (1989) "Oxidation of airborne C_2-chlorocarbons to trichloroacetic and dichloroacetic acid" *Fresenius Zeitschrift für Anal. Chem. 333*, 713.

[27] Frank, H. (1989) „Neuartige Waldschäden und luftgetragene Chlorkohlenwasserstoffe" *Z. Umweltchem. Ökotox. 4, 7-11.*

[28] Frank, H.; Vincon, A.; Reiss, J.; Scholl, H. (1990a) „Trichloroacetic Acid in the Foliage of Forest Trees." *HRC-Journal of High Resolution Chromatography 13*, 733-736.

[29] Frank, H.; Vincon, A.; Reiss, J. (1990b) „Montane Baumschäden durch das Herbizid Trichloressigsaeure" *Z. Umweltchem. Ökotox. 2*, 208-214.

[30] Frank, H. (1991) „Airborne Chlorocarbons, Photooxidants, and Forest Decline." *Ambio 20*, 13-18.

[31] Frank, W.; Neves, H. J. C.; Frank, H. (1991) „Levels of Airborne Halocarbons at Urban and Mountain Forest Sites in Germany and at the Atlantic Coast" *Chemosphere 23*, 609-626.

[32] Frank, H.; Scholl, H.; Sutinen, S.; Norokorpi, Y. (1992a) „Trichloroacetic acid in conifer needles in Finland" *Annales Botanici Fennici 29*, 263-267.

[33] Frank, H.; Scholl, H.; Sutinen, S.; Norokorpi, Y. (1992b) „Trichloroacetic acid, a ubiquitous herbicide in Finnish forest trees." In Tikkanen, E.; Varmola, M.; Katermaa, T.(Eds.) *Arctic Centre Publications,* 259-261, University of Lapland, Rovaniemi, 1992.

[34] Frank, H., Scholl, H., Renschen, D., Rether, B., Laouedj, A., Norokorpi, Y. (1994) „Haloacetic Acids, Phytotoxic Secondary Air Pollutants." *Environ. Sci. Pollut. Res. 1*, 4-14.

[35] Frank, H.; Renschen, D.; Klein, A.; Scholl, H. (1995) „Trace Analysis of Airborne Haloacetates" *HRC-Journal of High Resolution Chromatography 18*, 83-88.

[36] Gage, R.S., Aronoff, S. (1960) „Translocation III. Experiments with Carbon 14, Chlorine 36, and Hydrogen 3" *Plant Physiol.* 35, 53-64

[37] Gribble, G.W. (2003) „The diversity of naturally produced organohalogenes" *Chemosphere 52*, 289-298.

[38] Gryndler, M., Rohlenová, J., Kopecký, J., Matucha, M. (2007) *„Soil microbial community at elevated chloride concentration" Chemosphere*, accepted.

[39] Heal, M. R.; Dickey, C. A.; Cape, J. N.; Heal, K. (2003) „The Routes and Kinetics of Trichloroacetic Acid Uptake and Elimination in Sitka Spruce (*Picea sitchensis*) Saplings via Atmospheric Deposition Pathways" *Atmospheric Environment 37*, 4447-4452.

[40] Hjelm, O., Borén, H., Öberg, G. (1996) „Analysis of halogenated organic compounds in coniferous forest soil from a *lepista nuda* (wood blewitt) fairy ring" *Chemosphere* 32, 1719-1728.

[41] Hoekstra, E. J.; de Leer, E. W. B. (1995) "Organohalogens: the natural alternatives." *Chem. Ber. 31*, 127-131.

[42] Hoekstra, E. J.; Lassen, P.; van Leeuwen, J. G. E.; de Leer, E. W. B.; Carlsen, L. (1995) "Formation of organic chlorine compounds of low molecular weight in the chloroperoxidase-mediated reaction between chloride and humic material." (Grimvall, A.; de Leer, E. W. B., Eds.) *Naturally-produced organohalogens.* 149-158. Dordrecht, Kluwer Academic Publishers.

[43] Hoekstra, E. J.; Verhagen, F. J. M.; Field, J. A.; de Leer, E. W. B.; Brinkman, U. A. T. (1998a) „Natural production of chloroform by fungi." *Phytochemistry 49*, 91-97.

[44] Hoekstra, E. J.; de Leer, E. W. B.; Brinkman, U. A. T. (1998b) „Natural formation of chloroform and brominated trihalomethanes in soil" *Environ. Sci. Technol. 32*, 3724-3729.

[45] Hoekstra, E. J.; Juuti, S. (1999) „New Directions: Exchange of comments on "The origins and occurrence of trichloroacetic acid". *Atmospheric Environment 33*, 4526-4527.

[46] Hoekstra, E. J. (2003) „Review of concentrations and chemistry of trichloroacetate in the environment" *Chemosphere 52*, 355-370.

[47] Hoekstra, E. J., de Leer, E. W. B., Brinkman, U. A. Th. (1999) „Findings supporting the natural formation of trichloroacetic acid in soil" *Chemosphere 38*, 2875-2883.

[48] Johansson E., Ebena, G., Sandén, P., Svensson, T., Öberg, G. (2001) Organic and inorganic chlorine in Swedish spruce forest soil: influence of nitrogen. *Geoderma 101*, 1 - 13

[49] Juuti, S.; Norokorpi, Y.; Ruuskanen, J. (1994) „Trichloroacetic acid in conifer needles in area of kraft pulp mill plume" *Organohalogen Compounds 20*, 289-292.

[50] Juuti, S.; Norokorpi, Y.; Ruuskanen, J. (1995) „Trichloroacetic-Acid (TCA) in Pine Needles Caused By Atmospheric Emissions of Kraft Pulp-Mills" *Chemosphere 30*, 439-448.

[51] Juuti, S., Norokorpi, Y., Helle, T., Ruuskanen, J. (1996) "Trichloroacetic acid in conifer needles and arboreal lichens in forest environments." *The Science of the Total Environment 180*, 117-124.

[52] Juuti, S. (1997) *Trichloroacetic Acid in Forest Environment.* PhD Thesis , Univ. Kuopio.

[53] Kayama, M., Quoreshi, A.M., Kitaoka, S., Kitahashi, Y., Sakamoto, Y., Maruyama, Y., Kitao, M., Koike, T. (2003) "Effects of deeicing salt on the vitality and health of two spruce species, *Picea abies* Karst., and *Picea Glehnii* Masters planted along roadsides in nothern Japan" *Environmental Pollution 124*, 127-137.

[54] Keppler, F., Eiden, R., Niedan, V., Pracht, J, Schöler, H.F. (2000) " Halocarbons produced by natural oxidation processes during degradation of organic matter" *Nature 403*, 298-301.

[55] Kubátová, A., Matucha, M., Bubner, M. (1996a) „Application of 13C-Labelled Polychlorinated Biphenyl Congener 153 as Internal Standard in the Gas Chromatographic-Mass Spectrometric Analysis of Polychlorinated Biphenyls", *J. Chromatogr. A 750*, 245-251

[56] Kubátová, A., Matucha, M., Ševčík, J.G.K. (1996b) „Application of Correlation Analysis for Identification of Polychlorinated Biphenyls", *J. Chromatogr. A 752,* 197-207

[57] Kubátová, A., Matucha, M., Erbanová, P., Novotný, Č., Vlasáková, V., Šašek, V. (1998) Investigation of PCB Biodegradation Using ^{14}C-Labelling. *Isotopes in Environmental and Health Studies 34*, 325-334

[58] Lange, C.; Weißflog, L.; Strasser, R. J.; Krueger, G. H. J.; Pfennigsdorff, A. (2004) „Phytotoxic effects of trichloroacetic acid on Scots pine and birch determined by chl a fluorescence and the JIP-test“ *South African Journal of Botany 70*, 683-694.

[59] Laturnus, F., Fahimi, I., Gryndler, M., Hartmann, A., Heal, M. R., Matucha, M., Schöler, H.-F., Schroll, R., Svensson, T. (2005) „Natural Formation and Degradation of Chloroacetic Acids and Volatile Organochlorine Compounds in Forest Soil – Present Knowledge and Future Perspectives“ *Environ. Sci. Pollut. Res. 12*, 233-244.

[60] Laturnus, F., Matucha, M. (2007) „Chloride - a pre-cursor in the formation of VOCls by forest plants?” *J. Environ. Radioactivity,* doi:10.1016/j.jenvrad2007.07.008, accepted.

[61] Lee, R.T., Shaw, G., Wadey, P., Wang, X. (2001) „Specific association of 36Cl with low molecular weight humic substances in soils“ *Chemosphere 43*, 1063-1070.

[62] Lignell, R., Heinonentanski, H., Uusirauva, A. (1984) „Degradation of Trichloroacetic-Acid (TCA) in Soil.“ *Acta Agr. Scand. 34*,3-8.

[63] Matucha, M., Bubner, M., Kubátová, A., Erbanová, P., Uhlířová, H., Novotný, Č., Šašek, V. (1998) Some Applications of Isotopic Labelling in Ecological Research. Acta Universitatis Carolinae Environmentalica *12*, 49-60

[64] Matucha, M., Uhlířová, H., Fuksová, K., Forczek, S. T. (2000) „Halogenderiváty C_2-uhlovodíků a poškození lesa: účinek kyseliny trichloroctové na smrk ztepilý“(“Halogenderivatives of C_2-hydrocarbons and forest damage”) (In Czech) *Chem. Listy 94*, 830.

[65] Matucha, M., Uhlířová, H., Bubner M. (2001) “Investigation of Uptake, Translocation and Fate of Trichloroacetic Acid in Norway Spruce (Picea Abies /L./ Karst.) Using ^{14}C-Labelling“ *Chemosphere 44*, 217-222.

[66] Matucha, M., Uhlířová, H. (2002) „Těkavé chlorované uhlovodíky a úhyn lesa“ („Volatile chlorinated hydrocarbons and forest decline“), *Biologické listy 67*, 161-176.

[67] Matucha, M., Forczek, S.T., Gryndler, M., Uhlířová, H., Fuksová, K., Schröder, P., (2003) „Trichloroacetic Acid in Norway Spruce/Soil-System. I. Biodegradation in Soil“, *Chemosphere 50*, 303-309.

[68] Matucha, M., Gryndler, M., Forczek, S.T., Uhlířová, H., Fuksová, K., Schröder, P., (2003) „Chloroacetic acids in environmental processes“, *Environ. Chem. Letters, 1*,127-130.

[69] Matucha, M., Gryndler, M., Uhlířová, H., Fuksová, K., Rohlenová, J., Forczek, S.T., Schröder, P. (2004) “Microbiological aspects of determination of trichloroacetic acid in soil” *Folia Microbiologica 49*, 117-122.

[70] Matucha, M., Rohlenová, J˙, Forczek, S.T., Uhlířová, H., Gryndler, M., Fuksová, K., Schröder, P. (2006) „Determination of Trichloroacetic Acid in Environmental Studies Using Carbon 14 and Chlorine 36” *Chemosphere 63*, 1924-1932.

[71] Matucha, M., Gryndler, M., Schröder, P. Forczek, S.T., Uhlířová, H., Fuksová, K., Rohlenová, J˙, (2007a) „Chloroacetic Acids - Degradation Intermediates of Organic Matter in Forest Soil“ *Soil Biol. Biochem. 39*, 382-385.

[72] Matucha, M., Gryndler, Forczek, S.T., M., Schröder, P., Bastviken, D., Rohlenová, J, Uhlířová, H., Fuksová, K. (2007b) „A Chlorine-36 And Carbon-14 Study of the Role of Chlorine in the Forest Ecosystem“ *J. Labelled Cpd. Radioph. 50*, 437-439. doi: 10.1002/jlcr.1180

[73] McCulloch, A. (2002) „Trichloroacetic acid in the environment“ *Chemosphere 47*, 667-686.

[74] McGrath, D. 1976. Factors that influence the persistence of TCA in soil. *Weed Res 16*, 131-137.

[75] Niedan, V., Pavasars, I., Öberg, G. (2000) „Chloroperoxidase-mediated chlorination of aromatic groups in fulvic acids“ *Chemosphere 41*, 779-785.

[76] Norokorpi, Y.; Frank, H. (1993) „Effect of Stand Density On Damage to Birch (Betula pubescens) Caused By Phytotoxic Air-Pollutants“ *Annales Botanici Fennici 30*, 181-187.

[77] Norokorpi, Y., Frank, H. (1995) "Trichloroacetic Acid As a Phytotoxic Air Pollutant and the Dose-Response Relationship For Defoliation of Scots Pine" *Science of the Total Environment 160/161*, 459-463. Nyberg, L., Rodhe, A., Bishop, K. (1999) „Water transit times and flow paths from two line injection of 3h and ^{36}Cl in mikrocatchment at Gardsjön, Sweden." Hydrol. Process. *13,* 1557-1575

[78] Öberg, G. (1998) „Chloride and Organic Chlorine in Soil“ Acta Hydrochim. *Hydrobiol. 26*, 137-144.

[79] Öberg, G., Groen, C. (1998) „Sources of Organic Halogens in Spruce Forest Soil“ *Environ. Sci. Technol. 32*, 1573-1579.

[80] Öberg, G. (2002) „The natural chlorine cycle – fitting the scattered pieces.“ *Appl. Microbiol. Biotechnol. 58*, 565-581.

[81] Öberg, G. (2003) "The Biogeochemistry of Chlorine in Soil" in The Handbook of Environmental Chemistry, Vol. 3, Part P, (Ed. Gribble, G.) *Springer-Verlag,* pp 43-62.

[82] Öberg, G., Sandén, P. (2005) „Retention of chloride and cycling of organic matter-bound chlorine“ *Hydrological Processes 19*, 2123-2136.

[83] Öberg, G., Holm, M., Sandén, P., Svensson, T., Parikka, M. (2005) „The role of organic matter-bound chlorine in the chlorine cycle: a case study of the Stubbetorp catchment, Sweden“, *Biogeochemistry 75*, 241-269.

[84] O´Neill, P. (1998) „Environmental Chemistry“, 3rd Edition, *Blackie Academic & Professionals,* London.

[85] Peters, R. J. B. (2003) „Chloroacetic acids in European soils and vegetation.“ *J. Environ. Monitor. 5*, 275-280.

[86] Plümacher, J., Renner, I.,Schröder, P.(1993) „Volatile chlorinated hydrocarbons and trichloroacetic acid in conifer needles“ In: Schröder, P., Frank, H., Ether, B. (Eds.) *„Volatile Organic Pollutants“* Levels, Fate and Ecotoxicological Impacts. Wiss. Verlag Dr. W. Maraun, Frankfurt/M., pp. 37-51.

[87] Plümacher, J., Schröder, P. (1994) „Accumulation and Fate of C1/C2-Chlorocarbons and Trichloroacetic Acid in Spruce Needles from an Austrian Mountain Site“ *Chemosphere 29*, 2467-2476.

[88] Rohlenová, J., Matucha, M., Gryndler, M., Forczek, S.T. (2007) „Microbial chlorination of organic matter in forest soil." prepared for publication

[89] Rodstedth, M., Stahlberg, C., Sandén, P., Öberg, G. (2003) Chloride imbalances in soil lysimeters. *Chemosphere 52*, 381-389

[90] Salomon, K.R. (1996) „Chlorine in the bleeching of pulp and paper“ *Pure Appl. Chem. 68*, 1721-1730.

[91] Schöler, H. F., Keppler, F., Fahimi, I. J. and Niedan, V. W.(2003) „Fluxes of Trichloroacetic Acid between Atmosphere, Biota, Soil and Groundwater.“ *Chemosphere 52*, 339-354.

[92] Schröder, P.; Wolf, A. E. (1996) Characterization of glutathione S-transferases in needles of Norway spruce trees from a forest decline stand. *Tree Physiology 16*, 503-508.

[93] Schröder, P.; Juuti, S.; Roy, S.; Sandermann, H.; Sutinen, S. (1997) Exposure to chlorinated acetic acids: Responses of peroxidase and glutathione S-transferase activity in pine needles. *Environ.. Sci. Pollut. Res. 4*, 163-171.

[94] Schröder, P., Plümacher, J. (1998) „Evaluation of the impact of volatile chlorinated hydrocarbons on forest trees based on air and needle measurement data and the detoxification kapacity of spruce needles. *Environ. Sci. Pllut. Res. Spec.* Issue *1*, 38-45.

[95] Schröder, P., Matucha, M., Forczek, S.T., Uhlířová, H., Fuksová, K., Albrechtová, J. (2003) "Uptake, Translocation and Fate of Trichloroacetic Acid in Norway Spruce/Soil System" *Chemosphere 52*, 437-442.

[96] Sidebottom, H., Franklin, J. (1996) "The Atmospheric Fate and Impact of Hydrochlorofluorocarbons and Chlorinated Solvents" *Pure & Appl. Chem. 68*, 1757-1769.

[97] Silk, P.J., Lonergan, G.C., Arsenault, T.L., Boyle, C.D. (1997) "Evidence of natural organochlorine formation in peat bogs", *Chemosphere 35*, 2865-2880

[98] Steinberg, C.E.W., Kamara, S., Prokhotskaya, V.Y., Manusadžianas, L., Karasova, T.A., Timofyeev, M.A., Jie, Z., Paul, A., Meinelt, T., Farjalla, V.F., Matsuo, A.Y.O., Burison, B.K., Menzel, R. (2006) „Dissolved humic substance-ecological driving forces from the individual to the ecosystem level?" *Freshwater Biology 51*, 1189-1210.

[99] Stevenson, F.J. (1981) *Humus chemistry*. Wiley, Chichester, UK.

[100] Stidson, R., Dickey, C. A., Cape, J. N., Heal, K. V., and Heal, M. R.(2004a) „Fluxes and Reservoirs of Trichloroacetic Acid at a Forest and Moorland Catchment." *Environ. Sci. Technol. 38*,1639-1647.

[101] Stidson, R. T., Heal, K. V., Dickey, C. A., Cape, J. N., Heal, M. R. (2004b) „Trichloroacetic acid fluxes through a conifer forest canopy" *Environ.Pollut. 132*, 73-84.

[102] Stork A, Witte R, Fuhr F (1997) „(CO_2)-C-14 measurement in air: Literature review and a new sensitive method" *Environ. Sci. & Technol. 31*: 949-955.

[103] Sutinen, S., Juuti, S., Koivisto, L., Turunen, M., Ruuskanen, J. (1995) "The Uptake of and Structural Changes Induced by Trichloroacetic Acid in the Needles of Scots Pine Seedlings" *J. Exper. Botany 46*, 1223-1231.

[104] Sutinen, S., Juuti, S., Ryyppo, A. (1997)"Long-Term Exposure of Scots Pine Seedlings to Monochloroacetic and Trichloroacetic Acid: Effects on the Needles and Growth" *Annales Botanici Fennici 34*, 265-273.

[105] Svensson, T., Laturnus, F., Sandén, P., Öberg, G. (2007) "Chloroform in runoff water-a two year study in a small catchment in Southeast Sweden" *Biogeochemistry 82*, 139-151.

[106] Šašek, V., Volfová, O., Erbanová, P., Vyas, B.R.M., Matucha, M. (1993) Degradation of PCBs by White Rot Fungi, Methylotropic and Hydrocarbon Utilizing Yeasts and Bacteria. *Biotechnol. Lett. 15,* 521-526

[107] Uhlířová, H., Pasuthová, J., Matucha, M. (1995) "Air Pollutions and Forests. II. Chlorinated Compounds." (in Czech), *Zprávy lesnického výzkumu* (*Reports from the Forest Research) 40*, 30-35.

[108] Uhlířová, H., Matucha, M., Kretzschmar, M., and Bubner, M. (1996) „Aufnahme und Verteilung von Trichloressigsaeure in Trieben Norwegischer Fichte" (Uptake and Distribution of Trichloroacetic Acid in Norway Spruce (*Picea abies/L.Karst*) Shoots - in German). *Z. Umweltchem. Ökotox. 8*,138-142.

[109] Uhlířová, H., Matucha, M., Forczek, S. T. (2002) "Účinky chloroctových kyselin na smrk ztepilý (*Picea abies /L./ Karst.*)" (in Czech), *Zprávy lesnického výzkumu* (*Reports from the Forest Research) 47*, 16-20.

[110] Verhagen, F.J.M., Swarts, H.J., Wijnberg, J.B.P.A., Field, J.A. (1998) „Organohalogen production is a ubiquitous capacity among basidiomycetes" *Chemosphere 37*, 2091-2104.

[111] Vyas, B.R.M., Šašek, V., Matucha, M., Bubner, M. (1994a) „Degradation of 3,3′,4,4′-Tetrachlorobiphenyl by Selected White Rot Fungi" Chemosphere *28*, 1127-1134

[112] Vyas, B.R.M., Bakovski, S., Šašek, V., Matucha, M. (1994b) „Degradation of Anthracene by Selected White Rot Fungi", FEMS Microbiology Ecology *14*, 65-70

[113] Weissflog, L.; Manz, M.; Popp, P.; Elansky, N.; Arabov, A.; Putz, E.; Schüürmann, G. (1999) „Airborne trichloroacetic acid and its deposition in the catchment area of the Caspian Sea" *Environmental Pollution 104*, 359-364.

[114] Weißflog, L., Pfennigsdorff, A., Martinez-Pastur, G., Puliafito, E., Figueroa, D., Elansky, N., Nikonov, V., Putz, E., Krueger, G., Kellner, K. (2001) „Trichloroacetic acid in the vegetation of polluted and remote areas of both hemispheres – Part I. Its formation, uptake and geographical distribution" *Atmospheric Environment 35*, 4511-4521.

[115] Weißflog, L., Krüger, G., Elansky, N., Putz, E., Pfennigsdorff, A., Seyfarth, K-U., Nüchter, M., Lange, C., Kotte, K. (2003) „Input of trichloroacetic acid into the vegetation of various climate zones - measurements on several continents" *Chemosphere 52*, 434-449.

[116] Weißflog, L., Elansky, N., Putz, E., Krüger, G., Lange, C., Lisitzina, L., Pfennigsdorff, A. (2004) „Trichloroacetic acid in the vegetation of polluted and remote areas of both hemispheres - Part II: Salt lakes as novel sources of natural chlorohydrocarbons." *Atmos.Environ. 38*, 4197-4204.

[117] Weißflog, L., Lange, C.A., Pfennigsdorff, A., Kotte, K., Elansky, N., Lisitzina, L., Putz, E., Krueger, G., (2005) „Sediments of salt lakes as a new source of volatile highly chlorinated C1/C2 hydrocarbons" *Geophysical Research Letters 32*, L01401.

[118] Weissflog, L., Krueger, G., Forczek, S.T., Lange, Ch. A., Kotte, K., Pfennigsdorff, A., Rohlenová, J., Fuksová, K., Uhlířová, H., Matucha, M., Schröder, P. (2007) "Oxidative Biodegradation of Tetrachloroethene in Needles Of Norway Spruce (Picea Abies L.)" *South Afric. J. Botany 73*, 89-96.

[119] Winterton, N. (2000) „Chlorine: the only green element – towards a wider acceptance of its role in natural cycles" *Green Chemistry 2*, 173-225.

[120] Zachař, P., Novotný, Č, Vozňáková, Z., Matucha, M., Tesařová, E., Sýkora, D., Kubátová, A., Popl, M., Šašek, V. (1996) „Physical Factors Negatively Affecting Evaluation of Long-Term Biodegradation Experiments of Polychlorinated Biphenyls", *Chemosphere 33,* 2411-2421

INDEX

B

C

D

E

F

G

H

I

J

K

L

M

N

Q

R

S

T

U

V

W

X

Y

Z